THE PHILOSOPHY OF LOGICAL MECHANISM

SYNTHESE LIBRARY

STUDIES IN EPISTEMOLOGY,

LOGIC, METHODOLOGY, AND PHILOSOPHY OF SCIENCE

VOLUME 206

Arthur W. Burks, Fall 1978
Photo by Bob Kalmbach

THE PHILOSOPHY OF LOGICAL MECHANISM

Essays in Honor of Arthur W. Burks
With his responses

With a bibliography of Works by Arthur W. Burks

Edited by

MERRILEE H. SALMON˙

University of Pittsburgh

KLUWER ACADEMIC PUBLISHERS
DORDRECHT / BOSTON / LONDON

Library of Congress Cataloging-in-Publication Data

The Philosophy of logical mechanism : essays in honor of Arthur W.
  Burks, with his responses / edited by Merrilee H. Salmon.
      p.   cm. -- (Synthese library ; v. 206)
    Bibliography: p.
    Includes index.
    ISBN-13:978-94-010-6933-5   e-ISBN-13:978-94-009-0987-8
    DOI: 10.1007/978-94-009-0987-8

    1. Electronic digital computers--History. 2. Science--Philosophy.
  3. Burks, Arthur W. (Arthur Walter), 1915-   . I. Burks, Arthur W.
  (Arthur Walter), 1915-   . II. Salmon, Merrilee H. III. Series.
  QA76.5.P3998 1989
  005.13'1--dc20   89-11238
                                                                  CIP

ISBN-13:978-94-010-6933-5

Published by Kluwer Academic Publishers,
P.O. Box 17, 3300 AA Dordrecht, The Netherlands.

Kluwer Academic Publishers incorporates
the publishing programmes of
D. Reidel, Martinus Nijhoff, Dr W. Junk and MTP Press.

Sold and distributed in the U.S.A. and Canada
by Kluwer Academic Publishers,
101 Philip Drive, Norwell, MA 02061, U.S.A.

In all other countries, sold and distributed
by Kluwer Academic Publishers Group.,
P.O. Box 322, 3300 AH Dordrecht, The Netherlands.

printed on acid-free paper

TABLE OF CONTENTS

PREFACE

This work is divided into two parts. Part I contains sixteen critical essays by prominent philosophers and computer scientists. Their papers offer insightful, well-argued contemporary views of a broad range of topics that lie at the heart of philosophy in the second half of the twentieth century: semantics and ontology, induction, the nature of probability, the foundations of science, scientific objectivity, the theory of naming, the logic of conditionals, simulation modeling, the relation between minds and machines, and the nature of rules that guide behavior. In this volume honoring Arthur W. Burks, the philosophical breadth of his work is thus manifested in the diverse aspects of that work chosen for discussion and development by the contributors to his Festschrift.

Part II consists of a book-length essay by Burks in which he lays out his philosophy of logical mechanism while responding to the papers in Part I. In doing so, he provides a unified and coherent context for the range of problems raised in Part I, and he highlights interesting relationships among the topics that might otherwise have gone unnoticed.

Part II is followed by a bibliography of Burks's published works. This list of his publications provides ample evidence that technical virtuosity is not only compatible with serious work on traditional philosophical problems but also enriches that work. Burks's contributions to philosophy emphatically refute the often-heard complaint that contemporary philosophers, particularly philosophers of science, are all narrow specialists who have abandoned the traditional problems that draw most people to the study of philosophy in the first place.

Burks, who is probably best known to philosophers for his major study of the foundations of science, *Chance, Cause, Reason*, has also distinguished himself in the field of computer science. While maintaining an active role in the Philosophy Department at the University of Michigan, he served for some years as head of the Computer Science Department. His work in this field began with his involvement in the development of the ENIAC, the first general-purpose electronic computer. Together with his wife Alice he has recently published *The First Electronic Computer: The Atanasnoff Story*, a history of the controversy surrounding the invention of modern high-speed computers. He was one of the first contemporary philosophers to write about the philosophical implications of using computers as models of nature and models of the human mind, and this continuing interest remains central to his philosophy of logical mechanism.

The philosophy of logical mechanism represents a modern development of the atomistic view that originated with the Greeks. Burks, acknowledging that early forms of mechanistic atomism have not been able to provide an adequate account of "the most unique human abilities: explicit reasoning, intentional goal-directedness, consciousness, and the inner point of view," attempts to supplement the traditional mechanistic theory with concepts drawn from logic, computers, and evolutionary genetics to provide a reductivist account of the "holistic-coherent systems of man, society, and science." In his account, which analyzes these holistic-coherent systems as hierarchical-feedback systems, Burks takes seriously the importance of such specifically human capacities as consciousness, self-consciousness, and free choice. His work enriches our understanding of these features instead of attempting to explain them away by denying our uniqueness.

While some of the papers in this collection contain technical material, most of them are accessible to a general reader with an interest in current issues in philosophy of science, philosophy of language, or artificial intelligence. An attractive feature of this volume is the way it brings together in a coherent fashion so many of the issues that concern specialists in the three fields just mentioned.

My sincere thanks are extended to Jaakko Hintikka for suggesting this volume, to Arthur Burks and the authors of the critical essays contained herein, and to Charlotte Ashby for compiling the Indices.

Merrilee H. Salmon
History & Philosophy of
Science
University of Pittsburgh

PART I

CRITICAL ESSAYS

FREDERICK SUPPE

IS SCIENCE REALLY INDUCTIVE?

Arthur W. Burks has made major contributions to Peirce scholarship, computer science, automata theory, modal logic, and the philosophy of science; his contributions to any one of these areas would constitute a productive career most academics would envy. Disparate as these fields are, in Burks' approach to them there are deep interconnections. And in his monumental work *Chance, Cause, Reason*[1] Burks manages to exploit these interconnections to provide a richly detailed account of "the ultimate nature of the knowledge acquired by the empirical sciences." (p. 651) It does so by presenting his Presupposition Theory of Induction, a theory that provides a rich foundational analysis of "standard inductive logic" – which, according to Burks, "is the system of rules of inductive inference actually used and aspired to by the practicing scientist." (p. 654)

Underlying Burks' analysis of the nature of scientific evidence is the view that scientific laws and theories are confirmed on the basis of data acquired via observation and experiment, and that such confirmation essentially involves inductive reasoning.[2] Burks really doesn't argue for this supposition, rather taking it as a given. When he began writing *Chance, Cause, Reason* in 1960 the inductive-confirmation view was so commonplace it scarcely seemed to warrant defense. But since then a major revolution has occurred within philosophy of science, one which centrally has involved historical analyses of the place of rationality in the growth of scientific knowledge; increasingly one finds philosophers such as Kuhn, Feyerabend, Laudan, Shapere, Lakatos, *et al.*, as well as historians such as Holton, presenting models of scientific reasoning which implicitly or explicitly reject the view that the nature of scientific evidence essentially involves using observational and experimental data to refute or else inductively confirm laws or theories on the basis of repeated observations using the canons of standard inductive logic. Thus it is no surprise that I have heard some philosophers of science

3

Merrilee H. Salmon (ed.), *The Philosophy of Logical Mechanism*, 3–27,
© 1990 *Kluwer Academic Publishers*.

express the opinion that *"Chance, Cause, Reason* would have been a great book if it had been published twenty years ago."

I

To my mind the failure to defend the inductive-confirmation view against its modern-day detractors is a serious weakness of *Chance, Cause, Reason*. To be sure there are discussions in it which do lay the basis for such a defense. For example, Feyerabend's rejection of the inductive confirmation view rests in part in his agreement with Popper that Hume's arguments against the justification of induction preclude induction being a vehicle of scientific knowledge. In Chapters 3 and 10 Burks extensively discusses Hume's problem and presents a neo-Kantian, neo-Keynsian defense of standard inductive logic. And, his separation of the logic of empirical inquiry from inductive logic provides the basis for a reply to others such as Lakatos, Laudan, and Shapere – viz, what they're investigating through historical analysis is the logic of empirical inquiry, which he acknowledges as legitimate and important; but he's investigating something else – inductive logic. Such a reply would not be convincing to me. For these authors seem committed to the idea that any separation of discovery and confirmation is artificial – that scientific discovery, experiment, observation, testing, theory development, modification, and evaluation all go hand in hand in ways that ultimately do not fit the inductive testing and confirmation model. This is a serious challenge which, if correct, cuts to the very heart of Burks' enterprise – reducing it to a brilliant piece of technical philosophy which is utterly irrelevant to real science. And in today's philosophy of science that is not only unfortunate, it is a mortal sin.

To sharpen the issue, consider the following passages:

Let us call the system of rules of inductive inference actually used and aspired to by the practicing scientists *standard inductive logic*. This will include the calculus of inductive probability together with rules for analogy, the method of varying causally relevant qualities, and induction by simple enumeration. (p. 103)

Now in fact we all employ standard inductive logic and no one has ever seriously contemplated using any other. (p. 130)

The challenge being offered by recent work on the growth of scientific knowledge is to question whether we in fact *do* employ standard inductive logic in science (second quote), whether in fact there is *a* system of rules of inductive inference actually used and aspired to by practicing scientists which meets Burks' characterization of standard inductive logic (first quote).

Historically there is reason to doubt whether science is inductive in the ways Burks claims. In Section 2.5.4., Burks briefly analyzes several historical episodes – including Galileo and The Young/Fresnel controversies over light – using Bayes' theorem. Since the calculus of inductive probabilities postdates these episodes, and since the argumentation in these controversies did not invoke inductive probabilities (or any probabilistic precursers), the scientists involved could not have been using standard inductive logic – contrary to what Burks seems to be implying. Now it may be that what Burks means to claim is this: These episodes evidence good scientific reasoning. Even though they didn't possess an articulate version of standard inductive logic, they behaved in accordance with its dictates, and so implicitly employed standard inductive logic.[3] Thus it is not inappropriate to explicitly invoke standard inductive logic in analyzing the reasoning employed in these historical episodes. However, such a reply presupposes that standard inductive logic captures good scientific reasoning, and that is precisely what is being questioned.

We should be suspicious of attempts to read the history of science through inductive eyes. For there is a long history of attempting to reconstruct scientific achievements to make them conform to contemporary inductive canons. The inductive approach to science is a normative philosophical approach to doing science that predates any significant scientific achievements guided by it. That is, it was advanced not as an analysis of how good science has been done, but rather as a program for doing science in a new way. In 1543 Nicholas Copernicus published his *De revolutionibus* which, while starting a scientific and intellectual revolution, was a work as far removed from the inductive-confirmation view of science as one can imagine. Quite likely he borrowed his celestial model from Aristarchus of Samos (*fl* ca. 281 BC), and his arguments for preferring his heleocentric view rested centrally on the claim it was more faithful to Aristotle than was the Ptolemaic geocentric theory. Repeated testing and inductive confirmation did not play any significant role. Those who were convinced by Copernicus realized that much of the rest of science would fall (especially Aristotelean physics) and saw science as having to start all over again. Descartes' method of doubt was his first step towards developing a rationalistic methodology of science which, it was claimed, would insure that the new science was accurate and correct. Francis Bacon's *Novum Organon* (1620) offered a different analysis, claiming that the reason ancient science had gone wrong was its reliance on "first principles." To avoid these, and ensure correct scientific theories, one ought to root science in observation and inductive generalization. To this end Bacon developed two new methods of induction which included in rudimentary form what Burks calls "the method of varying causally relevant qualities." Although Bacon advocated such an inductive approach to doing science, he did no significant scientific work himself.

During much of the 17th century the most impressive scientific achievements were Cartesian, and do not appear at all to fit the inductive-confirmation model. Kepler's discovery of the planetary orbits and the development of his three laws in his *Astronomia Nova* (1609) although often cited by inductivists as a paradigm instance of science being done within their model in fact rather badly fits the model.[4] While it is controversial the extent to which Galileo engaged in observation and experimentation, the consensus seems to be that his work generally does not conform closely to the inductive confirmation model. Indeed, in The *Dialogo*, when asked whether he had actually performed the important experiment of dropping a rock from a moving ship's mast, Salviatti (speaking for Galileo) replies, "Without experiment, I am sure the effect will happen as I tell you, because it must happen that way...". (p. 145). Hardly an inductive confirmation view!

Only towards the end of the 17th and the beginning of the 18th centuries do we find important scientists purportedly claiming to work in accordance with the inductive method. E.g., Newton in his *Optiks* (1704) writes that "although the arguing from Experiments and Observations by Induction be no Demonstration of General Conclusions, yet it is the best way of arguing which the Nature of Things admits of " (p. 404). But, as the historical controversy over *Hypotheses non fingo* in The General Scholium of the second edition (1713) of his *Principia* (first edition, 1687) shows, it is very unclear how strongly Newton did adhere to an inductivist approach to science. We also find a decline in the influence of rationalism and a growing allegiance to empiricism among philosophers beginning towards the end of the 17th century. For example, Locke's *Essay* (1690) is in part an inductivist vindication of the scientific work of his friend Boyle and of Newton. And with this growing championing of empiricism and its inductivism, we find systematic attempts to recast earlier science in an inductivist mold. One of the more scandalous, but not untypical, attempts was the alteration of Galileo's writings by Thomas Salisbury while translating them into the vernacular so as to make Galileo conform (in translation at least) to empiricist and inductivist normative demands; e.g., in his 1661 translation of the *Dialogio*, in the passage quoted above Salisbury drops the phrase *senza sperienza* ("without experiment" or "without recourse to experience")[5] thus defusing the true anti-empiricist thrust of the *Dialogues'* passage.

With the resurgance of empiricism in the 19th century at the hands of Herschel, Mill, and Whewell, we again find attempts to read the history of science through inductive eyes – often distorting the history in the process. Particularly noteworthy here is William Whewell's extensive three volume *History of the Inductive Sciences* (1837) which, while an impressive historical work still worth consulting, does shape the history of science to conform to the inductivist views he develops (supposedly based on his *History*) in his 1840 *Philosophy of The Inductive Sciences*.

Again, with the rise of Logical Empiricism this century, we find a new resurgance of inductivism. But it is a very transformed inductivism. Whereas induction in the 18th and 19th century consisted of non-probabilistic methods of inference from specific instances to causal generalizations, at the hands of Reichenbach, Keynes, and Carnap the enterprise becomes fundamentally probabilistic. Hume's arguments against the possibility of justifying induction no longer applied, although probabilistic analogues did emerge to plague, e.g., Reichenbach's self-corrective method for justifying induction. With the Logical Empiricists claims abound that scientific laws and theories are inductively confirmed (now in accordance with the probabilistic inductive logics they are developing, not the earlier inductions of the 17th-19th centuries). Little attempt is made to defend these claims on historical grounds, although one finds casual references to Kepler, Semmelweiss, *et al.*

The following points emerge from this sketchy discussion. The inductive view of science initially was promulgated as a prescriptive account about how science ought to be done. That prescriptive account has on occasion taken on some of the trappings of an ideology. Like most other ideologies it has looked to spectacular historical episodes which are seen as exemplifying and vindicating the correctness of the ideol-

ogy, frequently engaging in revisionist history to make the case for
legitimacy and correctness. Even when Logical Empiricism radically
altered the nature of induction by making it probabilistic, the old
claims that science was in fact inductive survived and were perpetuated
without serious reexamination. Thus there are good historical grounds
for being suspicious of the claim that standard inductive logic "is the
system of rules of inductive inference actually used and aspired to by
the practicing scientist" (pp. 103, 654).

II

But grounds for suspicion do not make the suspicions correct, so let us
turn to a detailed examination of whether science *does* reason in accordance with standard inductive logic. Immediately we face a
problem: Burks' characterization of standard inductive logic quoted
above is none too specific, essentially being an ostensive definition
("what is actually used and aspired to by practicing scientists") augmented by a gloss. Nevertheless, piecing together various comments
Burks makes, we know it involves using induction by simple enumeration; and since this makes the same probabilistic predictions of future
events as does Bayes' method for finding unknown probabilities (p.
126), it involves use of Bayes' method. It also involves the use of
repeated instance-confirmation to inductively confirm laws and other
causally necessary generalizations, as well as the use of inductive probabilities in other contexts such as determining empirical probabilities,
and rules for analogy.

Outside of natural history, where one would most expect to find
science proceeding inductively is in the behavioral sciences – especially
behavioral psychology and sociology. For much of this century these
fields have subscribed to the methodology of operationalism (a particular version of Logical Empiricism/Logical Positivism) and inductivism. But even though they subscribe to an inductivist methodology,
it does not in fact follow that they proceed in accordance with it. Indeed, the fact that virtually no operational definitions encountered in
the behavioral science literature meet the methodological requirements
"officially" subscribed to should prompt us to investigate whether, in
fact, the behavioral sciences do employ standard inductive logic.

It certainly is the case that probabilistic inference and reasoning per-
meates the behavioral science literature. But rarely does one find the

use cf instance-confirmation to confirm generalizations involving causal necessity. More typically one finds investigations of statistically significant measure differences between a subject and a control population, regression and correlational analyses to determine statistical (but not causal) regularities, analyses of variance to explain the sources (but not causes) of differences among subject and control groups, and statistical tests of hypotheses which result in the rejection or acceptance of hypotheses at some confidence level. Occasionally the hypotheses tested are causal, but there is a widespread skepticism that "cause" is inappropriate since the statistics can't really determine causes. Only with the recent development of path analysis does one find much in the way of attempts to determine causal influence, and then it is a statistical sense of causality (not involving Burks' causal necessity) as are other statistical investigations such as those surrounding the "causes" of cancer. On the surface, then, the behavioral sciences do not seem to exemplify standard inductive logic.

This first blush assessment surely is incorrect. While Burks stresses the confirmation or rejection of general hypotheses involving causal necessity, other hypotheses not involving causal necessity do come under his conception of standard inductive logic. And these include at least some of the statistical hypothesis testing found in the behavioral sciences. Burks in *Chance, Cause, Reason* never discusses his views about statistical theories of causality. But I am relatively confident that Burks would view the statistical theories of causality developed by e.g., Suppes and Giere as being complicated empirical probability claims. More generally, the statistical methods typically found in the behavioral sciences all seem to involve the use of statistics to establish (at some level of confidence) empirical probability claims. And since in Chapter 8 Burks has analyzed in detail empirical probabilities and how reasoning about them falls within the purview of standard inductive logic, we can conclude that most statistical or probabilistic reasoning in the behavioral sciences does fall within standard inductive logic as analyzed by Burks. To be sure, Burks does not address or attempt to show this is the case, but this is totally in keeping with his view that "the logic of inductive argument studies the most general, universal, and fundamental types of inductive arguments, leaving the more detailed and specific forms to the individual sciences and statistics," (p. 17) and the resulting enterprise he has set for himself.

Although the behavioral sciences do provide many instances of science proceeding in accordance with standard inductive logic, it provides little evidence that science uses standard inductive logic to obtain instance-confirmation of generalizations involving causal necessity. Since for Burks this is a most central part of standard inductive logic (indeed the main thrust of *Chance, Cause, Reason* is to examine, analyze, and explain how this is possible), the behavioral sciences provide an unacceptably weak vindication of his claims for standard inductive logic being the means whereby science reasons about evidence.

Burks' own examples (cited above) suggest it is the physical sciences which he takes as best exemplifying standard inductive logic. We now examine this idea. Our starting points are his claims that "the rule of induction by simple enumeration makes the same probabilistic predictions of future events as does Bayes' method" (p. 126) and his Bayesian analysis of instance confirmation (Section 10.3.2); for these collectively indicate that the Bayesian model is at the very heart of how inductive scientific inference proceeds in the verification of laws and other generalizations involving causal necessity. On the Bayesian model one has a set of *logically possible*, mutually exclusive and jointly exhaustive hypotheses which one wants to choose between on the basis of repeated testing. One attaches prior probabilities to each (and the assignment may be arbitrary), tests them, then obtains posterior probabilities. Using the latter now as new prior probabilities, one iterates the procedure until one hypothesis emerges as the probabilistic victor; and one hypothesis must if one repeats the test long enough.

As a first challenge to this Bayesian view of instance-confirmation, let us look at Lakatos' research programme model of theory development; in doing so I do not intend to minimize either the serious problems it faces or the historical accuracy challenges besetting it.[6] Lakatos' basic idea is that the unit of scientific assessment is not the theory, but rather a supra-theoretical unit which he designates as a *research programme*. Research programmes contain a *problem shift* – a sequence of thories, each successor in the sequence being an augmentation or semantic reinterpretation of its immediate predecessor in the sequence. Allowable modifications in producing the sequence are constrained by what Lakatos terms the positive and negative heuristics. Collectively these heuristics and the problem shift constitute a *research programme*. As indicated in the work cited in note 6 above I have been fairly critical of this supposedly "historically-derived" model, and do not subscribe to its veracity. Nevertheless, despite all its defects in analysis, Lakatos does target an important aspect of theoretical development and the roles of experimentation and testing therein – viz, that when one is developing a theory one uses experimentation not to refute or confirm a hypothesis, but rather as a crucial ingredient in further refining a crude initial approximation of what is believed to be a promising theory into what, hopefully, will be an ever-increasingly-close approximation to an ultimately adequate theory. Initially this (roughly right-headed) perspective seems quite at odds with Burks' instance-confirmation or refutation view, and especially his Bayesian construal thereof.

Is Burks' analysis really so incompatible with such a Lakatosian view? Probably not. For minimally, on Lakatos' view, should a current theory in the problem shift prove inadequate (as, in neo-Popperian fashion, it is supposed to), presumably the positive and negative heuristics are supposed to provide some (unspecified) guidance for the revised next incarnation successor theory in the problem shift. Thus, one might plausibly construe the testing of the current theory in the problem-shift as being the choice, on the basis of observation and experiment, between the present incumbent in the problem shift and some anticipated alternative replacement. If so,[7] we plausibly can view theory development as involving repeated testing of successive theories in a problem shift as being a Bayesian decision between the current problem-shift incumbent, an anticipated replacement (or a few such), and a "none-of-the-above" alternative.[8] It is my opinion that much of the sort of development of theories Lakatos attempts to capture in his problem-shift model at best roughly typifies some roles experiment does play in testing, evaluating, and developing theories. And to the extent that the use of observation and experiment to produce the successor theory in a problem-shift is non-serendipitous and involves the testing of the current occupant theory against an anticipated replacement, where some semblance of prior probabilities can be attached to each, I think Burks' Bayesian' instance-confirmation model is basically applicable (although its specific application could be detailed much more if Burks chose to engage in "the more detailed and specific forms" (p. 17) of inductive argument which he eschews in his search for "the most general, universal, and fundamental types of inductive argument" (p. 17)). Thus, to the extent that Lakatos' research programmes' model enjoys legitimacy, it does not appear to be obviously incompatible with Burks' claims for standard inductive logic –

at least if we grant Burks the claim that he has defused the Humean anti-inductivist arguments which are at the core of Popper's philosophy of science and which, despite various distancing efforts, do influence Lakatos' view of the roles of observation, experimentation, and testing in science.

* * *

A much more important threat to Burks' Bayesian instance-confirmation view of science comes from the history of science-based philosophical analyses advanced by Shapere. A key point of Shapere's examination of rationality in the growth of scientific knowledge is the recognition that, contra the presuppositions of the Bayesian model, science does not resort to observation and experiment to decide against the *logically possible* alternatives to the favored hypothesis, but rather more selectively uses observation and experiment to test among only the favored hypotheses which are plausible.[9] This recognition, which I believe to be essentially correct, seems to me to constitute one of the deepest challenges I know of to the view that science fundamentally does reason and proceed in accordance with standard inductive logic.

To an extent, Burks is aware of the problem I've just raised. In his discussion of the Bayesian method he is aware that the logically possible alternatives to a favored hypothesis are potentially unlimited but that Bayes' method requires a specific finite set. To this end he engages in the practice of combining the seriously entertained alternatives (the plausible ones – as opposed to the logically possible ones) together with a "none of the above" alternative (with negligible prior probability) as a means of meeting the specific finitary option requirement (pp. 85ff, 510). In so doing he is tacitly acknowledging Shapere's point that we use observation and experiment to decide between the plausible, not the logically possible alternative hypotheses, laws, or theories.

Such tacit acknowledgement of Shapere's contention does not, however, salvage his claim that science works in accordance with standard inductive logic. Indeed, it is a near-fatal concession which seems to concede his case to the newly-sprouted anti-inductivists, in the process bringing attention to what is one of the more vexing problems which Burks addresses in *Chance, Cause, Reason* but does not fully resolve: the problem of prior probabilities (pp. 90–92, Section

8.3.4). The Bayesian method gives an excellent account of how, *given prior probabilities,* one can use testing to come up with improved (in the long run at least) estimates of new prior probabilities (viz, use the posterior probabilities from the prior ones in Bayesian iteration), but it tells one absolutely nothing about how to get the enterprise started. If one wishes to start a Bayesian inductive enterprise in a pristine manner, where does one obtain the *initial* prior probabilities? The Bayesian model gives no clue – other than to say that however capricious, uninformed, or misguided, it doesn't make any difference in the long run. But this doesn't cohere with Shapere's analysis. The very choice of which alternative hypotheses to treat seriously amounts to a choice of which logically possible alternative hypotheses enjoy significant prior probabilities on Burks model, and in the first instance the Bayesian model gives no guidance. In short, the Bayesian model says that, when we have logically prior plausibility judgments as to which alternative hypotheses are worth pursuing, we can start applying the Bayesian method – which is the heart of standard inductive logic. But to acknowledge this is to acknowledge that standard inductive logic is a hopefully impoverished model of actual scientific reasoning in the evaluation of hypotheses. For it is to concede that standard inductive logic presupposes a logically prior sort of reasoning – plausibility assessment. Differently put, we have a *dilemma*: If standard inductive logic is intended to provide an analysis of that plausibility reasoning, then we have a vicious regress where each iteration of the Bayesian method requires a logically prior application; hence it is impossible to ever get the Bayesian method going. Hence standard inductive logic is an inadequate model of scientific reasoning about evidence and the evaluation of hypotheses. If, on the other hand, standard inductive logic does not provide an analysis of that plausibility reasoning, standard inductive logic is a critically incomplete, hence an inadequate, model of scientific reasoning about evidence and the evaluation of hypotheses.

Burks is not unaware that there are problems concerning prior probabilities:

In each application of Bayes' theorem it is fair to ask: What is the source of the prior probabilities? Usually they derive from previous empirical situations. They might be posterior probabilities just renamed. More typically they come from general experience.... But this process of tracing back prior probabilities to earlier evidence must finally terminate, since the total available evidence is finite. And when it does terminate the same question arises: What is the epistemological basis of the prior probability assignment? (p. 91).

Burks goes on to suggest that "Since prior probabilities are expressed by atomic inductive probability statements, this problem is a special case of our first main question: What is the general nature of an atomic inductive probability statement?" (*ibid.*)

In Section 8.3.4. Burks returns to the issue. There his discussion is of how his dispositional theory of empirical probabilities and the foil positivistic theory of empirical probability differently treat prior probabilities. His conclusion is:

> The dispositional theory. . . .holds that prior probabilities are conditional and are subject to the norms of standard inductive logic. These norms are common to all men and are thus intersubjective rather than subjective (pp. 103, 136, 305, 326–327). Consequently, the dispositional theory makes prior probabilities intersubjective, in contrast to the positivistic theory that makes them subjective and relative to the individual. (p. 529)

The intersubjectivity of norms, as amplified at the locations he cites, consists in the claim that we all employ standard inductive logic and his analysis of atomic inductive probability statements via his pragmatic theory of inductive probability (summarized on pp. 305–306).

Does Burks' treatment of prior probabilities dissolve the dilemma we posed above for standard inductive logic? It is clear from the last quote that Burks is committed to embracing the first horn of our dilemma. Thus standard inductive logic must provide the plausibility assessment means for obtaining the prior probabilities needed to get the Bayesian method going. The most plausible account I can think of allowed by Burks' treatment is the following: Prior probabilities are either empirical or inductive probabilities. Either way, they ultimately are based upon atomic inductive probabilities. Atomic inductive probabilities have action implications. "An atomic inductive probability statement does and should express a disposition to act or 'bet' in certain ways under conditions of uncertainty." (p. 305; italics deleted.) The "certain ways of betting" consist in betting rationally in accordance with the canons of Burks' calculus of choice. Then on Burks' account one's rational beliefs produce betting dispositions which determine a system of atomic probability assessments – and these can be the source of prior probabilities needed to get the Bayesian method going.

This appears to be Burks' intended treatment of the problem of prior probabilities. Is it adequate to dissolve our dilemma? On the calculus of choice, we are faced with various decisions to be made in the face of uncertainty; rationally making those decisions induces atomic inductive probabilities. The decision making reflects a coherent belief and value structure. It is not unreasonable to interpret those decisions – which are logically prior to the probabilities –[10] as involving non-probabilistic plausibility assessments. If so, Burks is able to agree with us that the Bayesian method of standard inductive logic does presuppose a logically prior plausibility assessment logic. But his pragmatic theory of inductive probability incorporates such plausibility assessment via the calculus of choice. And since these are incorporated into his pragmatic theory of inductive probability, and the latter *inter alia* is part of his characterization of standard inductive logic, the first horn of the dilemma is incorrect: Suitably understood standard inductive logic does incorporate the sorts of pre-probabilistic reasoning needed to avoid the vicious regress.

Technically this enables Burks to escape the dilemma. But it doesn't vitiate the points I was trying to make via the dilemma – viz: That there is a form of reasoning involved in hypothesis assessment in science (plausibility assessment) which is more basic and fundamental than inductive probability assessment, and that an adequate model of scientific reasoning about evidence and hypothesis assessment must analyze and address it. Burks has done so only to a limited degree, in effect telling us that when it is done systematically in accordance with his calculus of choice, it will determine a set of atomic inductive probabilities which could be used to characterize what the scientist or other individual actually did. Thus it really is not the case that standard inductive logic constitutes "the system of rules of inductive inference *actually used*" (p. 103), but rather that standard inductive logic is a rational reconstruction of rational scientific inference. While it does allow for plausibility assessment, and tell us that it can be reconstructively characterized probabilistically, it tells us nothing about how those plausibility assessments are made or what their logic is (beyond being coherent enough to satisfy the calculus of choice). Thus as a rational reconstruction it is seriously incomplete since standard inductive logic can reconstruct nothing of the reasoning wherein one obtains the plausibility assessments which are reflected as the atomic inductive probabilities needed to get Bayes' method going.

* * *

A closely related difficulty is the following. On a Bayesian model it is tacitly supposed that the prior probabilities initially attached to hypotheses ordinarily are sufficiently far removed from the correct probabilities that a sizable number of observations and iterations of the Bayesian procedure will be needed to determine which hypothesis is the correct one. With surprising frequency this is not what happens: Rather hypotheses, laws, and theories are accepted on the basis of single, or just a few observations or experiments. E.g., General Relativity theory has been put to the test only a handful of times, and most of these came after the theory enjoyed general acceptance. (This in part reflects the fact that in the more theoretical portions of physics observation and experiment, while crucially involved, play a somewhat down-played role which is not reflected in the inductive-confirmation picture.) How can such limited testing as a basis for acceptance be accounted for on Burks' Bayesian model? The only plausible way is to suggest that in such cases the scientists assigned prior probabilities which were so close to correct that one or a few iterations of the procedure were sufficient. It we make this move, the question immediately arises: How were these prior probability assessments arrived at? As our previous discussion indicates, Burks' answer is to suggest they came from prior applications of the Bayesian method or else from general experience. I seriously doubt that historical analysis will vindicate the former possibility. E.g., key tenets of General Relativity theory had not been around to obtain prior probabilities via Bayesian means. Thus, it appears that the prior probability problem not only comes up at the very beginning of the scientific enterprise (e.g., in the move from prescience to science) but at very sophisticated stages as well. As to the idea that these prior probabilities come from "general experience," this is so vague that I don't know what to make of it. Given our discussion of prior probabilities above, the most plausible interpretation seems to be that it involves the non-probabilistic plausibility assessments which can be reconstructed as atomic prior probabilities. But, then, this is just to concede that in such one or two experiment/observation cases, it is the plausibility assessment, not the Bayesian-analyzed repeated testing and inductive confirmation, that is at the heart of reasoning about evidence and hypotheses. This increasingly strengthens the contention that scientific research and the evaluation of hypotheses centrally involves plausibility assessments which cannot adequately be captured by the Bayesian method or Burks account of prior probabilities and thus that Burks' inductive-confirmation model inadequately captures or analyzes the role of observation and experiment, and the ways of reasoning about them, that actually characterizes the evaluation of hypotheses in science.

*   *   *

Shapere's work lays the basis for one final challenge to Burks' claim that "*the* system of rules of inductive inference actually used and aspired to by the practicing scientists [is] *standard inductive logic*." (p. 103; italics added). This view implies that there is a fixed, stable, set of rules for reasoning about evidence and the evaluation of hypotheses adequate for and employed by all of science. One of the more important historical findings resulting from Shapere's work is that as sophisticated science proceeds it develops improved patterns of reasoning not previously employed in ways that are conditioned by the content of science, and that much of scientific progress consists in such development of increasingly subtle improved patterns of reasoning (e.g., ones which enable us to assess an hypothesis on the basis of one or a few observations or experiments). As he likes to put it, "We learn how to learn as we learn." Further, he sees no good philosophical grounds for supposing that there will be some ultimate, fixed set of reasoning canons for science; rather there is no reason to suppose this process of evolving new improved canons of reasoning is not open-ended and ongoing. Thus, there is good reason to reject the view that there is a fixed system of rules for scientific reasoning. And if we reject that, we must reject Burks' basic contention that standard inductive logic is that fixed system of rules.

*   *   *

Collectively, the above discussions lead me to conclude that (a) plausibility assessment plays a central role in scientific reasoning about evidence which standard inductive logic does not capture; (b) that standard inductive logic thus is not a good descriptive model of actual scientific reasoning about evidence and hypotheses; (c) because of the dilemma discussion given above, any inductive confirmation analysis of science must be based on a logically prior theory/model of plausibility assessment which must be non-probabilistic and non-inductive; and (d) there is no good reason to suppose that there is a fixed set of inference canons which will exhaust the range of successful reasoning patterns

which science has and will continue to develop. Thus standard inductive logic seems to be a seriously defective model of scientific reasoning
about evidence and hypotheses.

III

Given these defects in the standard inductive logic model, does it follow that *Chance, Cause, Reason* is beside the point and fails to make
any significant contribution to philosophy of science? I think not, and
in the remainder of this paper will try to make a case for the value and
importance of Burks' work on inductive logic.

Burks' various analyses and theories presented in *Chance, Cause,
Reason* are put forward as models. I have been arguing that standard
inductive logic (as analyzed by Burks) is not a good *descriptive* model
of actual scientific reasoning with respect to evidence and hypotheses.
But it does not follow that it is not a good model for other, nondescriptive, philosophical purposes. For example, I mentioned previously that science evolves increasingly efficient canons of reasoning
which allow the economical assessment of hypotheses on bodies of data
that would be insufficient on Burks' Bayesian inductive-confirmation
model (unless the prior probabilities were virtually on target). Shapere
and others have tried to analyze and present models of actual such
reasoning patterns evidenced in science.[11] The generation of such an
open-ended variety of reasoning patterns raises some interesting, new
philosophical problems concerning the evaluation, assessment, and explanation of such reasoning patterns. Specifically, not all reasoning patterns employed by science are or will be good reasoning patterns in the
sense that hypotheses which are accepted using them are very likely to
be correct. In the evaluation of inference patterns as good or bad, it is
not sufficient to just rely upon the historically-oriented philosophers of
science's intuitive judgments of what constitutes good reasoning. What
we would like to do is to be able to give philosophical explanations of
why particular patterns of reasoning developed by science are good
ones (in the sense given above) and others are not. I believe that
standard inductive logic as modelled by Burks can play an important
role in such philosophical assessments and explanations – in ways that
are not compromised by the descriptive inadequacy of these models.

To show how this is possible, I am going to have to involve myself in
a number of controversies within epistemology and philosophy of

science. I will rather dogmatically take positions on these, since this is not the appropriate place for me to defend my views on these. Rather my attempt will be to sketch a plausible but controversial epistemological view about scientific knowledge and show how Burks' analysis of standard inductive logic could be incorporated into it to provide the sorts of philosophical explanations of reasoning patterns we are concerned with. I do not mean to suggest that the philosophical value of *Chance, Cause, Reason* requires accepting the controversial epistemological position I sketch. Rather, my discussion is intended to be *illustrative* of the *kinds* of philosophical roles Burks' analysis can perform despite the inadequacy of standard inductive logic as a descriptive model of scientific reasoning. These caveats aside, let me turn to sketching an epistemological view.

* * *

Let us begin with the traditional suppositions that knowledge is justified true belief and that truth is to be construed in some correspondence sense. Let us further suppose that a causal and cognitive interaction of perceivers plays a crucial role in obtaining empirical knowledge of the world. Specifically, perceptual knowledge is obtained as follows: The human perceiver possesses receptors (eyes, ears, touch sensors, heat sensors, etc.), extensive neurological apparatus, various cognitive skills (including the ability to think thoughts in response to sensory stimulation and adopt doxastic propositional attitudes towards those thoughts) all of which are highly organized together via something akin to what Burks terms the "innate structure-program complex" (pp. 612, 632ff, 643–47) and Sellars terms the "brute organization" so as to be able to causally interact with the external world and respond cognitively and doxastically towards it. When a person perceptually comes to know something about the world (That A is B), he is caused to undergo a sensory experience as a result of looking (hearing, tasting, touching, etc.) at something under circumstances where he *couldn't have had* that experience unless A were B, as a result of having that sensory experience thinks a thought to the effect that A is B, and doxastically accepts that thought (i.e., believes that A is B). These are, I maintain, sufficient for perceptually coming to know that A is B, and enable us to obtain the ordinary kinds of common garden singular knowledge we believe humans routinely obtain from pre-school age on.[12]

While the above account does fit the knowledge as justified true belief view (the justification being undergoing the appropriate causal/cognitive/ doxastic processes as specified above), to avoid extreme skepticism it must deny another traditional tenet, the K-K thesis. This thesis says that

'S knows that P' entails 'S knows that S knows that P'

and usually is analyzed as being equivalent to

'S knows that P' entails! 'S knows that P is true and
S knows that S believes that P and
S knows that S is justified in believing that P'.

But on the above account it will be false that ordinary knowers in common-garden circumstances will know that they are justified in believing that P since such knowers (e.g., a six year old or a Socrates) will lack the requisite knowledge of the complex causal/cognitive/doxastic process which constitutes the belief's justification. Indeed such knowledge can only come at a relatively sophisticated scientific stage which ultimately must be parasitic on pre-scientific common-garden perceptual knowledge. So we reject the K-K thesis.[13]

A corollary to our rejection of the K-K thesis is that the evidence we adduce in defending a claim to know that P generally will be different from that which actually justifies the belief that P (the latter being the complex causal/cognitive/doxastic process S underwent, the former being the reasons one gives if the claim that P is challenged – e.g., "I saw it under clear observational circumstances"). Indeed, while beliefs always must be justified (in the way I've indicated) to be knowledge, we generally do not demand any justification be given for common-garden knowledge claims unless we have specific reasons for doubting them; rather, absent such specific doubt, we accept them at face value. I speculate that the reason we do is that, absent duplicity, common-garden knowledge claims have a high likelihood of being true – the reasons for this involving the close intertwining of perceptual processes with language acquisition wherein learning a language involves learning to apply it correctly to the world we perceive. That is, learning one's first language and learning to know are inextricably intertwined in ways that make our singular common-garden knowledge claims usually correct.

We also develop means for obtaining generalized empirical

knowledge. The empiricist view is that we obtain it inductively from our singular knowledge. I have no doubts that induction by simple enumeration or some intuitive precursor ("All crows are black because every crow I've seen, and I've seen a lot, was black.") plays an important and frequent role in the *defense* of common-garden generalized knowledge claims; and unlike singular common-garden knowledge claims we do not take general ones at face value and typically do demand some justification be given.[14] However, I seriously doubt that induction by simple enumeration or standard inductive logic constitutes the actual justification which makes the belief of a true generalization be knowledge. Rather, I think that the justification for generalized knowledge rests in the same complex causal/cognitive/doxastic processing that enables singular knowledge.[15] Whatever the means whereby our general beliefs are justified and thus constitute knowledge, standard inductive logic is incapable of supplying that justification. For as our prior probabilities discussion established, standard inductive logic presupposes a logically prior plausibility assessment which almost certainly is rooted in the complex causal/cognitive/doxastic processes of perception and empirical knowledge. Thus it seems to me that standard inductive logic is a suitable means for evaluating or defending knowledge claims, but does not provide the justification required for our true beliefs to constitute knowledge.[16]

Let us now turn from common-garden knowledge and knowledge claims to scientific ones. Precisely the same epistemic means are involved in such knowledge – perceivers interacting via a complex causal/cognitive/doxastic process with the external world. Only here the causal links between the objects we observe and our receptors are far more complex – e.g., photons emitted from the interior of the sun are caught by a satellite photon detector, which telemeters signals to earth-bound computers, where the signals are computer enhanced and printouts, optical displays, etc. are produced which then are looked at by human eyes which result in sensory experiences, thoughts, and the belief that *P* about the sun's interior, where the observational circumstances have been so contrived that we couldn't have had those sensory experiences unless *P* were true of the sun. Thus we obtain observational knowledge that *P* about the sun's interior. The knowledge is obtained in exactly the same way that I know my dog is sleeping on the sofa – only the subject-matter is more remote from me and the causal processes involved are far more complex.

Just because what we observe in science usually is so much more remote and accessible only by contrived highly complex causal processes, we have good reasons not to take singular scientific observational knowledge claims at face value in the way that we do common-garden ones. In general we require that scientific knowledge claims – be they singular observational claims or general ones involving causal necessity – be accompanied with some justification: Experimental designs and procedures are to be described, as are statistical and other evaluation procedures employed. Precisely what is to be described depends on the scientific discipline and its stage of development, and will vary among these. But the general requirement is that the justification demanded be such that, given the discipline's extant domain of knowledge, the background knowledge (e.g., about instrumentation techniques) it accepts, its methodological standards and its accepted canons of reasoning, it is highly likely the knowledge claim is correct.

I cannot stress too much the extreme variability between disciplines and over time in these factors which affect what justification is required by a scientific discipline. And the fallibility of these also must be stressed. Given a poor methodology, inaccurate background "knowledge," or bad canons of reasoning a science may fail to sanction correct knowledge claims[17] or, even worse, incorrect knowledge claims may be accepted. Here we need to return to our earlier observation that it is characteristic of sophisticated sciences to develop new and "improved" canons of reasoning which, frequently, are remarkably efficient in the paucity of data needed for a knowledge claim to pass methodological muster and be accepted – in ways that far exceed the data efficiency of the repeated instance-confirmation standard inductive logic. Other than these and standard inductive logic being alternative means to assess scientific knowledge claims and decide which ones to accept, are there any important connections between them?

Burks' statement that "the logic of inductive arguments studies the most general, universal, and fundamental types of inductive arguments, leaving the more detailed and specific forms to the individual sciences and statistics." (p. 17) suggests there is. Specifically, in light of our foregoing discussion it suggests that the more efficient canons of reasoning a science develops (which tend to be highly specific in their applicability) are the more specific forms Burks mentions, and that standard inductive logic augmented with a logic of plausibility assessment constitutes one general albeit inefficient method for assessing general knowledge claims involving causal necessity which could in principle be used whenever these more specialized canons are used. *Chance, Cause, Reason* does mount a quite persuasive defense of the latter claim provided one accepts Burks' presuppositions of induction – which I am willing to do.[18] Thus I am willing to accept the claim that given unlimited resources and computers capable of making manageable the massive amounts of data that would be required all of science in principle could be done using just standard inductive logic and the Bayesian inference model together with a logic of plausibility assessment.

Does this mean, then, that these special improved canons of reasoning science so frequently develops are just special case applications of standard inductive logic? No, for as I argued above it is plausibility assessment, not standard inductive logic, which bears the main evaluative burden in many of these canons. Adequate understanding of scientific reasoning about evidence will require investigating, and possibly developing a logic of, non-probabilistic plausibility reasoning. But absent such a logic there is a place for Burks' standard inductive logic – in the assessment of canons of scientific reasoning and explaining why good ones succeed and bad ones don't. Suppose a canon of reasoning says only $H_1, \ldots H_n$ are plausible. We can set up a Bayesian instance confirmation model with $H_1, \ldots, H_n, \sim (H_i \& \ldots \& H_n)$ as the candidate hypotheses. Then, possibly by computer manipulation, using existing and hypothetical data consistent with available knowledge, we vary the prior probabilities and see whether the instance-confirmation model yields the same assessments as does the canon of reasoning. If it does we have good reasons for accepting the canon as a good one, and have lain at least part of the basis for explaining why it is a good canon to follow. If they do not, several possibilities arise. First, it may be that $\sim (H_1 \& \ldots \& H_n)$ wins out. In this case, we know that the plausibility assessments provided by the canon are defective (in the sense of being incompatible with the available background and domain knowledge of the discipline) and lay a basis for explaining why. Second, it may be that different of the $H_1, \ldots, H_n$ pass muster under the Bayesian and the canon evaluations. Then we know the canon is incorrect and we have a starting basis for investigating and hopefully explaining why the canon fails. Third, since finitary constraints require truncating the

Bayesian process short of the confirmation-theorem's "in the limit," we have the possibility that for some prior probability assignments the Bayesian results are compatible with the canon and for others they are incompatible. In this case we will have to analyze, on other grounds, the reasonableness (plausibility) of the various prior probability assignments. Hopefully such assessment will enable us to restrict attention to just the plausible prior-probability assignment cases in such a manner that one of the other possibilities already discussed results.

Thus, if one accepts the sort of epistemological view I've been sketching here, there is an important philosophical place for standard inductive logic and Burks' analysis of it. It is in evaluating, explaining, and understanding the way general knowledge claims are justified and assessed, but not in the analysis of the role of evidence in obtaining general knowledge. It also constitutes a general means for justifying and defending knowledge claims, although it is not one that all of science in fact uses. One does not have to accept the epistemology I've outlined here to grant standard inductive logic and Burks' analysis of it these valuable philosophical roles. For the same roles will be available on virtually any epistemology which allows the justifications used to assess knowledge claims to be distinct from what constitutes the justification required for true belief to be knowledge. Of course standard inductive logic has a different role to play in epistemologies which make induction be the justification required for true belief to be general knowledge, but I am convinced that no such epistemology is defensible and compatible with the view that science as we know it is capable of, and routinely does, produce realistic knowledge of the world. Indeed, I am convinced that standard inductive logic will have the valuable philosophical roles I've been urging here in any viable epistemology.

IV

Is science really inductive? In this paper I've tried to argue that standard inductive logic is not a good descriptive model of science although some branches of science (e.g., behavioral sciences) do reason in accord with it; thus there is an important sense in which Burks' claim that standard inductive logic is the system of rules of inference actually used and aspired to by practicing scientists is incorrect. Moreover non-probabilistic plausibility assessment underlies the applicability of standard inductive logic, is logically prior and more basic to reasoning about evidence, and such plausibility assessment often plays a more important role in scientific reasoning about data than does standard inductive logic. Nevertheless, I believe standard inductive logic and Burks' analysis of it provides an important philosophical tool for assessing, analyzing, and explaining the successes and failures of the canons of reasoning which science does employ, and I have tried to show at some length how it can do this and have a valuable and important place in a comprehensive epistemology of science. Indeed, properly understood and deployed it is congenial with, and has a role to play, in my own epistemological efforts.

Professor Frederick Suppe
Department of Philosophy
University of Maryland

NOTES

1. Burks (1977). Unless otherwise indicated, all page references are to this volume.

2. 'Inductive' here and throughout this paper usually is meant not in the looser common sense of 'non-deductive' but rather in the stricter, technical sense of inductive logic wherein probability theory plays a central role.

3. Such a construal is especially plausible since in Chapter 5 Burks analayzes inductive probabilities in terms of beliefs and actions – willingness to bet in the face of uncertainty – in such a way that inductive probabilities are manifestations of behavior, not vice versa.

4. For defense of this claim, see Suppe (1977), p. 707, note 241.

5. Cohen (1977) p. 340. Section 5 of that paper contains an extended discussion of falsification of texts in translation to make scientists conform to approved philosophies of science.

6. Cf. Lakatos (1970). For a detailed summary and critical evaluation of Lakatos' views see Suppe (1977), pp. 659–670, *passim.*

7. Admittedly, this suggestion leaves no room for insightful new opportunities resulting from "surprising discoveries" in the testing which suggest new moves for designing the next theory to enter the Lakatosian problem shift. However, the positive and negative heuristics Lakatos allows are so constipative that, on his analysis, this seems virtually precluded. In any case, pursuing the line followed below, on a Burks's Bayesian analysis this can be handled by construing the "none-of-the-above" clause as gaining an increased probability – and so can be accommodated – provided the "none-of-the above" hypothesis is given a non-zero prior probability.

8. Meeting the mutually-exclusive and jointly exhaustive competing hypotheses requirement for Bayes' method typically is feasible only by listing a set of candidate alterna-

tives, together with the non-zero-probability denial of the disjunction of these (a "none-of-the-above" clause). Burks himself invokes this gambit in (1977) pp. 85ff.

9. Much of the above is based upon unpublished work of Shapere's. For a fairly comprehensive (but now dated) published account of Shapere's relevant work, see my (1977) pp. 682–704; my present account is based on works of Shapere discussed therein.

10. One feature of Burks' pragmatic theory and the related personalistic theories is that probabilities are reflections of behavioral dispositions and beliefs rather than the traditional view of basing beliefs and decisions upon available probability assessments.

11. Cf., e.g., Suppe (1977), pp. 541, 545–546, 553, 697.

12. This view is similar to that of F. Dretske (1969) and A. Goldman (1967). I believe it can be developed in ways that avoid criticisms that have been made against their versions. Inference can play a role in obtaining empirical knowledge, and a full account would have to allow for inference-aided perceptual knowledge; for simplicity I ignore this possibility here. See my (1989), chapter 12.

13. For expanded discussion of the K-K thesis issue, and some of the other epistemological moves I am making, see my (1977) pp. 716–728, portions of my (1974), and part IV of my (1989).

14. Unless, of course, the generalization is common knowledge. Before it becomes common knowledge, justification would be demanded.

15. Specifically I believe that we can know generalizations involving causal necessity on the basis of singular perceptual episodes. The argument in support of this goes roughly as follows: We are able to see that objects possess dispositional properties (e.g., salt is soluable) on the basis of suitable single perceptual episodes. But dispositions involve causal necessity, so we are able to perceive through single instances that certain things are causally necessary. The account of singular perceptual knowledge sketched above can be embellished to accommodate this ability perceptually to obtain dispositional knowledge, and essentially the same means of accommodation can be extended to generalizations involving causal necessity. However, detailed development of this claim is too extensive to give here, and so I will not make acceptance of it crucial to my main discussion.

16. Note that Burks's approach of rooting standard inductive logic in our dispositions towards action (willingness to bet in the face of uncertainty) as conditioned and constrained by our innate structure-program complexes is quite congenial to the views I'm speculatively propounding here.

17. Under positivistic-dominated behaviorist social science methodologies such as Skinner's this was the case for several decades after World War II.

18. Returning to my earlier speculations about the intertwined connections between perception, knowledge, and language acquisition, I further speculate that these proceses would not be possible unless something akin to Burks's presuppositions of induction were correct. Further, I agree with him that if they were not true, science as we know it would be impossible.

REFERENCES

Burks, A. W.: 1977, *Chance, Cause, Reason*, University of Chicago Press, Chicago.

Cohen, I. B.: 1977, 'History and the philosopher of science', in Suppe (1977) pp. 308–349.

Dretske, F.: 1969, *Seeing and Knowing*, University of Chicago Press, Chicago.

Galileo Galilei: 1953, *Dialogue Concerning the Two Chief World Systems*, trans, Stillman Drake, University of California Press, Berkeley and Los Angeles.

Goldman, A.: 1967, 'A causal theory of knowledge, *Journal of Philosophy* 64, pp. 357–372.

Lakatos, I.: 1970, 'Falsification and the methodology of scientific research programmes', in Lakatos and Musgrave (1970), pp. 91–196.

Lakatos, I. and A. Musgrave (eds.): 1970, *Criticism and the Growth of Scientific Knowledge*, Cambridge University Press, Cambridge.

Leinfellner, V. and E. Kohler (eds.): 1974, *Developments in the Methodology of Social Science*, Reidel, Dordrecht.

Newton, I.: 1952, *Optics*, Dover, New York.

Suppe, F.: 1954, 'Theories and phenomena' in Leinfellner and Kohler (1974), pp. 45–92.

Suppe, F.: 1977, *Structure of Scientific Theories*, 2nd ed., University of Illinois Press, Urbana.

Suppe, F.: 1989, *The Semantic Conception of Theories and Scientific Realism*, University of Illinois Press, Urbana.

L. JONATHAN COHEN

BOLZANO'S THEORY OF INDUCTION

I

Bolzano's *Wissenschaftslehre* was published in 1837, although most of it seems to have been written during the decade 1820–1830. John Stuart Mill's *System of Logic* was published in 1843, but had been in gestation or preparation since 1825. Neither author seems to have exercised any influence on the other, and in their views about the fundamental nature of logical and mathematical reasoning they notoriously represented very different trends. Bolzano sought to direct philosophers' attention away from mental processes towards relationships between ideas in themselves and between propositions in themselves, while Mill's logic insisted on a study of the mental process which takes place whenever we reason, of the conditions on which this process depends, and of the steps of which it consists. But in their views about the methodology of natural science the divergences are much more fine-grained. Both assign a central role to the search for causes and both discuss the same basic procedures for the discovery of these. It is just that Bolzano shows a greater sensitivity than Mill does to the inherent difficulties of the enterprise.

Let us look a little more closely at the matter. According to Mill's Method of Agreement (Mill, 1843, ch. III, ch. VIII). 'If two or more instances of the phenomenon under investigation have only one circumstance in common, the circumstance in which alone all the instances agree is the cause (or effect) of the given phenomenon.' According to the Method of Difference 'If an instance in which the phenomenon under investigation occurs, and an instance in which it does not occur, have every circumstance in common save one, that one occurring only in the former; the circumstance in which alone the two instances differ is the effect, or the cause, or an indispensable part of the cause, of the

29

Merrilee H. Salmon (ed.), *The Philosophy of Logical Mechanism*, 29–40,
© 1990 *Kluwer Academic Publishers*.

phenomenon.' Both principles were long established in the inductive logic tradition: precursors of them occur, even before Francis Bacon's *Novum Organum*, in the writings of certain medieval philosophers such as Robert Grosseteste, Albertus Magnus, Duns Scotus and William of Ockham.[1] Not surprisingly, therefore, we find Bolzano accepting that any events which all occur whenever E takes place, and which are always followed by E so that they do not occur when E does not take place, are either a partial or a complete cause of E.

Again, according to Mill's Method of Concomitant Variations 'whatever phenomenon varies in any manner whenever another phenomenon varies in some particular manner, is either a cause or an effect of that phenomenon, or is connected with it through some fact of causation.' So too we find Bolzano remarking correspondingly: 'It is reasonable to suppose that the phenomena which increase and decrease concomitantly with the effect belong to its cause, unless, indeed, we discover that they are its effect; the latter case could be discovered if we found that these phenomena occur somewhat later than the event we wish to explain.'[2] Mill's example was the sun's gravitational pull on the earth, Bolzano's the sun's heating of a spot on the earth.

Finally there is the problem of partial causes. Mill dealt with this in his Method of Residues: 'Subduct from any phenomenon such part as is known by previous inductions to be the effect of certain antecedents, and the residue of the phenomenon is the effect of the remaining antecedents.' Similarly Bolzano argues that the safest way of determining which parts of a complex causal condition are responsible for which aspects of the effect is to trace the causal relations between these parts and aspects when they occur separately (1972, p. 380). But Bolzano's formulation here is – wisely enough – more cautious than Mill's. Bolzano's formulation allows for the fact that a complex causal process whereby C_1 & C_2 & ... & C_n produce the effect E_1 & E_2 & ... & E_n may not always be analysable into a set of simpler processes whereby each C_i causes each E_i. For example, consider E_1 & E_2 & ... & E_n as the normal functioning of a motor-car and C_1 & C_2 & ... & C_n as the motor-car's resources of energy. We may know that the lights and wipers work off the battery, but it certainly does not follow, as Mill's Method of Residues implies, that the residue here, which includes the movements of the pistons in the cylinders, works off the petrol supply alone.

II

A more important difference between Mill and Bolzano, however, is in their expectations with regard to the certainty that should result from this methodology. Mill formulates each of his canons, or methods, in such a way as to imply that it produces conclusive results. No doubt this was because he envisaged the possibility of throwing the whole course of any inductive argument into a series of syllogisms, where the ultimate syllogism has some principle of causal uniformity as its major premiss (1843, ch. III, ch. III). But in this respect Mill diverged from the mainstream of the Baconian tradition. Bacon himself envisaged no connection between inductive and syllogistic reasoning, and explicitly 'rejected' the latter. He thought instead that his method involved setting up degrees of certainty, and his description of how to apply the method in a particular case introduces gradation in two ways – in regard to the variety of features listed in the tables of presence and absence and in regard to the level reached in the pyramid of axioms.[3] Of course, Bacon thought that conclusive certainty was attainable in the end, but only after a long and difficult progress through many intervening degrees of probability. And in this connection admirers of his inductive logic tended to stress that some level of probability which falls short of absolute certainty is all that is normally attainable. At any rate that was the view of the scientists Hooke and Boyle, as well as of the philosophers Glanvill and Hume.[4] Even in Mill's own day J. F. W. Herschel (1830, p. 155), who set an engraving of Bacon's bust upon the title page of his *A Preliminary Discourse on the Study of Natural Philosophy*, insisted on treating the success of any particular inductive investigation as a matter of degree. Experiments, he thought, become more valuable, and their results clearer, in proportion as they possess the quality of agreeing in all their circumstances but one, since the question put to nature becomes thereby more pointed, and its answer more decisive. So R. L. Ellis (1879, I, p. 23), Bacon's most influential editor, was wrong to regard absolute certainty as the distinguishing feature of Baconian induction, and Hacking (1975, p. 76) was equally wrong to assert that 'there is little room in [Bacon's] conceptual scheme for a working concept of probability'. Indeed the very same Latin phrase 'gradus certitudinis', which Bacon uses in the preface to his *Novum Organum* in order to describe the successive stages envisaged by his methodology, is also used by James Bernoulli in his *Ars*

Conjectandi as a definition of the concept of probability (1713, p. 211). It is as if Mill's *System of Logic* had supplied Ellis with his conception of what inductive methodology aims to achieve and thereafter he was blind to any signs that Bacon himself had held a different view.

Bolzano's view on this issue is quite clear. Two factors should affect the confidence with which we expect an observed regularity to continue. One is the number of observed cases, the other is the extent of their variety. What he says (1972, p. 378) is that, if A, B. C, ... are not only to be a partial, but a complete cause of E, then

> The larger the number of cases in which this occurs, and the greater the differences in all other respects, the greater the confidence with which we may assume that the object A or one of the several A, B, C, ... is a complete cause of E. For, the greater the difference between the remaining circumstances under which E took place ... the smaller the danger that some other, hidden object A' or B' was a partial cause in the production of E. Thus somebody can be all the more confident that he gets a headache from drinking coffee, the more often he realizes that he gets this headache when he drinks coffee and the greater the differences in the rest of his activities on these days.

Indeed, Bolzano is quite explicitly concerned to emphasise (1972, pp. 383–4) that the greater reliability of a supposedly 'established' experimental truth, as compared with a hypothesis which we maintain only 'with probability', rests only on the greater number of observations. So Bolzano seems not to have believed, as Bacon did, that conclusive certainty is eventually obtainable in regard to natural uniformities. Rather, the approach to certainty, as evidence of the right kind increases, is an asymptotic one. And at root the reason for this seems to be that Bolzano held the number of circumstances which accompany every event to be infinite, at least if we suppose that he meant an infinite number of mutually *independent* circumstances. For, if every event is accompanied by an infinite number of mutually independent circumstances, then, even if we assumed that every kind of occurrence is correlated with a characteristic kind of cause, it would still be the case that no finite series of observations or experiments could ever suffice to eliminate all but one possible hypothesis in relation to some particular kind of occurrence that is under investigation. Bacon, on the other hand, held[5] that Nature contains only a finite number of 'forms' for experimental scientists to investigate, and this tenet is echoed in J. M. Keynes's principle of limited independent variety (1921, p. 259 ff.).

Does the truth lie with Bacon here or with Bolzano? Perhaps neither has grasped the whole of it. No doubt the logical possibility always remains, in any particular area of investigation, that an infinite variety of circumstances may be relevant to the phenomena in question. But in practice we can only specify a finite list of these in a hypothesis; and, though any such hypothesis is always open to correction or amplification, an investigation that is carried out on its basis can be complete, or incomplete, according to the standard that it sets. So, relatively to that standard, we can have a conception of conclusive certainty, or of mere probability. But no such standard can itself claim absolute validity.[6]

III

Some further interesting differences now emerge between Bolzano's views about induction and those of his English contemporaries.

In the first place it is noteworthy that, whereas Herschel claimed that experiments become more valuable in proportion as they agree in all their circumstances but one, Bolzano argued that the greater the differences between the circumstances in which A is accompanied by E within a series of cases, the greater the confidence that A is causally connected with E. Herschel seems to be arguing that experiments on a particular issue should differ from one another as little as possible, Bolzano that they should differ as much as possible. How is this divergence of opinion to be explained or elucidated?

It is clear how Herschel's thesis can be justified. Suppose that we conduct two experiments in which A is produced, and in both E ensues, while circumstances C and D are both present in the first experiment and both absent in the second. Then we learn thereby that neither the joint presence of C and D, nor their joint absence, suffices to prevent E's occurrence when A occurs. But conceivably C has some neutralising effect and we might find that, if C and D were varied independently, we obtained E under the conditions C & D, C & not–D, and not–C & not–D, but not under the condition not–C & D. For example, if what was at stake was the therapeutic efficacy of a particular drug in relation to a certain kind of disease, and C was constituted by the patient's being over the age of 70 while D was constituted by the patient's having had the same disease on a previous occasion.

In the light of cases such as this it is obvious that, for any set of independently variable factors which we control in our experimental trials, we maximise the amount of information to be extracted from the trials if we try out every possible combination of the relevant circumstances rather than just some of these combinations. Under those conditions each of the various trials will agree with many of the others in all their circumstances but one, so far as the variables actually controlled in the trials are concerned. Of course, there may well be other relevant variables, which are not controlled in a particular series of experiments either because of the cost in time or money that would be involved or because their relevance is as yet unknown. For example, as well as a patient's age and medical history we might need to take into account such factors as his diet, his occupation, the climate of his place of residence, etc. So we cannot yet say without qualification that the various trials will agree with many of the others in all their circumstances but one: there might accidentally be additional differences, since differences might exist in respect of the as yet uncontrolled variables. But again it is obvious that, the more we take account of other relevant variables in the construction of our experiments, the more confidence we can have in any hypothesis that survives them all; and, if as each further relevant variable is taken into account we continue to test out all possible combinations of relevant circumstances, we are bound to get closer and closer to a situation in which it can be said without qualification that each trial really agrees with many of the others in every relevant circumstance save one. For example, if there are n relevant variables, and each of these has r relevant variations, then the maximum possible number of different combinations of relevant circumstances is r^n, and each of these will differ from each of $n(r-1)$ others in respect of only one relevant circumstance.[7]

But, if this is the insight with which we should credit Herschel, what is to be said about Bolzano's claim that the reliability of a hypothesis increases directly, not inversely, as the differences between the circumstances of the various successful tests on it increase? If we interpret Bolzano's claim as just being the opposite of Herschel's, the claim is evidently false. Nor would it serve to suppose that Bolzano was here concerned just to maximise the range of difference within each single variable, e.g. just to test out a new drug on the very old and the very young: this would be a very risky strategy in many cases. Rather, what is worth emphasising is that the range of relevant variables should be a wide one. If we vary very different kinds of known factor, we can have a better hope of catching any hidden factor, through some causal interdependence that it may have with one or other possible combination of the known factors. For example, if we vary age and sex as well as medical history, our experiments might reveal a special effect of the drug that is confined to some women in the middle age-group and that turns out, on an even more refined analysis, to be due to pregnancy. Admittedly, on this interpretation of Bolzano's remark, he did not do well to illustrate it in terms of the width of a coffee-drinker's range of other activities as being important for whether or not the headache comes from drinking coffee. But perhaps that was a badly chosen example in any case. After all, sitting indoors all day and climbing a mountain over 12,000 feet are extremely different activities, and both might well bring on a headache. Of course, they may both bring on the headache because of the same underlying cause, viz. deficiency of oxygen. But that just helps to show why it is that mere width of observable difference is a bad objective to aim at when one varies the circumstances of an experiment. What really counts is the width of the range of circumstance-types that one varies. It is not just the other activities of the coffee-drinker that need to be varied, for example, but also the composition of the air that he breathes, and so on. If, despite his unfortunate example, Bolzano really had this in mind then what he says is both correct and compatible with what Herschel says. In short, the more the experimental tests on a particular hypothesis come to agree with many of the others in all their circumstances but one (because they vary all possible combinations of circumstance for more and more relevant variables), the greater the variety of actual differences that are covered by the series of tests.

IV

If inductive reliability is to be conceived of as a matter of degree, a question arises as to the nature of the function that grades this reliability. In particular, is it a probability-function of some kind that conforms in its logic to principles derivable within the classical calculus of chance?

On this important question Bolzano's answer seems to be an affirmative one. He tells us, (1972, p. 380ff.), for instance, that if a given effect E is probably caused by one or more of A, B, C, ..., then if the number of cases in which A, say, was found to be operating is not larger than half of the total number of observations, then the probability of the assumption that E was caused by A is not large enough to warrant the assertion that it really was that way. But, subject to that qualification, we must favour that assumption which has the highest degree of probability over all the others.

Here Bolzano seems to be favouring what Carnap called 'the straight rule' for evaluating the reliability of a hypothesis. This is a rule that David Hume too had endorsed, when he said that if a person finds by long observation that of twenty ships which go to sea only nineteen return, then $\frac{19}{20}$ measures the probability with which he believes that a ship which goes to sea in the future will return (1888, Bk. I, pt. III, sec. XII). But unfortunately Bolzano did not explain – neither did Hume – how such a measure is to be integrated with the measure of the confidence that is due to the total number of relevant observations that have been made and to the variety of differences in relevant circumstances between those observations.

Statisticians now have various techniques for coping with part of this problem, viz. the size of the sample observed. For instance, this may be reflected in the Neyman-type confidence limits within which the probability is said to hold. But the credit due to the variety of experimental circumstances seems to resist exact measurement. Attempts to construct a measure, as in Carnap's or Hintikka's work, rely very heavily on prior assumptions about the language of science. At some point all expressions of a given linguistic category have to be put on a level with one another in order to generate the required prior probabilities. It is as if, at the appropriate level of complexity, all circumstances in nature have to be deemed equally important. Yet in the actual conduct of scientific reasoning some evidential circumstances, or some families of predicates, may be deemed to be much more important than others in relation to hypotheses of a particular kind. A patient's previous medical history turns out to be more important than the colour of his eyes in relation to most hypotheses about the safety of a drug for medical purposes.

Moreover, where causal hypotheses are concerned, any attempt to distribute appropriately unequal weights between different circumstances encounters a very serious difficulty. The weight to be assigned to a given combination of circumstances cannot be regarded as a mere function of the weights assigned separately to each of these circumstances. Sometimes two such factors may have an explosive potential in combination: sometimes they may cancel each other out. For example, in testing the safety of a drug the volume of the dose and the body-weight of the patient may constitute a much more important combination of relevant factors than do the patient's sex and occupation, even if dosage, body-weight, sex and occupation are all of equal importance on their own. Yet it seems that the weights assigned separately to each of the constituent factors ought to be the ones which affect our evaluations of hypotheses on given test-results, if hypotheses are to be considered more and more reliable to the extent that the tests which they pass incorporate more and more relevant variables. Contrariwise, if the cumulative value of controlling more and more of the relevant features cannot be reflected by any uniform mathematical operation on their supposed weights, then it looks as though this cumulative value is not a measurable quantity and, a fortiori, does not admit of evaluation by a probability-function.

Bolzano did, however, make one interesting suggestion in regard to a probabilistic measure for eliminative induction. He assumes that each possible hypothesis about the cause of a phenomenon under investigation may be decomposed into a number of atomic and independent components, and that 'the degree of probability' of each of these components may be roughly estimated. Then the hypothesis that shows the greatest value for the product of the probabilities of its various components is the most probable of all, with a preference being given – other things being equal – for the hypothesis with the smallest number of atomic components. Thus, if the phenomenon E can be explained in two ways, namely by the accidental conjunction of three objects a, b, c, and also by the accidental conjunction of four objects d, e, f, g, and if these seven objects are all of equal probability, then the first explanation must be preferred to the second. Moreover, Bolzano held, this point is closely connected with the problem of simplicity. People sometimes suggest, he said (1972, p. 384) that the reason why a simpler hypothesis is preferable to a more complex one is to be sought in a presupposition that nature always follows the simplest rules. In Bolzano's opinion however we do not need a special presupposition to justify this preference, since it follows from the rules of the probability calculus itself. Where each of a set of propositions is equiprobable with, but independent of, each of the others, the probability of a conjunction of two or more of them will be equal to the product of the probabilities of its conjuncts and thus its probability will vary directly with its simplicity, if its simplicity may be thought of as varying inversely with the number of its conjuncts.

To those familiar with Popper's arguments for identifying the simplicity of a hypothesis with its *im*probability, this doctrine of Bolzano's may at first sight appear rather paradoxical. But in fact it does not necessarily conflict with Popper's conception of simplicity. The point is that Bolzano and Popper differ substantially from one another in regard to the nature of the hypotheses that they are considering. Bolzano is considering, as a statement of the cause of a given event E, a hypothesis of the form 'A & B & C', whereas Popper is considering, for that kind of situation, a hypothesis of the form 'If A & B & C, then E'; and there is no reason at all to suppose that a high probability for a particular proposition p is inconsistent with a low probability for a conditional ('If p then q') that has p as antecedent. Indeed, it is clear that 'If A, then E' makes a bolder claim in Popperian terms than does 'If A & B, then E' since, other things being equal, it has more falsifiers. Specifically, while both 'A & B & not–E' and also 'A & not–B & not–E' would falsify 'If A, then E', only the former of these two statements would falsify 'If A & B, then E'. So, viewed like this, Bolzano and Popper are actually saying the same thing.

But, if that is so, Bolzano's method of evaluating eliminative induction suffers from the same weakness as Popper's measure of what he called (1959, p. 251ff.) 'corroboration'. It assumes that experiments have already taken us as far as they can towards eliminating potential explanations and that we now have to choose between the remaining, uneliminated hypotheses. In Popperian terms, our tests have been as severe as possible and we now have to choose between the hypotheses that remain unfalsified. What has not been taken into account here is the precise level of severity that was attained in the tests, i.e. the number and importance of the relevant variables that were controlled in the experiments. So the Bolzano-Popper criterion of simplicity might be effective as a method of comparing the values of different hypotheses in the light of experimental evidence if it can safety be assumed that every series of experimental tests is always conducted at precisely the same level of thoroughness. But this assumption is in fact a rather risky one, since considerations of time or available resources frequently force us to be content with what we know to be sub-optimal tests, and in any case new knowledge about relevant variables may create new standards of optimality in this connection, i.e. new standards of thoroughness for experimental tests. It follows that Bolzano's interesting idea here about the relationship between probability and simplicity does not enable him to construct a working measure of inductive support where support is gauged by the variety of relevant evidential circumstances, and it remains very doubtful whether any probabilistic proposal for such a measure could ever be satisfactory.[8]

L. Jonathan Cohen
The Queen's College
Oxford University

NOTES

[1.] Cf. Weinberg, (1965), p. 121ff.
[2.] B. Bolzano, *Theory of Science*, ed. R. George, 1972, p. 379. Cf. F. Bacon, *Novum Organum*, 16, bk. II, sec. XIII.
[3.] Cf. Cohen (1980), pp. 219–231.
[4.] For references see Cohen (1980). Mill did discuss probability in this connection: Cohen (1988).
[5.] Cf. 'Of the Advancement of Learning', in *The Philosophical Works of Francis Bacon*, ed. J. M. Robertson, 1905, p. 95.
[6.] For a fuller discussion of this issue see Cohen, (1977), p. 345ff.
[7.] The details of all this are worked out in L. Jonathan Cohen. (1970), p. 35ff; cf. also (1977), p. 129ff., and especially pp. 143–4 for a complication that may affect Herschel's method adversely in certain cases.
[8.] See further Cohen (1977), p. 188ff.

REFERENCES

Bacon, F.: 1879, *The Works of Francis Bacon*, ed. J. Spedding, R. Ellis and D. N. Heath, Longman, London.
Bacon, F.: 1905, *The Philosophical Works of Francis Bacon*, ed. J. M. Robertson, G. Routledge and Sons, London.
Bernoulli, J.: 1713, *Ars Conjectandi,* Thurnisiorum, Basle.
Bolzano, B.: 1972, *Theory of Science* (ed. R. George), University of California Press, Berkeley and Los Angeles.
Cohen, L. J.: 1970, *The Implications of Induction*, Methuen, London.
Cohen, L. J.: 1977, *The Probable and the Provable*, Oxford University Press, Oxford.
Cohen, L. J.: 1980, 'Some historical remarks on the Baconian conception of probability', *Journal of the History of Ideas* XLI, pp. 219–231.
Cohen, L. J.: 1988, 'Who introduced mathematical probability into the philosopy of induction?' in E. Scheibe (ed.) *The Role of Experience in Science*, de Gruyter, Berlin, pp. 75–81.
Ellis, R. L.: 1879, General preface to Bacon's philosophical works, in Bacon (1879).
Hacking, I.: 1975, *The Emergence of Probability*, Cambridge University Press, Cambridge.

Herschel, J. F. W.: 1830, *A Preliminary Discourse on the Study of Natural Philosophy*, Longmans, London.

Hume, D.: 1888 *A Treatise of Human Nature*, ed. L. A. Selby-Bigge, Oxford University Press, London.

Keynes, J. M.: 1921, *A Treatise on Probability*, Macmillan, London.

Mill, J. S.: 1843, *A System of Logic*, Longmans, London.

Popper, K. R.: 1959, *The Logic of Scientific Discovery*, Harper and Row, New York.

Weinberg, J. R.: 1965, *Abstraction, Relation and Indication*, University of Wisconsin Press, Madison.

BERNARD P. ZEIGLER

CELLULAR SPACE MODELS: NEW FORMALISM*
FOR SIMULATION AND SCIENCE

CELLULAR AUTOMATA AND THE LAWS OF NATURE

Arthur Burks, when explicating what is meant by a "Law of Nature" states that it has three main features: uniqueness, modality, and uniformity. (Burks, 1977) (p. 425) Very briefly, "uniqueness" means that the law makes unequivocal assertions, "modality", that it holds not only for actual, but also possible, situations, and "uniformity", that it applies uniformly to all points in space and time. Proceeding to formalize these concepts, Burks presents cellular automata as model systems in which such laws of this nature are readily comprehended. (Burks, 1977, p. 562). A cellular automaton[1] is characterized by specifying three parameters: a set of states S, a neighborhood N, and a transition function T. The interpretation of the triple $<S,N,T>$ is as follows:

One imagines a checkerboard stretching out towards infinity in both north-south and east-west axes. At each square is located a cell with state set S. The neighbors of a cell located at square (i,j) (where as will be apparent "neighbors" is intended in an informational sense) are simply determined from the neighborhood N: in fact, N is a finite ordered set of integer pairs and the neighbors of this cell are located at the squares obtained by adding (i,j) to each pair of N (vector addition). Now imagine a global state of this system pertaining at some time instant t, i.e., assign to each cell a state from the set S. Then the system will move to a succeeding global state at time instant t+1 which is determined as follows: Simultaneously, for each cell apply the transition functionn T to the states of its neighbors and let the result be the state of the cell at t+1.

Within such a framework, Burks identifies a law of nature as asserting a property of the transition function. Indeed, further placing a contiguity requirement on such a law – that it assert a Humean direct spatio-temporal cause/effect relation – it can be identified with a statement of the form:

If the states of the neighbors of a cell satisfy P at time t then the state of the cell will satisfy Q at time t+1.

41

Merrilee H. Salmon (ed.), *The Philosophy of Logical Mechanism*, 41–64,
© 1990 *Kluwer Academic Publishers*.

A somewhat less general way to capture this idea, is that a law of nature reveals what the transition function T is for one or more of its arguments.

Let us see how cellular automata universes exemplify the three features of natural laws. "Uniqueness" is embodied in the transition relation being a function, i.e., specifying a unique next state for a cell as determined by the present states of its neighbors. In the system theoretic context this feature is called *local determinism*, and also implies *global determinism*: every global state of the system has a unique successor.

"Modality" is exemplified by the fact that there are many possible initial global states of the system, each initiating a state trajectory which can be postulated to be the actual history of our universe. Only one can in fact represent the actual universe while the others represent "causally possible" universes, i.e., state histories that could have happened according to the laws of nature – the transition function – but did not because the initial state was as it was. Burks formalizes this "could have happened" (counterfactual) feature by means of the non paradoxical causal implication operator of modal logic:

*) For all cells c and times t, the neighbor states of c satisfy P at t nonparadoxically causally implies that the state of c satisfies Q at t+1.

Such an implication asserts a transition relation which holds for all causally possible universes including the postulated actual one.

"Uniformity" is embodied in the universal quantification over space and time of the foregoing scheme. More specifically, every cell no matter where located, has the same state set and the same transition function. And while neighborhoods are not literally the same, they are all "isomorphic" in the sense that any one is a translation of any other (or indeed a translation of the set N which turns out to be the neighborhood of the origin (0,0)). Thus a cellular automaton specifies a system which is both discrete and invariant in both time and space.

While cellular automata exemplify the features of natural laws, it does not follow that they are the only formal models which do so, or indeed, that these features are mandatory for any formalism to be useful for science. In this article we shall examine some of the questions raised by the foregoing observations. We shall introduce a class of cellular models called discrete event cell spaces which also exhibit some features of natural laws but which are more convenient to employ than the cellular automata in many simulations of natural systems. What constitutes "convenience" in this context is interesting in itself and throws some light on the larger philosophical issue of adequacy of formalisms for expressing real world relationships. A second kind of issue raised arises from our demonstration that in the discrete event cell space global, but not local determinism, holds. Accordingly, laws of nature must be phrased with elliptical, rather than (non-elliptical) causal implication. This result opens up a whole set of questions concerning the role of laws of nature in model construction and behavior prediction.

DISCRETE EVENT CELL SPACE MODELS

Discrete event cell space models preserve the spatial discreteness of cellular automata as well as their space and time invariance. However, the time base is no longer discrete but is continuous, that is to say, there is no intrinsic time step for such models. Events, i.e., cellular state transitions, may occur at irregularly spaced intervals, not necessarily synchronized to the beat of a clock as in the cellular automaton case. These events are all scheduled to occur as a consequence of the actions of cells: a cell may schedule events to occur to itself as well as to its neighbors, and these events may in turn schedule other events, and so on.

Next event models form a subclass of the more general discrete event models – other subclasses of the latter are the activity scanning and process interaction models.[2] Activity scanning and process interaction cell space models can be defined by combining the cell space structure with the appropriate dynamic structure, as we now shall do for the next event models.

A *Next Event Cell Space* (NEVS) is specified by a quadruple $<S,N,T,$ SELECT$>$, where the first three parameters bear a resemblance to those of the cellular automaton, but their interpretation is quite different. As before, one imagines an infinite checkerboard, with each square containing a cell. However, the state of any cell is now a pair (s,σ) where s is an element of S and σ is a non negative real number (it will be useful to allow σ to take on the value of infinity ∞ as well). A cell in state (s,o) will remain in "sequential" state s for a time σ before undergoing a self induced transition. However in the meantime its state may be altered as a result of some other cell's action. Thus s and σ are called the *sequential state* and *time left* components of the total state (s,σ). If cell c is in state (s,σ) at time t, this can expressed as:

> *)   Cell c is in sequential state s and is scheduled for a transition in time σ (or at time $t+\sigma$).

As before the neighborhood of a cell is computed by adding its coordinates to the prototype neighborhood N (a finite set of pairs as before). When a transition event occurs to a cell, it and and its neighbors undergo an immediate state change as dictated by the transition function T. Thus, in contrast to the automaton case, T takes a list of pairs (s,σ) – the total states of the cell and its neighbors just before the event – and produces a list of such pairs – giving the total states of these cells just after the transition.

In contrast to the cellular automaton, the updating of an NEVS model takes place at irregularly spaced computation instants. Let t_i be such an instant, then its successor t_{i+1} is computed as follows:

> Imagine the global state at time t_i to be a list of pairs (s,σ) containing the total states of all the cells at this time. Let σ^* be the minimum of all the σ's on the list. Then $t_{i+1} = t_i + \sigma^*$.

The global state at time t_{i+1} is computed as follows: Call the cells whose σ's at t_i were equal to σ^* the IMMINENT cells. These cells are all scheduled for a transition event at t_{i+1}. If there are more than one, apply the SELECT function to choose one of the IMMINENT cells call it c^*. (Thus the SELECT function provides the desired tie breaking rules when more than one transition event is scheduled for activation at the same time.) The chosen cell c^* carries out the transition function as indicated above resulting in a new global state which is related to that pertaining at t_i as follows:

The total states of the neighbors of c^* are as dictated by T; the total state of every other cell is converted from (s,σ) to $(s,\sigma-\sigma^*)$.

EXAMPLES OF DISCRETE EVENT CELL SPACE MODELS

Some examples of next event cell space models will be given to illustrate their operation and versatility. They will also serve to point out the differences in expressive capability relative to cellular automata.

Example 1:   motion of a single body

Let us begin simply by expressing the motion of a single object in a NEVS. The object is to move to the right, at a speed of one square every MOVETIME seconds. An appropriate NEVS is defined as follows:

Every cell has two states, one to indicate the presence of the object and the other its absence: thus $S = \{0,1\}$. The absence indicator 0 is a *passive* state i.e., a cell in this state will never become active of its own accord. This is represented by the total state pair $(0,\infty)$ which indicates that the time left for a cell in sequential state 0 to change state is inf. On the other hand, the presence indicator 1 is an *active* state (opposite of a passive state); indeed, it is the source of motion in this example.

To get the object to move we must arrange for a cell in state 1 to cause its right adjacent (physical) neighbor to enter state 1 after staying in this state for MOVE_TIME seconds. Thus for the neighborhood N we require only the two integer pairs $\{(0,0), (1,0)\}$ – adding (i,j) to each we obtain $\{(i,j), (i+1,j)\}$ the co-ordinates of the cell at (i,j) and its right neighbor.

In the light of the above comments, the transition function can be expressed as follows:

> *)   set your own sequential state to 0
> and passivate (set your own time-left to ∞)
> set right neighbor state to 1 and
> schedule a transition event there in MOVE_TIME
> (set neighbor's time_left to MOVE_TIME)

The SELECT function is arbitrary in this example since at most one event is scheduled at any time.

Starting from an initial global state in which all but one cell are in state $(0,\infty)$, the exception being in state $(1, \text{MOVE\_TIME})$, successive global states are generated every MOVE_TIME seconds in which a 1 is seen to travel horizontally to the right.

Example 2:   Traffic congestion

Note that the above model does not specify what happens in the case of collisions – when two or more objects want to move into the same square. The present example exemplifies this phenomenon. Consider

cars which move to the right at a constant speed when possible (trav-
elling in an east bound lane for example). When the car immediately
ahead obstructs progress a car seeks to pass preferably to the right, but
also to the left if the first is not possible. The patterns of traffic congestion which develop can be studied with NEVS models of which the following is prototypic:

The state set $S=\{0,1\}$ to represent absence and presence of a car as
in Example 1. We put in the neighborhood N all cells immediately
above, below and to the right of the center cell to allow for the possibility that a car may move in these directions if direct progress is
blocked. Let N be ordered in the order of preferred motion as illustrated in Figure 1. Then the transition function T can be expressed
as follows:

*)   Find the first unoccupied neighbor cell
    (scanning in the preference order)
    set this cell to state 1 and schedule a
    transition event there in MOVE_TIME
    set your state to 0 and passivate.
    leave all other cells unaffected
    (let their sequential states
    states remain unchanged and subtract
    your time left (σ^*) from theirs)[3]
    if no unoccupied neighbor exists
    reschedule yourself in time CHECK_AGAIN_TIME
    leave all other cells unaffected.

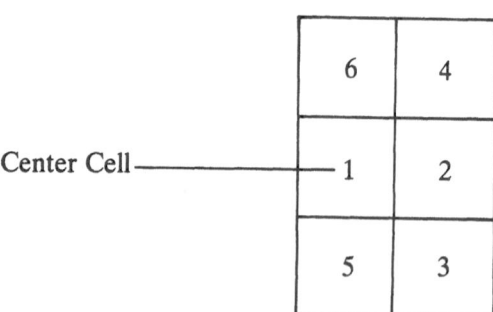

Fig. 1.   Neighborhood for traffic congestion model.

The SELECT will determine which of two or more cars contending for the same square at the same time will move into it. Thus, the order in which IMMINENT cells are selected *does* make a difference in this example. SELECTION, based on "right of way" principles for example, therefore incorporates essential model hypotheses in this case.

An initial global state in this model consists of putting a finite set of cells in sequential state 1 with a time_left σ between 0 and MOVE_TIME; all other cells are set to the passive $(0,\infty)$ state. The pattern then generally moves right with rearrangements as cars take diversionary manuevers to bypass obstructions.

Note that to make the simulation more realistic, MOVE_TIME and CHECK_AGAIN_TIME can be made to be random variables so that cars move at randomly assigned rather than constant speeds.[4]

Example 3: Plant growth: gravity signal

This example shows how next event models can handle "action at a distance" in a very natural manner. Consider embedding a plant in a cell space as illustrated in Figure 2. Growth is made to occur at the tip of the plant by sprouting a new cell every GROWTH_TIME hours. The extra weight is transmitted to the stem and base cells by a gravity signal which travels instantaneously downward. Thus at the same computation instant at which the tip extends itself the weights on the other cells are simultaneously increased – and this happens no matter how long the plant becomes. The following NEVS will exhibit this behavior:

Let the state set S be $\{0,1,2,3, \ldots\}$ where 0 is a passive environmental state and $i > 0$ is a plant cell supporting a total weight of i units (itself plus i-1 cells above it). The neighborhood N consists of the center cell and the cells above and below it. The transition function T is expressed as follows:

> *) If you are the tip (center cell sequential state = 1)
>   set the upper neighbor to state 1 and
>   schedule a transition event to
>   occur there in GROWTH_TIME

if the lower neighbor is a plant cell (state=i > 0)
add 1 to its state and schedule a
transition event to occur there in 0 time
set yourself to state 2 and passivate

Otherwise (you are a plant cell)
  if the lower neighbor is a plant cell
    add 1 to its state and schedule a
    transition to occur there in 0 time.

Note that if the activated cell is a tip then it carries out three actions:
sprout a new tip upward, start a gravity signal downward and transform itself to a passive plant cell; if the activated cell is a plant cell it
passes on the gravity signal (unless it is a base cell known by the fact
that its lower neighbor is an environmental cell).[5]

When started in a global state such as illustrated in Figure 2, this
NEVS will display plant extension with weight accumulation in "real
time" (see later comparison with cellular automata).

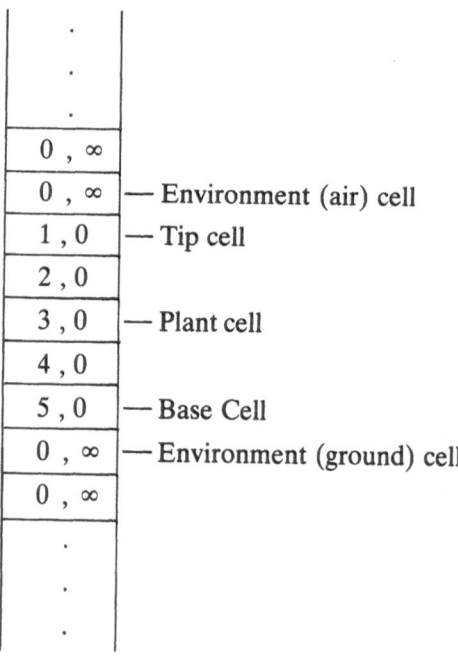

Fig. 2. Embedding of a plant in a next event cell space.

Example 4: Reaction systems: ecological, chemical, etc.

This example illustrates how collisions which result in changes of state are naturally expressed in the NEVS framework. Such collisions occur for example, when two molecules collide and combine chemically. The example however, concerns ecological interactions of this nature. Each cell is to represent a "patch", a hospitable area isolated from other patches by inhospitable terrain. Prey species colonize such patches and are predated upon by predators. Both species tend to remain localized to a patch until forced by shortage of resources to migrate. How far migrants can get to is species dependent (depending on such factors as speed of travel, ability to survive without food, etc.) The following prototype is drawn from the literature (Zeigler, 1977; Sampson and Dubreuil, 1979):

The sequential states are as follows: EMTPY, PREY (patch colonized by prey, population not yet at equilibrium), PREY' (prey population at equilibrium, sends out migrants), PRED (predators have immigrated to cell in PREY or PREY' state), and PRED' (predator population has reached maximum, sends out migrants, and dies out).

The neighborhood is the union of NPREY and NPRED, cells reachable from the center cell by prey and predator migrants respectively.

The transition function is expressed as follows:
*) if you are colonized by prey
(state = PREY or PREY')
set yourself into state PREY' and
reschedule an event
to occur here in INTER_MIGRATION_TIME
for each cell in NPREY:
if the cell is EMPTY, then
convert it to PREY
and schedule an event to occur
there in PREY_GROWTH_TIME.

if you have been attacked by predators
(state = PRED)
set yourself into PRED' and
reschedule an event
to occur here in PRED_EXTINCTION_TIME

       for each cell in NPRED'
         if the cell is in PREY or PREY', then
           convert it to PRED and
            schedule an event
              to occur there in
                PRED_GROWTH_TIME

         if you are a doomed predator
                  cell
               (state = PRED')
         set yourself to EMPTY and
         passivate.

In actual studies, colonization and migration are made stochastic and the space is limited to a finite region. From an initial global state consisting of EMPTY, PREY and PRED cells, the evolution of the system is followed to see if global persistence of predators is maintained despite guaranteed local extinction (Zeigler, 1977).

Example 5: Networks of computing devices

Von Neumann originally used the cellular automaton framework to embed networks of logical elements with the purpose of exhibiting systems with self reproductive ability (Von Neumann, 1966). Subsequently, this framework under the rubric "tesselation or iterative arrays" has been extensively employed for hardware design. Extending the same principles to the discrete event cell space, one can readily model networks of asynchronous elements for questions of logical correctness, coordination control and reliability. Such networks may model computer systems at all levels: whether the components be electronic elements, hardware devices, or full scale computers connected by communications links.

On the natural computing side, cellular automata have been used to model networks of neurons and the extension to discrete event models is natural (Barto, 1975 and citations therein).

CELL SPACE FORMALISMS: THE ADEQUACY ISSUES

We raised the issue of adequacy of model formalisms for expressing real world relationships. This is a multi-sided question which we cannot hope to answer fully in this article. But we can explore some of its aspects. One of these is a formal one: the relative expressive power of a formalism. We shall first comment upon this "hard" criterion as a backdrop for consideration of the "softer" questions such as convenience, versatility, naturalness, etc.

EXPRESSIVE POWER: A "HARD" CRITERION

By the expressive power of a formalism is meant the class of systems it can specify (Zeigler, 1977b). We can compare powers of two formalisms by asking whether every system specified by a model in one can also be specified by a model in the other. In this form, automata and discrete event systems are incomparable since the time bases are distinct. Thus we are led to the more sophisticated formulation: can every system specified in one be "simulated" by a system specified in the other? Now the formal concept of "simulation" is expressed in a hierarchy of system preservation relations (articulated in Zeigler 1976). At the lowest level, we require only that one system reproduce the external behavior of another in order to grant it simulation capability; at the higher levels, we require that the first represent more and more of the internal structure of the second to be accorded such status. Thus as we ascend the hierarchy, simulation related systems will be "closer" together in structure and behavior. Conversely, as we descend, systems so related may be more and more different in their detailed operation. A consequence of this fact is that as we loosen our definition of simulation, we increase the expressive power of formalism but at the expense of burdening the observer with the task of making the translation required to interpret the behavior of the simulator as revealing that of the simulatee.

With this as prologue, let us compare the powers of the cellular automata and NEVS formalisms. We shall restrict our attention to computable models i.e., models whose behavior can be computed by a (idealized) digital computer.[6] Then it is easily seen that the cellular automata can simulate any NEVS in the sense of computing its behavior. The argument is simple: to be computable is more particularly, to be computable by a Turing machine. Now any Turing machine can be embedded in a cellular automaton space (using Von Neumann's construction or the very straight forward approach of (Smith, 1972). Composing the two simulations we get that any next event cell space model can be simulated by some cellular automaton.

Of course, the concept of simulation employed in this result is a rather weak one (lies at the lower level of the hierarchy) and states little more than that a cellular automaton can be used as a computer to generate the behavior of a NEVS. It leaves the burden of writing and interpreting this simulation to the user and so says nothing of the ability of the automaton formalism to directly express NEVS like models.

In fact, this ability is demonstrably limited when the simulation relation is tightened (hierarchical level increased) by requiring the simulation to be in "real time". This requires that the simulator reproduce the behavior of the simulatee in phase with it (allowing the simulator no intermediate computation[7]). Let us employ such a relation in which we require that each cell state of the simulator be interpretable as a cell state of the simulatee (we allow the simulator cells to have any (including infinite) number of states). Under such a readout map we require that the simulator produce the same output trajectory as the simulatee when started in corresponding global states.

Using the above simulation relation we can show that every cellular automaton can be simulated by some NEVS but that the converse does not hold. Thus the next event formalism is the more powerful in terms of direct model expression.[8] We sketch the proof of this result:

First the positive part. Let $<S,N,T>$ specify a cellular automaton. We specify a NEVS $<S',N',T',SELECT>$ as follows:

S' is composed of two copies of S, one to hold the present state of a cell after its next state is computed. Also there is a third component to S' which is a phase indicator: in phase 1, the next state of a cell is computed. in phase 2, the next state is transferred into the present state location and the inverse neighbors are activated (see below).

N' is the union of the neighborhood N and the inverse neighborhood −N. The inverse neighborhood consists of the set of cells directly influenced by the center cell: those cells to whose neighborhood the center cell belongs. The inverse neighborhood of the origin can easily shown to be −N i.e., the negatives of each of the pairs of N (see Zeigler, 1976, Chapter 4).

The transition function T' is expressed as follows:

*) in phase 1:
  apply T to the present states of the cells

in N and store the result as the next state
of the automaton cell
set the phase indicator to 2 and reschedule
an event here in 0.5
in phase 2:
transfer the next state to the present state
set the phase indicator to 1 and
reschedule an event here in 0.5
if you truly changed state
(next state not = present state):
schedule an event to occur in
time 0.5 at all cells in –N
which are passive

The SELECT function is arbitrary: the result will not depend on the order in which scheduling ties are broken.

The readout map in this simulation just consists of paying attention to the present state component of S' and ignoring the rest. Initialize the NEVS to a global state in which all cells are in phase 1 and have time-left either 0.5 (representing initially non blank cells and their inverse neighbors) or ∞ (all other blank cells). It will then output a trajectory which reproduces the global state transitions of the cellular automaton every 1 time unit.[9]

Now for the negative part: By virtue of its finite neighborhood the cellular automaton eschews "action at a distance" (Burks, 1977). By allowing instantaneously propagating signals, as in the plant growth example, the NEVS formalism does not impose this restriction.[10] In particular the plant model can be shown not to be simulatable by any cellular automaton in the real time sense. To see this consider a plant configuration which has extended beyond the neighborhood of the base cell. To simulate the plant model, the base cell readout must increase by 1 at the next time step (for simplicity let GROWTH_TIME be unity). But now consider the same configuration in which the growth tip is removed. Then it can be readily checked that the NEVS model will not change state (no gravity signal will be sent down to the base). But in the purported simulation the base cell readout must still increase by 1 since its neighborhood configuration has not been affected by the removal of the growth tip.

In short, a cellular automaton realization of the plant model would require an infinite neighborhood to instantaneously pass the weight information generated at the arbitrarily extending tip to the base and stem cells. A finite neighborhood would not do since however large it is, a plant configuration can be chosen so that the tip lies beyond the neighborhood of the base cell. Starting from such a state, if the base cell of the automaton replicates the behavior of the base cell of the NEVS, it cannot do so starting from the same configuration with the stem cell removed.

Holland (1970) developed an extension of the cell space concept which removes the action-at-a-distance limitation as well. His scheme lays down wires capable of passing instantaneous signals in such a way that no cycles or infinite paths are generated. In our scheme such conditions are possible and we shall discuss this issue in a moment. Holland's cellular space can be regarded as a cellular automaton in which the neighborhoods of cells can grow as the computation develops. This kind of cellular space can simulate the plant model (with unity GROWTH_TIME) since the neighborhood of the base can be extended to keep up with the tip as it moves out from the base.

As just indicated, in Holland's space the neighborhoods of cells, though not fixed, are always finite, and the model remains well defined as it moves through any global state trajectory. This is not the case for the NEVS. Indeed, there may be trajectories ending up in a condition in which the model is ill defined in the sense that the clock underlying the simulation cannot advance. Moreover, it is undecidable whether or not such a condition can be reached during a simulation.

As an example of such a situation, imagine the plant model modified so that the gravity signal emanating from the tip propogates away from the base rather than toward it. While this signal propogates, the clock is not advanced. Since this signal never reaches a stopping point (that the base cell provides in the original plant model), the clock never advances and the trajectory is ill defined. A model in which such a condition cannot occur is called *legitimate* and is fully characterized by Zeigler (1976, Chapter 9).

That legitimacy is not effectively decidable is readily demonstrated with the following construction. Take any Turing Machine and embed it into a one dimensional NEVS in such a way that each cell has a copy of the finite control, a tape square, and a marker for indicating the presence of the head at the square. In the initial global state, each cell is passive ($\sigma = \infty$) except the one "containing" the head which is imminent ($\sigma = 0$). At each step of the computation, the head moves to an adjacent neighbor, activating it (setting $\sigma = 0$) and passivating itself (setting $\sigma = \infty$). If a final state is reached, the active cell passivates itself and a global equilibrium state is entered (in which all cells are passive). Note that during the course of the computation, the clock is not advanced, (each computation step takes 0 time) and to tell whether it would ever be advanced requires solving the halting problem for Turing Machines, a recursively unsolvable problem.

A corollary of the above demonstration, is that it is not possible to effectively compute a bound on the spread of information in an NEVS during one "time step". Although the NEVS does not have fixed time steps as does the cellular automaton, we can take a time step to mean a non zero advance of the clock. Formally, we define a *transitory* global state to be one for which the time advance is 0. Then a time step occurs from a global state, if the trajectory evolving from it moves through a finite sequence of transitory states to a non transitory state. Suppose that given a global state we could effectively circle a finite region in which all activity will be limited during the next time step, i.e., any change· in state will be limited to cells within this region during the sequence of transitory states. Then by embedding Turing Machines in the above manner, we could effectively bound the space used by such machines, a problem which is known not to be recursively solvable.

In sum, using the real time simulation relation defined previously, every cellular automaton can be simulated by an NEVS but not conversely. (The relation of the NEVS to Holland's model is still an open question). The reason underlying this asymmetry in expressive power is that there is no bound on the distance that information can propogate in the NEVS in one "time step" while this propogation is strictly limited in a cellular automaton (including Holland's version). More precisely, it is not possible to effectively bound the region in which activity will happen during a "time step" nor even to decide whether such a "time step" will occur for arbitrary NEVS. Extra expressive power gained at the expense of increased undecidability is a well known phenomenon in mathematical logic and computer science.

CELL SPACE MODELS: THE "SOFT" ADEQUACY CRITERIA

The expressive power of a formalism throws light on, but does not close the question of its adequacy. We can see for example, that the two formalisms under discussion are computationally equivalent (a loose notion of simulation) but differ quite drastically when more direct model representation is required – the cellular automaton being incapable of exhibiting immediate signal propagation in real time. And although any cellular automaton can be directly represented in next event form, the representation $<S',N',T', SELECT>$ contains much additional apparatus not found in the original specification $<S,N,T>$. Thus if one had already developed a simple cell space model, one would have to complicate its description in order to convert it to next event form. Moreover, the result would obscure the parameters of the original specification and would be more difficult to modify because of the interdependencies of the components of the new parameters (N' for example consists of N and –N: if a change is made in N a corresponding change must be made in –N). The same considerations hold, in even more striking form, when expressing simple NEVS models as cellular automata. To make this point we shall consider expressing the models given in Examples 1–4 as cellular automata.

Expression of motion

Consider Example 1 modelling the motion of a single body. To represent this in a cellular automaton, we require a neighborhood consisting of the center cell and its *left* adjacent cell (rather than the right neighbor as in the NEVS case). The reason is that motion consists of two parts: leaving one cell and entering the next. Since all state transitions are made independently and simultaneously by automaton cells, both cells affected by the motion must do some computation. The cell containing the object must release it (change from state 1 to 0) and the cell receiving the object must accept it – look to the *left* to see if the neighbor is a 1 and if so change itself to state 1. Thus what is a single action in the NEVS case, is redundantly expressed in two actions in the automaton case.

Handling of collisions

The redundancy of computation required is greatly compounded when collisions must be resolved as in the traffic congestion model in Example 2. Consider that to compute its next state, the automaton cell must independently decide whether any car it contains will move right and also whether any other car will move into it. To see whether its car can move, it must look not only for unoccupied cells in its own preference order but also check that an unoccupied cell is not "desired" by a car with higher priority (otherwise both cars will attempt to move into the same cell). To see whether a car will move into a cell, the cell must check whether any car which can move into it, will prefer to do so (a car below will move up only if all of its other possible moves are blocked). The reader can convince himself that this computation will require a neighborhood of approximately 20 cells (!) and is, of course, heavily redundant – the same checks being done independently by many cells.

Choice of time step

The actual value of the time step does not appear explicitly in the automaton formalism. However, when discrete time models are formulated for real systems, the choice of this value is always an issue. As discussed by Barto (1975) cellular automata can be viewed as providing the formal basis for the usual representation of partial differential equation models for computer simulation. Such models are based on continuous time and space variables which are discretized for numerical integration. The time step and spatial unit must be chosen carefully to conform to the rates of propagation expressed in the original model.

Similar considerations apply when the underlying model is a discrete event type. The most direct way to simulate a NEVS by a cellular automaton is to let the automaton cell state be a pair (s, σ) where now σ is a non negative integer or ∞. The transition function sees to it that σ is treated as a timeleft variable by decrementing it by 1 at every time step; when 0 is reached, a new total state pair is determined according to the transition function of the NEVS. Of course, the cell must also check at each time step for cells reaching $\sigma = 0$ for changes that they would cause to it.

The question is: what real value should the integer σ represent? Too small a discretization will result in much unnecessary recomputation of global states that essentially don't change; too large a value risks missing of events which would have occurred in the original. We can conceptualize this problem by limiting consideration to recurring cycles of a propagating nature (such as the motion of Examples 1 and 2 or the signals of Example 3) or of a local nature (such as that of Example 4). Then if the cycle durations are constant the best step size is given by their greatest common divisor. However, in even the simplest of models, these durations may not be constant due to being random variables or due to interactions which determine the continuation of the cycle (this is especially evident in Example 4 where the cycle must await successive prey and predator migrations in order to proceed).

In conclusion, no general procedure exists for selecting the step size underlying cellular automaton representation of NEVS models. Even if such a procedure were employed, the step size would have to be recalculated, and the timeleft variables appropriately rescaled, for every change in the NEVS parameters. Thus a cellular automaton realization of models such as those of Examples 1-4 would be either inefficient or inflexible (or both).[10]

WORLD VIEW OF FORMALISMS

Let us summarize the above findings in terms of the "world view" of the formalisms. The cellular automaton cell space is one in which each cell updates its state at each time step. The updating is done simultaneously and independently by the cells. The cell space thus expects a state change to occur at every cell even though a true change may not in fact take place. To obtain the information necessary for computing its next state, a cell interrogates the cells of its neighborhood – which must be sufficiently extensive to provide this information. Since cells cannot immediately act on one another, a cell must predict what effect other cells would have had on it in the more interactive NEVS, and this may require large neighborhoods and redundant computation. Since all cells use the information current at the same date (i.e., the previous global state) to compute their next state, the result is independent of any sequential order imposed by a sequential simulator.

The discrete event cell space on the other hand, considers cells to be capable of acting upon their neighborhood as well as themselves. Cells undertake their activity in one-at-a-time fashion so that the result is deterministic but also may be dependent on the order in which simultaneously scheduled cells are serialized by tie breaking rules. Updating of states occurs only via events which need not be uniformly distributed in time or space. The information required to schedule its transition event is contained within the cell state (in the timeleft component).

One's perception of the real system being modelled (presumably reflecting its intrinsic nature) will determine whether a formalism provides an appropriate "world view" for it. My own perception is that for most situations, the discrete event formalism provides the more natural world view. That its direct expressive power is probably the greater provides support for this ultimately subjective assessment of adequacy but does not clinch it. One could for example reject instantaneously propagating signals as valid explanatory mechanisms – relying on the accepted postulate of the finite velocity of light – and so maintain that the extra expressive power is not of use. (An answer is that often the speed of light is so great compared to other rates of interest that it is best modelled as infinite; a cellular automaton which represented a wide range of signal speeds would suffer from the time step problem discussed above.) Further support for my preference for the discrete event view is supplied by the fact that even when discrete event models are expressible in the automaton formalism, the resulting descriptions seem to be unnecessarily complex, redundant and not easily modifiable. Again, it could be countered that the reverse representation of cellular automata models in the NEVS framework involves a similar inelegance – but the *degree* of inelegance is significant.

Thus my belief is that the discrete event formalism is the more adequate in terms of range of expressive power, simplicity of expression and flexibility. In any case, the wider issue of what constitutes adequacy for a modelling formalism has been elucidated as a consequence of this discussion.

DISCRETE EVENT CELL SPACES AND LAWS OF NATURE

Accepting that the NEVS is at least as adequate a formalism for modelling real systems as is the cellular automaton begs the consideration of laws of nature in this context. As we have indicated Burks explicates a law of nature as exhibiting the "uniqueness", "modality", "uniformity", and later the "contiguous causality" features. He employs the automaton cell space to formalize these features and thus identifies a law of nature with a statement concerning the transition function of such a space.

In a parallel manner we propose that a law of nature can be identified as a statement concerning the transition function of a discrete event cell space. To make this proposal credible, we must show that it satisfies the same criteria that Burks employs in his development, viz. that appropriate statements concerning the transition function turn out to have the features of natural laws just listed. The procedure seems to break into two parts: 1) show that the cell space has the formal properties corresponding to "uniqueness", etc. and 2) exhibit a scheme for making statements about the transition function which is consistent with these properties.

The first is readily accomplished. The NEVS was set up to be time and space invariant – the formal equivalent of "uniformity". There is a concept of global state trajectory which parallels that of the cellular automaton and which is the basis for the "modality" feature. The "uniqueness" feature is not so straight forwardly handled however. It is true that there is a transition function which uniquely determines what the cell will do when it is activated. It is also true that the NEVS is globally deterministic: the time and nature of the next global state transition are uniquely determined by the current global state. However, local determinism *does not* hold. It is not in general possible, given time t and later time t', to specify a finite set of cells C, such that the state of the cells in C at time t uniquely determines the state of a center cell at time t'. Suppose this were possible in the case of the plant model for example. Let us choose the base cell as the center cell. Let t' = t + GROWTH_TIME. Let C be finite set of cells such that the states of these cells at t uniquely determine the state of the base cell at t'. Choose a sufficiently long plant configuration whose tip lies outside C. Then, as in the proof that a cellular automaton cannot simulate the plant model, the states of the cells in C are the same whether or not the tip is removed from the plant, yet the state of the base cell is different at t' in each case. This clearly contradicts the local determinacy hypothesis for C.

The absence of local determinacy in the NEVS is in marked contrast with its presence in the cellular automaton (a fact which is easily demonstrated: to predict what the state of a center cell will be in n time steps, it is only necessary to know the states of the cells within the extended neighborhood N + ... N (n times) of the cell.) Indeed, the distinction is due to the same phenomena discussed earlier: in a NEVS there is no bound on the distance that information can travel in a fixed time interval: no matter how far we extend out from a given cell, there are cells beyond this region which can send signals to this cell which arrive within the given time interval. (Recall the discussion concerning the inability of the cellular automata to simulate the NEVS in real time.)

What form of law is formulateable in such a space? Our proposal that the law be a statement about the transition function must take into account the absence of local determinism. In other words it must be hedged! Consider the following scheme:

> *) if the state of a cell c satisfies P at time t then there is a time t' later (computed from P) such that if the states of the neighbors (including c) satisfy P' just prior to t' they will satisfy Q just after t' ... provided that no other events affect the cell and its neighbors in the period [t,t'].

It is easy enough to verify that this is a disguised way of saying that knowing the total state of a cell (in particular the value of σ) one can predict the time of the next event at the cell and using the transition function one can predict the states of the cell and its neighbors just after the event ... provided that there has been no external inter-ference (the model in Example 4 evidences such external interference: in the PREY' state a cell will predictably send out migrants ... unless there has been a predator invasion).

Indeed, Burks (1977) provides the appropriate logical formulation for a hedged law: elliptical causal implication. This form of causal im-plication is the same as nonparadoxical causal implication except that unspecified antecedent conditions must obtain in addition to the stated antecedent condition in order that the consequent be causally implied. Burks gives the example:

> An electron in a magnetic field is deflected is formalized as
> Me elliptically causally implies De

where Me=an electron in a magnetic field and De=is deflected. The unspecified additional antecedent condition is that there be no other forces acting on it.

Thus in our formulation of *) above, we may drop the proviso and employ elliptical causation:

> *) For every cell c and time t, the state of c satisfies P at t *elliptically causally implies* that there is a time t' such that if the states of c and its neighbors satisfy P' just

prior to t' then the states of c and its neighbors satisfy Q just after t'.

In this way we arrive at a formalization of law of nature which is appropriate to discrete event cell spaces and which seems more like actual laws formulated by scientists (being intrinsically elliptical in form) than the formalization given by Burks for the cellular automaton.

CONCLUSION

We have considered the adequacy of discrete event cell spaces *vis à vis* cellular automata as modelling formalisms and as vehicles for expressing "laws of nature" as characterized by Arthur Burks. Many issues are raised by such a discussion:

Concerning adequacy: What are the dimensions and criteria for the assessment of formalism adequacy? What is the role of expressive power, a "hard" criterion much studied in formal treatises. Can "soft" criteria such as convenience, flexibility, etc., be usefully quantified.

Concerning "laws of nature": Is the concept relative to a modelling formalism? Our results suggest that each formalism may carry with it a concept of "law of nature" that is most congenial to it. Absence of local determinism in discrete event cell spaces and the replacement of unrestricted causal implication by elliptical implication seem to reflect the limitations of real world science, but do the stronger laws phraseable in the automaton spaces better characterize our beliefs concerning lawful regularities in nature? In other words, should laws predict unequivocally or merely help to specify the structure of a model? Is there a necessary tradeoff between the convenience of a formalism and its strength in terms of admitting strong forms of "laws of nature" (as seems to be true in relation to the cellular automata and discrete event cell spaces)?

These kinds of issues offer meat for a better understanding the computer and systems science bases of modelling and simulation as well as the yet more fundamental investigations of the philosophy of science. Stimulated by the thoughts of Arthur Burks, it is no surprise that they penetrate far beyond the usual disciplinary wells to where the true fundaments lie.

Professor Bernard P. Zeigler
University of Arizona,
Electrical & Computer Engineering

NOTES

*A shortened version of this work appeared under the title "Discrete Event Models for Cell Space Simulation", *International Journal of Theoretical Physics,* Vol. 21, Nos: 6/7, 1982.

1. More precisely, we are resstricting our discussion to two-dimensional, deterministic cellular automata. The reader wishing more background in the subject may consult Burks (1970), Barto (1975), and Zeigler (1976, Ch. 4).

2. The discrete event formalism and its subformalisms express the models implemented in discrete event computer simulation languages. For background in this area, the reader is referred to Zeigler (1976, Ch. 9).

3. This is to update the time-left values appropriately as with all non-neighboring cells in the space.

4. Zeigler (1976, Ch. 6) shows how this is done formally.

5. Studies of plant growth using discrete event models may be found in Hogeweg (1980). Actually, it is interesting that a further extension of the cell space formalism is necessary to conveniently express growth in the interior of a plant – the neighborhood must be allowed to change.

6. For cellular automata, a cell state is "quiescent" if the global state in which all cells are assigned this state is an equilibrium state. A computable (or algorithmic) cellular automaton is one having at least one quiescent state called the "blank" state and whose initial global states are restricted to those in which only a finite number of cells are not in the blank state (Burks, 1977, pp. 566–67). In the NEVS case, the role of the blank state is played by a passive state (recall earlier definition). A computable NEVS has such a blank state and is initialized only from global states having a finite number of nonblank cells. We also require that all real numbers used must be representable by fixed precision rationals as they would be in a digital computer.

7. In the case of cellular automata one can even allow a fixed number of intermediate state transitions without increasing the power of the class, since all this does is effectively expand the neighborhood of the simulator, a free parameter in the cellular automaton class.

8. However, as a consequence of this observation and the previous one, the unrestricted simulation capability of the two is the same, both being computation universal.

9. This construction is of more than theoretical interest: it parallels an efficient strategy for digital simulation of cellular automata (Zeigler, 1976, Ch. 4).

10. It should be noted that the same difficulties *do not* apply to digital simulation of discrete event models (as they do to simulation of differential equation models). The reason is that the simulation strategies employed by such discrete event languages as GPSS, SIMSCRIPT, and SIMULA are based on event-driven rather than fixed step time-advance.

REFERENCES

Barto, A. G. (1975), *Cellular Automata as Models of Natural Systems,* Doctoral Diss., CCS Dept., University of Michigan, Ann Arbor.

Burks, A. W. (1970), (ed.), *Essays on Cellular Automata,* U. of Illinois Press, Urbana, Ill.

Burks, A. W. (1977), *Cause, Chance, and Reason,* U. of Chicago Press, Chicago, Ill.

Holland, J. H. (1970), "A Universal Computer Capable of Executing an Arbitrary Number of Subprograms Simultaneously", In: *Essays on Cellular Automata*, A.W. Burks (ed), U. of Ill. Press.

Herman, G. T., and G. Rosenberg (1975), *Developmental Systems and Languages*, North Holland, Amsterdam.

Hogeweg, P. H. (1980), "Locally Synchronised Developmental Systems", *Intnl. J. of General Systems*, Vol. 6, pp. 57-73.

Lindenmeyer, A. (1968), "Mathematical Models for Cellular Interactions in Development", *J. Theo. Biol.*, 18, pp. 280–312.

Sampson, J. R. and M. Dubreuil (1979), "Design of An Interactive Simulation System for Biological Modelling", In: *Methodology in Systems Modelling and Simulation*, B. P. Zeigler et. al. (eds.), North Holland Press, Amsterdam.

Smith, A. R. (1972), *Simple Computation Universal Cellular Spaces*, Jnl. Assoc. Computing Mach. Vol 18 p. 339–353.

Zeigler, B. P. (1976), *Theory of Modelling and Simulation*, J. Wiley and Sons, New York.

Zeigler, B. P. (1977), "Persistence and Patchiness of Preditor-Prey Systems Induced by Discrete Event Population Exchange Mechanisms", *J. Theo. Biol.*, Vol. 67, pp. 687–714.

Zeigler, B. P. (1977b), "Systems Theoretical Description: a vehicle reconciling diverse modelling concepts", In: *Proc. Intl. Conf. Applied General Systems Research*, (ed: G. J. Klir), von Nostrand.

MANLEY THOMPSON

SOME REFLECTIONS ON LOGICAL TRUTH AS A PRIORI

In the first part of this paper I outline briefly what is included under "logical truth" in Arthur Burks' *Chance, Cause, Reason*[1] and locate my topic under this heading as it is used by Professor Burks. In part II I indicate the sort of approach I believe has to be followed and the sort of issues that have to be faced if logical truths are to be explicated as a priori. I argue that in order to explicate them a priori one needs "un-naturalized" as opposed to a Quinean "naturalized" epistemology. I end this part of the paper by noting a difficulty that confronts any such explication. In part III I examine the possibility of escaping the difficulty by following Burks' proposal that we explicate the notion of an a priori concept "in terms of ideas from automata theory." In conclusion I urge that this proposal hardly offers an escape but rather a flight to naturalized epistemology.

I

At the outset of *Chance, Cause, Reason* Professor Burks explains that by a "logically true-or-false statement" he means one whose "truth-value ... *can be* established, in principle, by defining, intuiting, reflecting, calculating, and reasoning alone, without the use of observation or experiment" (p. 6). An "empirically true-or-false statement," in contrast, "cannot be confirmed or disconfirmed without the use of observation and/or experiment, but defining, reflecting, calculating, and reasoning may not be required." For convenience, the two sorts of statements are referred to simply as "logical" and "empirical." Later in the book, Burks refines the contrast between the two ways of fixing truth-values. The activities involved in fixing the truth-value of a logical statement may all be included under "reasoning" when the latter is taken "in the core sense of deductive rationality" (p. 316). A logical statement, then, "is one whose truth-value in principle, can be decided by deductive reasoning alone without the use of observation and experiment and without appeal to the rules of inductive reasoning."

65

Merrilee H. Salmon (ed.), *The Philosophy of Logical Mechanism*, 65–82,
© 1990 *Kluwer Academic Publishers.*

The notion of a *rule of reasoning* is central in this explication of logical and empirical truth. In commenting on this notion, Burks begins with "some very broad concepts of reason and logic" (p. 15). "Rules, procedures, algorithms, methods and recipes are all different, but they may be viewed in a common light, as a system of instructions (simple or complex) to be followed in a given type of situation. All of them may be covered by the word *rule*, if we widen it to include complex systems of rules providing for many cases and alrternatives ... *Logic* studies and evaluates the most general, common, and basic rules of reasoning, together with related matters." Logic then divides into two sorts, corresponding to two sorts of rules. The "logic of inquiry" studies "rules that are useful in solving problems, answering questions, and arriving at important results; it evaluates these rules in terms of applicability, simplicity, and utility" (p. 16). The "logic of argument," on the other hand, "treats rules for deriving conclusions from premises and relating evidence to hypotheses and theories, judging these rules to be valid or fallacious, correct or incorrect, reliable or unreliable."

In Burks' scheme, logic may also be divided with respect to kinds of *statements* rather than kinds of *rules*. One sort of logic deals with logical statements and another with empirical statements. Each sort has two branches. Logic dealing with logical statements divides into the logic of mathematical inquiry and deductive logic; logic dealing with empirical statements divides into the logic of empirical inquiry and inductive logic (p. 16). In this fourfold scheme, which Burks presents in the form of a table, the basic dichotomies are two types of activity, inquiry and argument, and two types of statement, logical and empirical. In terms of this scheme, my discussion of logical truth will be restricted to the activity Burks calls "argument" and to the type of statement he calls "logical." That is, I will be concerned with logical truth within the context of "deductive logic."

This restriction, I believe, accords with Burks' later characterization of a logically true-or-false statement as "one whose truth-value, in principle, can be decided by deductive reasoning alone without the use of observation and experiment and without appeal to the rules of inductive reasoning" (p. 316). By "deductive reasoning" here I assume Burks means the activity he called "argument" when he characterized deductive logic in terms of argument and logical statements. Although rules are not explicitly mentioned in this characterization, they are obviously presupposed. Argument as an activity is rule-governed, and deductive logic as argument is activity governed by rules of deductive reasoning.

The intimate connections Burks intends between "deductive argument," "rule of inference," "validity," and "logical truth" stand out clearly in his remarks: "We define *deductive argument* to cover rules governing deductive arguments as well as the arguments themselves ... we define a deductive argument (or rule of inference) to be *valid* if its corresponding conditional is *logically* true; otherwise it is *invalid*" (p. 22). Thus, in giving a deductive argument with premises $P_1, P_2, \ldots P_n$ and conclusion C we are also giving (at least implicitly) the rule: from P_1 & $P_2 \ldots$ & P_n to infer C. We say the rule is valid provided that it always takes us from true premises to a true conclusion. Its validity is assured if the corresponding conditional: 'If P_1 & $P_2 \ldots$ & P_n then C' is logically true. With this procedure "questions of validity are replaced by questions about logical truth" (p. 23). *What is logical truth* then becomes the basic question of deductive logic, at least when this logic is restricted to what I am going to consider as deductive logic.

I add this further restrictive remark because, as Burks notes, it "is difficult to draw the line between deductive logic and the rest of mathematics, especially since mathematical logic is the most important part of deductive logic" (p. 17). He observes that "very roughly" the distinction is that "Deductive logic treats the most foundational and general types of deductive arguments, while mathematics emphasizes those that depend on *the specific subject matter* of that branch of mathematics in which they occur" (my italics). He adds that "the logic of causal statements," which he considers later in the book, is also part of deductive logic. I restrict my consideration of deductive logic to that part of it which Burks very roughly distinguishes from the rest of mathematics because my central question is just *what is the subject matter of logical truth* when it pertains to only, in Burks' phrase, "the most foundational and general types of deductive arguments." I do not regard (and I do not think Burks regards) this restricted deductive logic as including the logic of causal statements. [2]

II

One answer to the subject-matter question is simply that logical truth is about everything. A paradigm of logical truth would be the statement that everything is self-identical. Logical truths on this view are then those that predicate of objects only what is true of any object whatsoever. But then what are the criteria for selecting predicates that are true of any object whatsoever? We might begin by noting that one may take any predicate that is not true of just any object whatsoever and construct a predicate that is true of any object whatsoever. For example, from the predicates 'is a dragon' and 'is a table' I may construct the predicates 'is a dragon or is other than a dragon' and 'is a table or is other than a table,' both of which are true of any object whatsoever. Reflection on examples of this sort suggests that we may specify the desired criteria solely in terms of a few basic concepts such as those symbolized by 'not', '=', and 'or'. A predicate true of everything – true of any object whatsoever – would then seem to be one constructed in an appropriate way from basic concepts. The construction may or may not contain a predicate that by itself is not true of everything, as 'is a dragon or is other than a dragon' contains 'is a dragon'. 'Is not other than itself but is other than anything other than itself', in contrast, contains only predicates true of everything.

If we follow the above suggestion we may decide to abandon any attempt to explicate logical truth metaphysically, i.e. as truth that is true because of the nature of being as such – being qua being. We are not to say that 'Everything is self-identical' is logically true because every being qua being is self-identical, because every being in every possible world is self-identical. On the contrary, in following the above suggestion we may decide to explicate logical truth epistemologically, i.e. as truth that is true because certain concepts are employed in a certain way. We may say rather that 'Everything is self-identical' is logically true because a predicate such as 'is not other than itself but is other than anything other than itself' is true of every conceivable object. If we say instead that it is true of every being qua being we commit ourselves to a metaphysical explanation based on claims about the nature of being rather than an epistemological explanation based on claims about the nature of conceivability.

The phrase 'the nature of conceivability' points to a serious problem for any epistemological explanation. With use of this phrase we seem restricted to a specific subject matter, psychological or linguistic. We seem headed toward an explanation of logical truth in which basic concepts such as *not* and *or* are accounted for by facts of human psychology generally, or perhaps more specifically by facts about the linguistic structure entrenched in our culture. In either case such an explanation seems open to the objection that it misses the basic insight with which a metaphysical explanation begins, viz. that a logical truth is true of any object whatsoever. An epistemological explanation appears to restrict logical truth to what is true of any object conceivable by beings with human psychological capacities, or, with even greater restriction, any object representable in a linguistic structure we can comprehend. With this restriction we seem forced to recognize the possibility of beings who, because they have extra-human psychological capacities or perhaps merely because they have developed a linguistic structure totally alien to our culture, have a set of logical truths entirely different from ours. The point is not just the claim that we must recognize alternative logics. We expect this claim to be accompanied by the presentation of at least one alternative, and when it is not we are not likely to take it seriously. The point here, however, is that we must recognize the possibility of alternatives even when the alternatives may be forever incomprehensible to us.

The problem, then, is that a recognition of incomprehensible alternatives hardly makes sense and yet with an epistemological explanation of logical truth we seem forced to such a recognition. We are indeed forced to it, it seems to me, if we restrict ourselves to epistemology à la Quine, to epistemology "naturalized" as a branch of natural science. Quine admits the possibility of creatures whose language is so alien to ours as to be untranslatable, so that we can never comprehend what they take to be true of objects.[3] We must recognize the possibility of alternatives to our present formulations of logical truth not only in the sense that we may always come to revise our formulations of logical as well as empirical truth, but also in the more radical sense that there may be revised formulations we will never discover because they are in principle incomprehensible to us.

In an effort to explain the possibility of comprehensible alternatives we may return to metaphysics. If we distinguish between objects as they are in themselves apart from any relation to us and objects as they can be conceived of or spoken about by us, we at the same time recognize the possibility of objects conceived of or spoken about in ways incomprehensible to us. But with this distinction, we do not merely return to metaphysics, we return to it in a way that makes metaphysics incoherent. In drawing the distinction we must conceive of or speak about objects as they are in themselves and then contrast such objects with *all* the objects we say we can ever conceive of or speak about.

Instead of trying to save an epistemological explanation of logical truth by attempting to explain the possibility of incomprehensible alternatives, we may take the opposite tack and deny the possibility. But in taking this course we must embrace an "unnaturalized" epistemology that distinguishes *a priori* from *a posteriori* truth. We take logical truth as a priori not only in the sense that it is true of every object of possible human experience (a condition purportedly satisfied by Kant's synthetic a priori), but also in the sense that it is true of every possible object – of any object whatsoever – (a condition purportedly satisfied by Kant's analytic a priori). We thus in effect fuse metaphysics and epistemology. Although we speak only of conditions of conceivability, we resolutely insist that these are also necessary conditions of being and deny that they have anything more to do with psychology than with any other natural science. This appears to be basically Wittgenstein's position in his *Tractatus*.[4] In stating conditions of conceivability one seems to be talking about oneself, about what one can and cannot conceive of, but in philosophy the self talked about is not the self of an individual person, which is the subject of psychology. "What brings the self into philosophy is the fact that 'the world is my world'" (5.641). The world that confronts the self is limited by the conditions under which the self can conceive of objects. But then the self as the determination of these conditions is not itself another object determined by them. It is not an object studied "in psychology" but rather "the metaphysical subject, the limit of the world – not a part of it" (*ibid*).

Two interrelated objections to this position spring to mind at once. "The world is my world" seems obviously true only if it is understood as referring to the world *for me*. But then "the world for you is your world" seems just as true. What we seek in metaphysics is not a determination of the conditions of my world, your world, or the world of any other individual, but of the world that is what it is no matter how it may be conceived by different individuals. In elevating the self to the status of metaphysical subject we seem to embrace solipsism. In the *Tractatus* this objection is answered by the contention that "solipsism, when its implications are followed out strictly, coincides with pure realism. The self of solipsism shrinks to a point without extension, and there remains the reality co-ordinated with it" (5.64). But this contention seems compatible with the claim that the self of each person shrinks to an extensionless point and there remains reality for that person. So understood the contention does not answer the objection that we cannot achieve what we seek in metaphysics merely by elevating the self to the status of metaphysical subject. We need to explain why the elevation obliterates differences among individual selves.

Even if this first objection is waived, there is the further objection that an individual's self determines the limit of the world for that individual differently at different times. What I find inconceivable today I may find perfectly conceivable tomorrow. I cannot find in my self what *must* be the limit of my world but only what is in fact the limit of my world at present. In an effort to answer this objection we may turn to the Kantian notion of an *a priori concept*. What I conceive of a priori is what I conceive of independently of all experience. My self determines the limit of my world differently at different times only to the extent that my self includes merely concepts I acquire from experience – a posteriori concepts. But if I can also find in my self a priori concepts and can show that without these concepts, which cannot come from experience, I could have no experience at all, then there is a sense in which I find in my self what must be the limit of my world. The a priori concepts would be those like *not* and *or*, concepts represented by what are sometimes called "logic words.". Insofar as I find that in order to have any experience I must form empirical judgments and that any judgment I can form has a structure determined by my a priori concepts, I find in my self – in my a priori concepts – the formal limit of my world. My world today may be different from my world yesterday in that I have acquired new empirical concepts that enable me to form new empirical judgments about objects in my world, but my world remains always the same in that whatever new judgments I may form they are all limited by the structure determined by my a priori concepts.[5]

The argument outlined above not only suggests a way of answering the objection that one's self limits one's world differently at different times, it also suggests a way of answering the objection charging solipsism. My self differs from yours in that it includes my experiences but not yours, yet if a priori concepts are independent of all experience, then selves restricted to a priori concepts are not distinguished by differences in experiences. You may of course have a priori concepts I don't have in the sense that you can express these concepts verbally and interrelate them in sentences while I cannot. For you may have studied logic while I haven't. But this is not to say that I don't have the concepts in the sense that any judgment I can form is limited by the structure they determine, even though I am unable to articulate the structure. Although I may be able to articulate the structure after I have had the experience of studying logic, it does not follow that the a priori concepts I thereby acquire are (at least for me) really empirical concepts. A priori concepts are independent of all experience in that they are the concepts in terms of which the structural conditions regulating the formation of all judgments are articulated. They are not independent of all experience in the sense that the ability to articulate them is acquired independently of all experience. But then the fact that you can and I cannot articulate a priori concepts does not imply that in the formation of all our judgments we are not restricted by the same structural conditions. Our respective selves as determined solely by these conditions (as restricted to a priori concepts) may be indistinguishable. If by "the metaphysical self" we mean this self indistinguishable as yours or mine (a transcendental as opposed to an empirical self), then we may each say "The world is my world" without implying solipsism.

We got to a consideration of the self, let us recall, in the course of considering different ways of specifying the subject matter of logical truth. We turned from the straightforward metaphysical view that the subject matter is everything and that logical truth is fixed by the structure of being qua being to the epistemological view that logical truth is fixed by the structure of thought or language. With this second view the subject matter becomes psychological or linguistic if we naturalize our epistemology, and with the naturalization we seem forced to recognize the possibility of alternative structures forever incomprehensible to us. To avoid this consequence we may forego naturalization and return to an epistemology that sharply distinguishes a priori from a posteriori truth. With this unnaturalized epistemology we seek to determine the structure of thought at a level of generality that transcends anything that distinguishes the thought of different individuals or of different cultures. We seek, in other words, to determine the structural conditions necessary for any thought – for any conception or judgment – at all. But then in response to the question *whose* thought are we considering we can only reply, the thought of the metaphysical subject, the transcendental self.

One may well object that in being forced to this reply we show that a turn to unnaturalized epistemology can add nothing to the straightforward metaphysical view we assumed was inadequate. If the metaphysical subject shrinks to an extensionless point and there remains the reality co-ordinated with it, what is the difference between 'the structure of thought qua thought' and 'the structure of being qua being'? Have we not merely exchanged one vague if not meaningless phrase for another?

The objection misses the point that with unnaturalized epistemology we forsake any attempt to specify a subject matter for logical truth. In saying that logical truth is fixed by the structure of thought we are saying, not that it is fixed by anything in the world, but that it is fixed by the structural conditions under which we can think of anything in the world. The point in introducing 'the metaphysical subject' and saying that it marks the limit of the world and is not a part of the world is to emphasize that in speaking of the structure of thought we are not speaking of the thought of a subject we can locate as an object in the world. While 'the structure of thought' and 'the structure of being' refer to the same structure, the second phrase is misleading insofar as it suggests a physically realized structure determinable, in principle, by observation of objects in the world. The first phrase, in contrast, may suggest a structure realized in the activity of thinking and determinable by thought reflecting on itself.

It is only as determined solely by this reflection that logical truth is known a priori. As Wittgenstein puts the point in the *Tractatus*, "What makes logic *a priori* is the *impossibility* of illogical thought" (5.4731). I find on reflection that I cannot think illogically. I may of course discover that I have contradicted myself, that I have come unintentionally to affirm both p and not–p. With this discovery I must either withdraw one of the affirmations and maintain the other or renounce both and rethink the question. If I do neither of these but insist on affirming both, I cannot succeed in thinking anything at all. With one affirmation I cancel whatever I might think with the other by itself. I cannot think illogically (contradictorily), I can only discover that in attempting to think I fail to do so if I contradict myself. With this discovery I become aware of rules I cannot avoid following if I am to think at all. In articulating these rules I form logical concepts such as *not* and *or*. These concepts are a priori in that they arise solely from the formulation of rules conformity to which is necessary for any thought at all. They are not to be confused with the a posteriori concepts that one learns by being told the rules for the use of particular words in a given language, e.g. the use of 'or' in English or the use of 'v' in a logical calculus.

Rules articulated solely with such a priori concepts express logical truth. We ordinarily speak of rules as conventions we may or may not choose to adopt. But in the case of logical rules we have no choice – they are not conventions. We unavoidably adopt these rules when we think at all, including the cases where we debate whether or not to adopt a convention. Once we adopt a convention we make certain statements true simply by virtue of that convention. We may say that such statements are analytic and true a priori, but they are so only in the derivative sense that their truth is determined by a convention we adopt and not by the way the world is independent of us. They are not a priori in the basic sense of being true by virtue of rules we cannot avoid adhering to if we are to think at all.

I distinguished two paragraphs back between a priori concepts that arise from thought reflecting on itself and a posteriori concepts that one learns by being told the rules for the use of particular words in a given language. Logical rules are articulated solely in terms of concepts of the former sort, linguistic rules are articulated mainly if not ex-clusively in terms of concepts of the latter sort. Yet this division of con-cepts seems difficult to maintain. What can 'thought reflecting on itself' possibly mean if not 'thought reflecting on the use of language'? That all thought proceeds within a system of signs and that a system of signs constitutes a language seem obvious platitudes. But then concepts that arise from thought reflecting on itself would seem to arise a posteriori from observations on the use of language. We may find, for example, that all observed languages have rules we cannot articulate without using the concept of negation, even though the language itself may be so impoverished (or so specialized) that it contains no sign of negation apart from, say, signs of assent and dissent. We may confidently assert on the basis of our observations that all languages have rules whose ar-ticulation requires the concept of negation. Users of any language, we may say, follow these rules whether or not they can articulate them in their language. This generalization about all languages, however, can-not be an a priori truth. It is based on observations and is subject to correction by future observations, even though at present we are ut-terly unable to say what the correction might be.

We now seem to have lost any claim to a priori truth and to have returned to naturalized epistemology in spite of ourselves. If we want to retain a priori truth we must reconsider the import of the phrase 'thought reflecting on itself'. Does it follow that if all thought proceeds within a language, thought reflecting on itself is thought reflecting on the use of language? We have already remarked that with unnaturalized epistemology we forsake any attempt to specify a subject matter for logical truth. Yet language is a subject matter, a system of signs, of sensible marks (usually written or spoken) embodying a structure. We determine the structure by observing the behavior of signs as they are used by speakers of a language. We articulate the structure by specifying rules speakers follow in their use of signs. Some of the sentences by which we express these rules may also be taken as sentences expressing logical truths, e.g. in assenting to a disjunction one is committed to assenting to at least one of the disjuncts. But a logical truth taken simply as the expression of a rule we observe to be universally followed in the use of language is not an a priori truth.

In order to have a priori truth we need some way of viewing the rules that makes the possibility of their discovery by observing the use of language incidental. We need, in other words, a special way of interpreting 'thought reflecting on itself'. Such a way is proposed by Burks in *Chance, Cause, Reason* and I turn now to a consideration of this proposal.

III

Burks proposes that "we give positive characterization of 'empirical concept' and 'a priori concept' in terms of ideas from automata theory" (p. 610). Analogizing man to a computer, we may characterize a priori concepts as those that derive solely from "man's innate structure-program complex," from his "innate information processing capacities," while empirical concepts, in contrast, always derive at least in part from the input man receives from his environment and belong to his "acquired structure-program complex" (p. 612). Although both complexes are physically realized in the human nervous system, Burks' proposal is not that we discover which concepts are a priori and which are empirical by observing the nervous system. Segregation of concepts by this means is not only imposssible with our present knowledge of neurology; its possibility is irrelevant to the question of a priori truth since knowledge acquired by observation is always a posteriori. Burks' proposal, on the contrary, concerns the exercise of certain mental activities.

"The abilities to abstract concepts from experience, to construct concepts from other concepts, and to reflect on one's experiences and to see what concepts are involved in them are," Burks assumes, "among man's innate information-processing capacities" (p. 612). When "one can directly experience instances of a concept and abstract the concept from these experiences, the concept is empirical." Burks does not say so explicitly, but presumably the awareness that a concept can be so obtained is itself a result of reflection on one's concept-forming activities. In any case, reflection is explicitly mentioned as the way in which one becomes aware that certain concepts are a priori. A concept may be a priori in the sense of being "a basic aspect or feature of the structure-program complex constituting a person's innate information processing capacities." One then "becomes aware of such an a priori concept by having experiences involving the concept and seeing on reflection that the concept is involved in these experiences" (p. 613). Concepts that can be constructed from a priori concepts alone are themselves a priori, while a concept with at least one empirical constituent is itself empirical.

This reflection that brings awareness of a priori concepts seems to be what we tried to capture by the phrase 'thought reflecting on itself'. While Burks speaks of reflecting on experiences and seeing that a certain concept is involved in them, he clearly means to distinguish 'involved in them' and 'abstracted from them'. For a concept abstracted from an experience is empirical and not a priori. I take Burks to mean that a concept is involved in but not abstracted from an experience when the concept characterizes the thought component as distinct from the sensory component of the experience. We may think of an experience as comprising a sensory matter organized by a thought or judgment. The rules or structural principles governing the organization remain constant while the sensory matter may vary with each experience. Whatever the sensory matter may be, there is the rule, for example, that the matter must be so organized that one and the same empirical concept does not, without qualification, both characterize and not characterize it. As we become aware of this rule by reflecting on our experiences we come to see that the concept of noncontradiction is involved in all our experiences and is not abstracted from a particular type of experience. In terms of the computer analogy, the concept belongs to our innate and not our acquired structure-program complex.

With this view of reflection we seem able to claim that the possibility of discovering logical rules by observing the use of language is incidental. Insofar as the rules are only discovered empirically by observing linguistic behavior they are not known as logical rules but merely as conventions which, as far as is known empirically, are universally adopted by language users. They are known as logical rules giving rise to a priori truth only as they are seen on reflection to be always involved in the thought by which sensory matter is organized into experience. We thus come back to Wittgenstein's dictum that what makes logic a priori is the impossibility of illogical thought. But with Burks' proposal we can now add that what gives rise to the impossibility is the nature of our innate structure-program complex. We do not have to stay with the vague phrase 'thought reflecting on itself'. We can say that on reflection we come to articulate features of our innate structure-program complex – features without which thought for us would be impossible – and that this capacity for reflection is itself one of our innate information-processing capacities.

But if this view of logical rules makes the possibility of their discovery by observing the uses of language incidental, it would also seem to make their discovery by reflection as a priori truths incidental. For if the rules are conditions determined by a structure physically realized in the human nervous system, they are not known as these conditions when they are known only by reflection as a priori truths. The knowledge of them as such truths, as rules we have to follow if we are to think at all, is incidental to, in the sense of being explained by, the knowledge of how our thought is limited by the structure of our nervous system. The fact that our knowledge of the latter sort is extremely meager need not prevent us from maintaining that we know from reflection that certain concepts are involved in but not abstracted from experience and that these involved concepts together with the rules we formulate by their means constitute a priori knowledge. What we now seem to be prevented from maintaining is that this a priori knowledge in any way represents either insights into the nature of being qua being or limitations imposed by a metaphysical subject on the structure of our world. We seem forced, however, to maintain that this a priori knowledge is only incidental knowledge of the structure of our nervous system.

One may reply that it is not incidental but is truly explanatory. Accepting the computer analogy, we may claim that with a priori concepts we gain knowledge of the formal structure of our innate information-processing capacities, while what we gain empirically from neurology is knowledge of the particular way in which this structure is physically realized. Our a priori knowledge explains what is realized. Without this knowledge we would not know what structure to look for in the nervous system, and if we happened to discover it independently we would know it only as a structure which, as far as we can tell empirically, is universally realized in the human nervous system. We would not know it as the physical realization of the structure of our innate information-processing capacities.

I do not think this reply is adequate to the charge that in viewing logical rules as determined by a structure physically realized in the human nervous system we make a priori knowledge of these rules incidental. We must address the question: What is our evidence that a priori knowledge we gain through reflection provides us with knowledge of a structure physically realized in our nervous system? Clearly, the evidence is empirical and of fairly recent origin. Logical rules were known in antiquity and associated with the use of language, while their association with the brain and nervous system remained unknown for centuries. But once the latter association is assumed, we seem forced to reassess our knowledge of logical rules. How do we know that discoveries in neurology will not show that our determination of logical rules by reflection as a priori truths was not wrong in at least some respects? With this question we face a dilemma. Either logical rules as a priori truths remain unassailable no matter what is discovered in natural science, or they remain subject to correction by discoveries in at least that part of natural science comprising neurology. With the first alternative, what is discovered in natural science, including neurology, is incidental to our a priori knowledge of logical truths. With the second alternative, our a priori knowledge of logical truths is incidental to our knowledge of them as determined by conditions realized in the structure of our nervous system.

We cannot escape between the horns of this dilemma by claiming that logical rules as a priori truths remain unassailable in their capacity of initially informing us of the structure to look for in the nervous system, while they become subject to correction as we acquire knowledge of this structure in its physical realization. With this claim we do not escape between the horns, we opt for the second horn outright. Truths subject to correction by empirical discoveries, even if only by discoveries now well in the future, are not a priori truths in the sense claimed in the first horn. In opting for the second horn we are back with naturalized epistemology and can dispense with any notion of a priori truth. From observations of behavior, linguistic and otherwise, we can discern all we need to know in the way of a structure to look for when we turn to the nervous system. Our entire project can be summed up in the statement that we seek in neurology to explain how the input received through sensory receptors is processed so as to yield a structured output we observe in human behavior. If we add that the output structure is one we can come to articulate in the form of a priori truths discernible by thought reflecting on itself, we are opting for the first horn of the dilemma and are speaking of a different project.

We do not combine the projects and escape between the horns when we claim that in articulating a priori truths we are characterizing features of our innate structure-program complex. In accepting the computer analogy we embrace naturalized epistemology. The complex that determines logical truth is to be explained ultimately in neurology as the structure of the human nervous system, and not in unnaturalized epistemology as limitations inposed by a metaphysical subject on the structure of our world, or outright in metaphysics as the structure of being qua being.

The computer analogy, as Burks uses it, renders acute the problem of recognizing incomprehensible alternatives to our most basic logical truths, a problem I suggested earlier is indigenous to naturalized epistemology. The innate structure-program complex realized in the human nervous system is the product of an evolutionary process. If what we humans accept as logical truths are fixed by this process, then we must recognize the possibility of species (perhaps on another planet) with an evolution so different from ours that what they accept as logical truths are incomprehensible to us. The problem is made acute by Burks' use of the computer analogy because he uses it in a way that segregates logical truths from others as being a priori and analytic rather than a posteriori and synthetic. The crucial point is Burks' assumption that one's ability "to reflect on one's experiences and see what concepts are involved in them" is not only part of "man's innate information-processing capacities," it also affords a priori knowledge (p. 612). Without this assumption the computer analogy provides no way of segregating truths into a priori and empirical. We investigate the human nervous system as we investigate any other physical reality. Although we have some prior access via both reflection and the observation of linguistic behavior to the structure realized in this physical reality, our prior access does not enable us to determine the structure with a priori certainty. The fact that in our investigation of the nervous system we seek to discover conditions that determine how we humans are forced to view our world does not force us into the paradoxical situation of having to recognize that what is a priori impossible may still be possible. We have to recognize only that certain truths whose denial we find incomprehensible because self-contradictory may not be so for beings radically unlike us, or indeed for our own decendents in some future state of evolution. This situation differs only in degree from those in which we find certain nonlogical truths (e.g. that two straight lines cannot enclose a space) so compelling that their denial at first seems incomprehensible.

IV

In considering Burks' proposal that a priori concepts are those that derive from man's innate structure-program complex and are discoverable by reflection on one's experience, I began with the remark that the proposal seemed to provide a view of logical rules that makes the possibility of their discovery by observing the use of language incidental. I must now conclude that insofar as the proposal succeeds in providing such a view it does so at a price. The price is acceptance of a naturalized epistemology according to which what the logical rules discoverable by observing linguistic bahavior are incidental to is the structure of the nervous system, for they are explained (when they are explained) by this structure. But then the logical structure of thought discoverable by reflection is likewise incidental to the structure of the nervous system. Burks' assumption that the capacity for this reflection is part of man's innate information processing capacities allows for the possibility of discovering the logical structure independently of observations of linguistic behavior and the nervous system. But insofar as the assumption is taken as allowing for the possibility of a priori truth it must also be taken as allowing for the possibility that the a priori impossible may be possible. Naturalized epistemology, as far as I can see, can escape this paradox only by going along with Quine and forsaking any absolute distinction between a priori and empirical, analytic and synthetic truths.

I have tried in part II of this paper to indicate the sort of approach I believe is required and the sort of issues that have to be faced when one wants to maintain that the most fundamental truths of logic are a priori in an absolute sense. The central question, as I see the issues, is whether or not fundamental truths of logic have a specific existential subject matter. The answer is negative with either a straightforward metaphysical approach that takes the subject matter to be everything, or an epistemological approach that takes it to be nothing in the world. Logical truth with either approach becomes a priori in an absolute sense. On the other hand, the answer is obviously affirmative if logical truths are simply the articulation of a formal structure discernible by observation of linguistic behavior and to be explained ultimately by the structure of the nervous system. With this approach logical truths do not differ in epistemic status from other truths about a specific existential subject matter. They cannot be segregated as a priori.

Saving a priori truth by accepting the revelations of reflection as unassailable also has its price. The reflection must be viewed as the act of a transcentental self or metaphysical subject aware of itself as comprising the limits of its world. Although this reflection existentially occurs in the world as acts of individual beings, and as such depends on physical conditions which, in our case at least, include a sufficiently complex nervous system, the knowledge of those conditions remains incidental to the reflection as transcendental. Discoveries in behavioral science and neurology no less than discoveries in other empirical investigations must be taken as governed by and not also as accounting for the a priori truths of logic. If the price for a priori truth revealed by reflection seems too high, the only alternative I can see is naturalized epistemology.[6] The proposal that "we give positive characterization of 'empirical concept' and 'a priori concept' in terms of ideas from automata theory" does not, it seems to me, provide a different alternative.

Manley Thompson
University of Chicago

NOTES

[1.] Arthur W. Burks, (1970). All page references in the text are to this book.

[2.] On this point see below, note 5.

[3.] Quine qualifies this statement in his most recent writings. "If there is a question in my mind whether a language might be so remote as to be largely untranslatable ... that question arises from the vagueness of the very notion of translation ... Translatability

is a flimsy notion, unfit to bear the weight of the theories of cultural incommen-
surability that Davidson effectively and justly criticizes." (1981, p. 42). While incomprehensible alternatives cannot be ruled out with naturalized epistemology, they do
not present a serious problem as long as logical truths are not taken as a priori. It is
only when logical truths are so taken that with naturalized epistemology we encounter
the paradox of having to recognize that what is for us a priori impossible may still be
possible. I consider this paradox below at the close of part III and in part IV.

4. Wittgenstein, (1963) References to this work are given in the text by the article number in parenthesis.

5. I restrict myself here and throughout this paper to those a priori concepts required for,
in Burks's phase, "the most foundational and general types of deductive arguments.' In
Kantian terms, this means restriction to the analytic a priori. I do not consider the synthetic a priori, which requires the introduction of space and time as a priori concepts. I
agree with Kant that only with this introduction can one argue for causality as an a
priori concept, though Burks does not seem to agree (cf. pp. 615 ff.). I do not want to
get into questions of modal logic.

6. I have argued elsewhere (1981) that the price of maintaining a priori truth in a
philosophically significant sense is acceptance of a radically first-person oriented point
of view that makes transcendental reflection intelligible.

REFERENCES

Burks, A. W.: 1977, *Chance, Cause, Reason*, University of Chicago Press, Chicago.
Quine, W. V.: 1981, *Theories and Things*, Harvard University Press, Cambridge, Mass.
Thompson, Manley: 1981, 'On a priori truth', *The Journal of Philosophy* 78, pp. 458–82.
Wittgenstein, L.: 1963, *Tractatus Logico-Philosophicus*, trans. D. F. Pears and B. F. Mc-
Guinness, The Humanities Press, New York.

MICHAEL A. ARBIB

SEMANTICS AND ONTOLOGY: ARTHUR BURKS AND THE COMPUTATIONAL PERSPECTIVE

1. INTRODUCTION

Arthur Burks has gained renown both as a philosopher and as a computer scientist. As a philosopher, he is known both as a Peirce scholar (Burks, 1958) and for his original investigations into meaning and induction (Burks, 1977). As a computer scientist, he is known both for his pioneering work on the stored-program concept (von Neumann, Burks and Goldstine, 1946; Burks, 1963) and for his studies of logical nets (Burks and Wright, 1953; Burks and Copi, 1956; Burks and Wang, 1957) and cellular automata (von Neumann, 1966; Burks, 1970).

Perhaps the paper which most deeply draws upon these varied interests – meaning and induction, the work of Peirce, the study of automata – is the paper "Ontological Categories and Language" (Burks, 1967), which was originally written in 1953. The aim of the present essay is to use this paper as the springboard for the exposition of a computational perspective on the relation between semantics and ontology.

This exposition will be enriched by reference to other writings by Burks, as well as to studies by various workers on the formal semantics of natural languages, artificial intelligence, and cognitive science.

To aid reference to Burks' work, we shall employ the following abbreviations:

EV = "Empiricism and Vagueness" (Burks, 1946)

OCL = "Ontological Categories and Language" (Burks, 1967)

CCR = "Chance, Cause and Reason" (Burks, 1977)

CSP = "Computer Science and Philosophy" (Burks, 1979)

EIEI = "Enumerative Induction versus Eliminative Induction" (Burks, 1980)

Merrilee H. Salmon (ed.), *The Philosophy of Logical Mechanism*, 83–97,

2. THE PITFALLS OF ONTOLOGY

Newcomers to philosophy are amused by the following example of the philosophy of language [Tarski, 1944]:

> (1) 'The snow is white' is true if and only if the snow is white.

The four words in quotes constitute a sentence of the English language whose truth we seek to establish, while the last four words of (1) comprise an ontological expression of the state of the world. The unintended humour comes from the fact that Tarski has used English not only as the language whose semantics we seek, but as the language in which states of the world are expressed.

We may decompose the problem further. We are creatures who have a variety of skills for making our way about the world. Some of these we readily express in words, some clumsily, some hardly at all. For example, I cannot describe any but the most distinctive face sufficiently well for you to select the person who bears it from a largish crowd. We thus have a mismatch between our language and the perceptual and behavioral abilities that give them meaning. We can thus make a genuine advance on (1) if we can give a precise account of these skills in some new technical language, and then relate the truth of a sentence to the exercise of those skills that give us confidence in asserting the sentence, or otherwise using it. But, of course, this does not solve the problem. The account of the skills must represent the internal state of man or machine (the language user) in interaction with the environment. We *represent* the environment. But how? A tree of such and such a shape and size? A finely digitized bas-relief of optical reflectances? A Newtonian description? Relativistic field theory? We have gone too far, and yet not far enough. Too far, because (save perhaps in discussions of space travel and nuclear power) a relativistic or quantum-theoretic representation of 'reality' (dangerous word) is more detailed than necessary to underpin the semantics of everyday discourse. Yet not far enough, because our current physics is still (and perhaps always will be) in a state of flux, and has no vocabulary for much that touches the human psyche.

In short, then, our ontology is an admittedly imperfect representation of that which is; and our semantics is not so much a translation as a relation – between sentences of the language and states-of-the-world expressed in the ontology-language.

In [OCL, Section 6], Burks offers an ontology which views the world as comprised of "objects" extended in space-time (though I suspect that no use is made of any relativistic implication of this term). Such an object can 'possess a universal', and if it does so, we say that the universal is exemplified in that space-time region. (Burks uses the term 'position' where I use 'region', but it is certainly part of my ontology that [non-void?] universals cannot be exemplified without spatio-temporal extent.) A crucial role is then played by the notion of a *continuant*, a notion which, for Burks, involves not only spatio-temporal contiguity but also causality. A continuant would seem to be the ontological basis for what we would call an object. Just which sequences of events $e_1, \ldots e_N$ located at space-times $s_1, \ldots, s_N$ (the division of the space-time history of the continuant into events is arbitrary) constitute a continuant must depend on our ontology of causal connections: Is the contiguity of each e_j to e_{j+1} such that no other object could have been substituted? If there are apparent changes in the object between e_j and e_{j+1}, is there a causal explanation which in some sense maintains the 'thread' of the object. This is an ontology which makes sense of the everyday reality of hard tables, but which we cannot even define at the wave-function level of quantum mechanics. I have suggested elsewhere [Arbib, 1967] that Wigner's [1961] argument for his claim that self-reproduction is virtually impossible in a quantum-mechanical system would also seem to imply that "it would be a miracle" if a chair could subbsist from moment to moment. (It is of interest, in the present context, that Wigner's analysis and my reply were both prompted by von Neumann's theory of self-reproducing automata, to which Burks has contributed so much [Burks, 1970; von Neumann, 1966].) The import of the argument is that we have little understanding of how to represent, at the quantum-mechanical level, the constraints that constitute the cohering of matter into an enduring macroscopic object with a high probability of maintaining at least some continuity of identity (of the kind perceptible by a human observer). In fact, even the basic laws of Newtonian mechanics cannot be 'derived' from quantum mechanics without the imposition of further 'laws' of statistical averaging, and the invocation of methods of approximation, such as the WKB method (Messiah [1961, pp. 231–241]).

In summary, then, Burks' ontology is of a world of objects which endure from their creation to their dissolution, and their creation, enduring, and dissolution are all to be seen as falling under causal connections. In view of the above remarks, it would seem that Burks could not agree with Quine's [1978] view that there is a single reality in which we may ground our ontology, namely that of modern physics, though he might well wish to see his ontology as an *aspect* of that reality, at a level fine enough to allow us to analyze what it is in the world that provides the basis for our knowledge and our semantics, yet not so fine that we cannot see the wood for the wave-functions.

But Burks' ontology goes further (or less far?) than this. Not only is a physical continuant to be seen as "reducible to a sequence of physical events connected by causal relations", but "these events are ontological-compounds whose ontological-elements are universals, regions of space-time, and causal potentialities" [OCL, p. 40]. Burks sees his ontology as very similar to that of Plato but with the following differences [OCL, pp. 40–41]: (i) He does not see universals as having a higher mode of reality than particulars. A particular is not a copy of the corresponding universal; rather, the particular has the universal as one of its ontological-elements. (ii) Particulars are not created by a Demiurge who uses the universals as patterns; thus Burks need not face Plato's question: How can a perfect universal be a constituent of an imperfect particular? (iii) Plato has trouble with the question of Parmenides: "How can the same universal be in two distinct particulars at the same time?" Burks has an easy answer – it is the nature of a universal to be capable of exemplification in many different regions of space at the same time. (iv) Peirce's category of causality plays an essential role in Burks' ontology, but not in Plato's. However, a different category of causality may be discerned in Plato. For Plato, the Idea functioned as an ideal pattern or goal, and hence as a Final Cause. It seems to Burks that, for Plato, final causes "do the work" of causal potentiality, determining (in a future to past way) many possible outcomes or possible universes. (cf. [Boden, 1972], who argues that purpose is compatible with and wholly dependent on mechanism, but that purposive accounts cannot be replaced by mechanistic ones without a real loss of explanatory power.)

3. THE RELATIVITY OF UNIVERSALS.

Burks uses 'brownness' and 'tableness' as examples of the universals to be appealed to in his ontology in presenting an event as an "ontological-compound whose ontological-elements are universals, regions of space-time, and causal potentialities". I believe that this is a mistake. One man's "brown" is another man's "dark olive". One woman's "table" is another's "workbench". For this reason, I will henceforth eschew the term 'universal' – with its inevitable implication of universality – and substitute the term 'concept'. This then leaves open the question as to whether certain concepts are or are not, in some sense, universal.

In the last section, I suggested that semantics should not be directly related to ontology, but that the mediation should be via a human's perceptual and motor skills. Let me, then, speak of language *per se* (e.g., English), skill-language, and ontology-language. I will argue that concepts such as 'brownness' and 'tableness' are to be seen as part of skill-language. Ontology-language can then talk of objects enduring in space and time. I agree with Burks that we need not use the language of quantum mechanics as our only ontology-language (though it *does* provide an extension of that language for certain phenomena – in the age of scientific revolutions, even ontology is dynamic). Rather, if we are to explain certain verbal abilities, or certain skills, we must use an ontology expressive enough to offer a real (if inevitably partial) explanation; not (2) but rather of the form (3):

(2) 'This is a brown table' is true because it exemplifies the universals 'brownness' and 'tableness'.

(3) 'This is a brown table' is true because it has reflectance properties which cause it to reflect white light with the following ranges of hue, saturation and intensity ... (which is why it is called 'brown'), and it has a horizontal surface ... (which is why it is called a 'table').

Note that in (3), the explanations of 'brownness' and 'tableness' could be refined to great levels of precision; that the choice of measurements and qualities invoked in the explanation will be defined by (and, in their achievement, help define) the ontology-language; and that different refinements may well lead to different extensions of these concepts. The 'universals', then, are those qualities that enter into the ontology-language. However, to a great extent, they will not have the Boolean, yes-or-no, quality often associated with the notion of universal (this is, or is not, a table), but rather have some aspect of measurement (not "This is a wavelength, yes or no?", but rather "What is the spectral decomposition of this light- bundle?").

I do not know how much of this would be acceptable to Burks. However, he would certainly agree that an important approach to characterizing concepts is to specify a machine which can recognize them, and that the embodiments of a concept in different brains or machines may well have different extensions. Burks invokes the pragmatic theory of Peirce to assert that "understanding a directly applicable symbol involves a capacity to use this symbol in experiential situations" [OLC, p. 30], and he suggests how one could construct an (unconscious) automaton capable of understanding symbols in this sense, e.g. by printing out the colour and geometrical description of a coloured square, circle or triangle presented to a light-sensitive input device [OCL, Section 5]. If we are to imagine a concept as neither 'innate' (in the human) nor 'hard-wired' (in the machine), then it must be learned. Burks shows convincingly that any concept derived from experience is, in principle, vague. Unless a concept can be unequivocally defined in terms of concepts which are unequivocal (using a language whose compositional semantics is itself unequivocal), then the acquisition of the concept must be based on the experience of a number of instances, both positive and negative (cf. Winston [1975] for a simple learning algorithm which embodies this principle):

"A single instance will not suffice, since it embodies many qualities, so that one presentation cannot determine which one is being denoted. These can be eliminated in the process of ostensive definition only by presenting other instances that lack them but to which the concept to be defined ostensively is applicable ... [Moreover, since, for example,] every correct application of the concept 'chair' is also a correct application of the concept 'furniture' ... [i]t is necessary that cases to which the word does *not* apply [also] be presented.... An ostensive definition is unavoidably limited to a finite number of presentations. Hence borderline cases may arise. If meaning is defined in terms of ostensive definition the possibility of vagueness is unavoidable, since it is impossible even in principle to define a term ostensively so that it will cover all possible situations" [EV, pp. 47B & 480].

Burks notes that his analysis shows that vagueness is a logical possibility, but not a metaphysical necessity, for any concept applicable to the real world. For example, our world may be such that we will never meet a borderline case which tests the limits of applicability of our concepts of 'dog' and 'cat'; or we might well agree on the classification of the dogs and cats of our domestic experience, and disagree as to whether an exotic newcomer was dog or cat or neither.

In later papers (cf. [CSP, pp. 408–409] and [EIEI, p. 174]), Burks has refined this notion of vagueness by means of a complexity argument. He holds that a concept is vague when the complexity of the structure of properties involved in instances of the concept in the domain of application exceeds the complexity of a pattern-recognition rule which attempts to classify instances as to whether or not they fall under the concept. In such a case, some instances of the domain will be too complex for the rule to analyze successfully. (One is reminded of Richard Gregory's [1967] positive answer to the question, "Will Seeing Machines have Illusions?") In our current terminology, we may say that we have a concept expressed in ontology-language which is to be learned; a concept expressed in skill-language which approximates it; and an almost inevitable mismatch. Such a perspective reinforces our warning against the danger of viewing such concepts as 'brownness' and 'tableness' as "ontological-elements". However, I suspect that Burks would not be happy with the way in which we have moved 'brownness' and 'tableness' from the universals of ontology to the concepts of skill, for he seems firmly wedded to the notion that these are universals *embodied* in the real world rather than concepts *ascribable* to the real world:

"The empiricist may deny that there are any genuinely vague concepts, but such a position contradicts experience. So he is driven either to conceptualism, a denial that there are universals embodied in things, or to that extreme form of realism which holds that participation in a universal by an object admits of degrees. ... [A]n empiricist can not consistently hold to the view that all universals embodied in the real world are precise universals" [EV, p. 483].

It would seem that I am a *conceptual realist*: I accept an ontology which views the world as made up of 'things' having spatial extent, optical reflectance, etc. ('universals' which may be quantifiable rather than Boolean); I hold that 'brownness' and 'tableness' are not universals of ontology but concepts acquirable by skill; and I hold that 'participation' in such a concept by an object admits of degrees, which depend on the observer and his relation with the object.

4. A GLIMPSE OF INDUCTION

Burks has developed a sophisticated theory of induction [CCR]. It is beyond the scope of this paper to analyze this theory. Instead, we briefly recall Burks' own ideas about the Peircean concepts that underlie his approach. Then in the next section, at long last, we shall return to the starting point of our investigation – the semantics of language.

Burks follows Peirce in holding that there is a strong analogy between the evolution of species, and the evolution of science:

"Peirce used the term 'habit' in this connection, employing it in a sense sufficiently general to cover the hereditary action patterns of the individuals of a species as well as learned habits. The concepts, rules, methods, procedures, etc., of social institutions ... are also included. In science the habits are the procedures and methods of enquiry ... as well as the accepted theories." [EIEI, pp. 177]

Burks then observes that Peirce intended his notion of habit to cover *any set of operative rules embodied in a system.*

"Peircean habits are both stable and adaptive and thus have the two characteristics essential to the construction of complex compounds by an evolutionary process. Thus Peircean habits are appropriate building blocks in a natural hierarchy ... Generalizations, laws, and theories established at one level become the background assumptions for those tested at the next level. For example, in conducting an experiment, scientists rely on the laws governing the apparatus. ... I suggest that our concepts have developed ... in a manner similar to the way that our genetic programs have evolved. Concept rules have been modified, compounded with other concept rules, tested for their utility in helping us adjust to the environment, with the most useful ones being retained at each stage. ... This is done at successive levels of the hierarchy, beginning with the common-sense concepts and generalizations, moving to scientific concepts and causal laws, and on up to the concepts and theories of highly advanced science." [EIEI, pp. 177–179, 181, 183]

I find myself in general agreement with this perspective and would just add a few comments:

First, Peirce's notion of 'habit' strikes me as highly similar to Piaget's notion of a schema (see, e.g., Part II of Beth and Piaget [1961]). Moreover, Burks' distinction between the habit as set of operative rules and the system that embodies it is exactly parallel to the distinction [Arbib, in press] between two senses of the Piagetian scheme, as corresponding to the input-output behavior of an automaton, and to the internal-state description of an automaton which exhibits that behavior, respectively. This internal-state description might be in terms of state-transition and output functions, or be more like a computer program together with its state of computation.

Second, as Burks outlines in [CSP], the carrying out of the above outline of induction in any detail will be a major problem in artificial intelligence and will provide a major meeting ground for computer science and philosophy. We need a theory not only of learning machines which embody adaptable habits, but also of control structures which can orchestrate these machines in 'coordinated control programs' to solve complex problems. (Hesse [1973] also takes a learning-machine perspective on the analysis of changing concepts by the philosophy of science.)

Third, this learning-machine view provides a new translation of the rationalist-empiricist controversy (of renewed topicality with the debate between Piaget, as neo-empiricist, and Chomsky, as rationalist, reported in Piattelli-Palmarini [1980]):

"Let us call the basic structure of the central nervous system together with any program stored in it initially 'man's innate structure program complex,' and let us assume some substantial correspondence between the operations of the human mind and the neural mechanisms of the central nervous system. ... The rationalist would hold that certain specific innate concepts (e.g., space-time, cause effect), rules (e.g., for grammar and inference), or principles (e.g., the uniformity of nature) exist in man's innate structure-program complex. The empiricist would hold that man's innate structure-program complex does not contain any such specific concepts, rules, or principles, but only a very general learning program." [CSP, Section 4]

What I would add to all this is that we not only require an account of learning machines *per se*, but also a theory of the means of communication which enable a community to attain consensus as to a set of instances of a concept by a process of mutual verification (Arbib [in press], Hesse [1979]).

5. THE SEMANTICS OF LANGUAGE

I started out, in Section 2, from the problem of ascribing meaning to a sentence, and soon came to consider the ontology offered by Burks in [OCL], a paper whose first paragraph asserts that "Since language is a tool for designating and describing reality, an adequate philosophy must provide a single categorial schema for both language and the non-linguistic part of reality." Burks asserts [OCL, p. 26] that the "manner in which a predicate represents a universal to a symbol user who understands that predicate involves a concept of this universal in the mind of the user. On this ground we take the meaning of a predicate to be a concept of a universal". I demurred somewhat, holding that the task of ontology was not to show continuants as embodying universals, but rather to provide a substrate for a theory of how a learning machine could come to ascribe concepts. But I agree with Burks as to the general outline of such a theory of learning machines as related to Peircean habits (or Piagetian schemas); with the filling-in of this outline posing a major challenge for cooperation between computer science and philosophy. This measure of agreement seems sufficient for us to continue.

There are several aspects to understanding . . . a language [in which empirical content can be expressed]. First is the ability to reason deductively in that language. Second is the capacity to reason inductively in that language. . . . A further ability is required: the ability to apply the symbols of the language to experiential situations, i.e., to decide the truth status [relative to the state of the world] of factual statements of the language." [OCL, p. 29]

"Meaningful-symbols may be divided into two classes, according to whether their meanings are derivable from their constituent symbols (and the rules of syntax governing their arrangement) or not; we shall call these *compound-symbols* and *element-symbols* respectively. . . . [T]he receiving of new information requires the comprehending of novel meanings [and this important use of language, the communication of new information, is made possible by a] process of deriving the meanings of compound-symbols from the meanings of their constituent element-symbols in accordance with the syntactical rules governing their arrangement." [OCL, p. 32]

This process, which Burks calls *the procedure of deriving meanings*, is known to linguists under the rubric of *compositional semantics*, where it is seen as a development of Tarski's model-theoretic semantics for logical languages. To get some flavor of Tarski's semantics, consider the following fragment, where $[|a|]$ is the truth-value ascribed to sentence a. If a and b have already been assigned truth values, *true* or *false*, then

$$[|a \, AND \, b|] = \begin{cases} true & \text{if} [|a|] = true \text{ and} [|b|] = true \\ false & \text{otherwise} \end{cases}$$

$$[|a \, OR \, b|] = \begin{cases} true & \text{if} [|a|] = true \text{ or} [|b|] = true, \text{ or both} \\ false & \text{otherwise} \end{cases}$$

$$[|NOT \, a|] = \begin{cases} true & \text{if} [|a|] = false \\ false & \text{if} [|a|] = true. \end{cases}$$

Burks gives such a model-theoretic semantics for his 'logic of causal statements' in [CCR, Section 6.3]. Again, it is beyond the scope of this paper to give a critique of this semantics or of its application in Burks' theory of induction. However, it does seem appropriate to note a strange gap in Burks' writing. Despite his early interest in "the procedure of derived meanings" in providing a semantics for natural language [OCL, written in 1953], and his explicit use of a model-theoretic semantics [CCR, 1977], he nowhere discusses *Montague grammar*, perhaps the most thoroughgoing attempt to provide a compositional semantics for English (Montague [1974], Partee [1976]). This is even more striking because both Montague and Burks have recourse to what is called a *possible world semantics*. Let me first sketch the way in which Montague grammar offers a truly articulated form of "the procedure of derived meanings" which is at most hinted at in [OCL], and then close with a critique of "possible world semantics" founded on the way in which it falls short of Burks' own call for an automaton-theoretic approach to induction (Section 4).

Briefly, Montague grammar uses a *categorial grammar* to represent the syntax. There are two *basic categories*, the category e whose semantic interpretation is a set $[|e|]$ of entities, and the category t of sentences whose semantic interpretation is the set $[|t|]$ of truth values. All other categories are derived categories of the form $<a, b>$, where a and b are either basic categories or previously derived categories. We have $[|<a, b>|] = [|a|] \rightarrow [|b|]$, i.e. any item of category $<a, b>$ has as its interpretation a map from $[|a|]$ to $[|b|]$. For example, intransitive verbs are assigned to the category $<e, t>$, for given any entity (the interpretation of the subject of a verb) we are to determine a truth value (the interpretation of the sentence so formed). Actually, the story is more complicated than this. Montague (mistakenly, in my opinion) does not actually have e as an explicit category in his grammar. Rather, e only occurs implicitly, as above where $<e, t>$ is the category of a predicate. Each subject is then to be assigned not to e but to the derived category $<<e, t>, t>$ – i.e. a subject is such that, given a predicate (of category $<e, t>$), it returns a sentence (of category t). Thus, 'John' is not to be interpreted by an element of the domain $[|e|]$, but rather as a map which assigns to each predicate, e.g., 'walk', a truth value j'[walk] which is *true* or·*false* according as to whether or not John walks in the world under consideration. (Note that a 'world' in this sense is a world at some given moment of time.) The virtue of this apparent complication (I shall not discuss its vices here) is that it enables us to treat uniformly entities such as 'John' and more complex subjects such as 'every man':

$$[|every man|] (P) = true$$

Just in case $[|(\forall x) (M(x) \rightarrow P(x))|] = true$ on the usual Tarskian semantics of the universal quantifier, M(x) being the formal expression for "x is a man". Barwise and Cooper [1981] further develop Montague's treatment of noun phrases, and show some of its implications for a theory of natural language. However, rather than further review current developments in Montague grammar, we now turn to an all-too-brief critique of model-theoretic semantics.

Succinctly, the problem with model-theoretic semantics is that it presents a world at precisely the level of aggregation of the language it is to interpret:

(4) 'Snow is white' is true if snow'[white] = *true*.

We are back to (1), but in logical notation! Of course, this is too glib. The achievement of Montague grammar is to provide a set of syntactic rules for parsing a sentence in such a way that we can indeed use the rules of syntactic composition to drive the semantic composition which lets us infer the truth value of the sentence from the interpretation of its constituents. Nonetheless, a 'world' in this framework is simply a set of value specifications of just such a kind as snow'[white] = *true*. This falls short of Burks' Peircean inspiration in at least three ways:

First, by not having a fine-level ontology (of the kind I argued for in Section 2), there is no room for truth values to be *computed* for a state of the world by a 'habit' or 'schema', rather than simply asserted.

Second, because a model must provide all the interpretations necessary to infer the truth value of every sentence, there is no room for the imparting of novel information which Burks [OCL, p. 32] has rightly seen as a fundamental role of language. Thus, even if we remain at the level of aggregation of model theory, we must use *partial* models, in which, for example, only certain subject-predicate pairs have already been assigned truth values. Then an assertion can be checked for consistency with the current 'data-base', and incorporated if it is novel and consistent. (Of course, this is only a small part of the story. For one of the first successful artificial intelligence contributions in this spirit, see Winograd [1972]).

For the third and final point, we must briefly recall how 'possible world semantics' handles the modalities of necessity, $\Box$, and possibility, $\Diamond$. Associated with each '"actual" world' W (a set of constituent interpretations) is a set P of other 'worlds' (each another set of constituent interpretations) which are '"possible" worlds' for that '"actual" world'. Then the statement q is necessarily true, i.e., $\Box q$ is true, in W just in case q is true in all the possible worlds in P; while $\Diamond q$ is true in W just in case q is true in at least one world of P.

What is missing here is the Peircean category of causality, and the crucial fact that we have only partial knowledge of W. My ontology says that if we were to have complete knowledge of a world, then truth, necessity and possibility would all be equivalent. It is only when we have partial knowledge (and we always do!) that necessity and possibility become interesting. I would say that "q is necessary in W" if "q is true in every completion of W", while "q is possible in W" if "q is true in at least one completion of W". We obtain a formulation skin to the classic one if we take each actual world to be incomplete, with the set P of all worlds possible relative to W being the set of all completions of W compatible with the causality embodied in our ontology.

6. CONCLUSION

Peircean 'habits' provide the bridge between ontology and the semantics of natural language. The further study of Peircean mechanisms of induction will require the development of a refined ontology of causality in tandem with new techniques in artificial intelligence. Arthur Burks has charted the path for a profound rapprochement between computer science and philosophy. It will be an arduous path to follow, but the challenges and insights will make the effort worthwhile.

Michael A. Arbib
Program in Neural,
Behavioral and Informational Sciences,
University of Southern California,
Los Angeles, CA 90089–0782, U.S.A.

REFERENCES

Arbib, M. A., 1967, Some comments on self-reproducing automata. In *Systems and Computer Science* (J. F. Hart and S. Takasu, Eds.), University of Toronto Press, pp. 42–59.

Arbib, M. A. (in press), A Piagetian perspective on the construction of language, *Synthese*.

Barwise, J. and Cooper, R., 1981, Generalized quantifiers and natural language, Linguistics and Philosophy 4: 159–219.

Beth, E. W. and Piaget, J., 1966, *Mathematical Epistemology and Psychology* (Translated from the French by W. Mays), D. Reidel Publishing Company.

Boden, M. A., 1972, *Purposive Explanation in Psychology*, Harvard University Press.

Burks, A. W., 1946, Empiricism and vagueness, *Journal of Philosophy* 43: 477–486.

Burks, A. W., 1949, Icon, index, and symbol, *Philosophy and Phenomenological Research* 9: 673–689.

Burks, A. W., 1951, A theory of proper names, *Philosophical Studies* 2: 36–45.

Burks, A. W. (Ed.), 1958, *Collected Papers of Charles Sanders Peirce*, Vol. VII, *Science and Philosophy* and Vol. VIII, *Reviews, Correspondence, and Bibliography*, Harvard University Press.

Burks, A. W., 1963, Programming and the theory of automata. In *Computer Programming and Formal Systems* (P. Brafford and D. Hirschberg, Eds.), Amsterdam: North-Holland, pp. 100–117.

Burks, A. W., 1967, Ontological categories and language, *Visva-Bharati Journal of Philosophy* 3: 25–46. Visva-Bharati (University), Santiniketan, West Bengal, India.

Burks, A. W. (Ed.), 1970, *Essays on Cellular Automata*, Urbana: Univ. of Illinois Press.

Burks, A. W., 1977, *Chance, Cause, Reason – An Inquiry into the Nature of Scientific Evidence*, Chicago: Univ. of Chicago Press.

Burks, A. W., 1979, Computer science and philosophy. In *Current Research in Philosophy of Science* (P. Asquith and H. Kyburg, Eds.), East Lansing, Michigan: Philosophy of Science Association, pp. 399–420.

Burks, A. W., 1980, Enumerative induction versus eliminative induction. In *Applications of Inductive Logic* (L. J. Cohen and M. Hesse, Eds.), Oxford University Press, pp. 172–189.

Burks, A. W. and Copi, I. M., 1956, The logical design of an idealized general-purpose computer, *Journal of the Franklin Institute* 261: 299–314 and 421–436.

Burks, A. W. and Wang, H., 1957, The logic of automata, *Journal of the Association for Computing Machinery* 4: 193–218 and 279–297.

Burks, A. W. and Wright, J. B., 1953, Theory of logical nets, *Proceedings of the Institute of Radio Engineers* 41: 1357–1365.

Gregory, R. L., 1967, Will seeing machines have illusions? In *Machine Intelligence 1* (N. L. Collins and D. Michie, Eds.), Oliver and Boyd, pp. 169–177.

Hesse, M., 1973, Models of theory change. In *Logic, Methodology and Philosophy of Science* (P. Suppes *et al.*, Eds.), North-Holland, pp. 379–391.

Hesse, M., 1979, Habermas' consensus theory of truth, *Philosophy of Science Association 1978*, Vol. ii (P. D. Asquith and I. Hacking, Eds.).

Messiah, A., 1964, *Quantum Mechanics*, Vol. I (Translated from the French by G. M. Temmer), North-Holland Publishing Company.

Montague, R., 1974, *Formal Philosophy* (Edited and with an introduction by R. Thomason), Yale University Press.

Partee, B. H., 1976, *Montague Grammar*, Academic Press.

Piatelli-Palmarini, M. (Ed.), 1980, *Language and Learning: The Debate between Jean Piaget and Noam Chomsky*. Harvard University Press.

Quine, W. V., 1978, Otherworldly (a review of *Ways of Worldmaking* by Nelson Goodman), *The New York Review of Books,* November 23, 1978, pp. 25–26.

Tarski, A., 1944, The semantic conception of truth. *Philosophy and Phenomenological Research*, 4: 341–376.

Wigner, E. P., 1961, The probability of the existence of a self-reproducing unit. In *The Logic of Personal Knowledge: Essays Presented to Michael Polanyi*, The Free Press, pp. 231–238.

Winograd, T., 1972, *Understanding Natural Language*, Academic Press.

Winston, P. H., 1975, Learning structural descriptions from examples. In *The Psychology of Computer Vision* (P. H. Winston, Ed.), McGraw-Hill, pp. 157–209.

von Neumann, J., 1966, *Theory of Self-Reproducing Automata* (Edited and completed by A. W. Burks), Urbana: Univ. of Illinois Press.

von Neumann, J., Burks, A. W. and Goldstine, H. H., 1946, *Preliminary Discussion of the Logical Design of an Electronic Computing Instrument*. Princeton Institute for Advanced Study (Second edition, 1947).

NOTE ADDED IN PROOF

The present chapter was written in 1982. A number of the themes addressed here are developed in M.A. Arbib and M.B. Hesse (1986), *The Construction of Reality*, Cambridge University Press, which is an expansion of the authors' 1983 Gifford Lectures.

STEVEN E. BOËR

NAMES AND ATTITUDES

INTRODUCTION

Arthur Burks was one of the first philosophers to recognize the inadequacy of the traditional "Description Theory" of proper names inherited from Frege and Russell; he was also one of the first to suggest a plausible revision of that theory (Burks [5]). His suggestion – that proper names be viewed as *indexical* definite descriptions – anticipated by nearly three decades the general drift of accounts of proper names such as those currently offered by Tyler Burge (in [4]) and Stephen Schiffer (in [22]). Description Theories, however, have recently come under frontal attack from proponents of so-called Causal-Historical Theories (e.g., Kripke [13] and Donnellan [10]), who urge a radically different account of the mechanisms of reference. Whatever one may think of this attack – and its success is by no means uncontested (cf. McKinsey [19]–[20] and Boër [2]) – Description Theories still have many enthusiastic supporters (including Burks in [6]). Nor do they lack ammunition for a counteroffensive. Their most powerful weapon is derived from the well-known tangle of problems attending the interpretation of names in propositional attitude contexts, which they increasingly point to as evidence against the Causal-Historical Theorists' contention that names "merely designate" and do not "express (descriptive) senses" (cf., e.g., Loar [18]).

As a convert to the Causal-Historical Theory, I shall attempt to supplement the frontal attack on Description Theories with domestic sabotage, by rebutting the charge that the Causal Theory cannot cope with the propositional attitudes. In so doing, I will be forced to make a number of assumptions, some of which will inevitably be unacceptable to certain proponents of the new faith, since acceptance of the Causal-Historical model is compatible with considerable divergence of opinion on other semantical matters. My aim, however, is merely to provide *a* plausible background against which one could embrace a Causal-

99

Merrilee H. Salmon (ed.), *The Philosophy of Logical Mechanism*, 99–130.
© 1990 *Kluwer Academic Publishers.*

Historical account of names while being untroubled by the data from propositional attitude contexts which are standardly urged against it.

1. THE PROBLEM POSED BY PROPOSITIONAL ATTITUDE CONTEXTS

As standardly presented, the Causal-Historical Theory of proper names has both a positive and a negative component. The positive component is the by now familiar account of dubbings and chains of transmission, which explains how – in the absence of "senses" – a particular name-token may be ultimately connected with a certain object as its referent. The negative component is the claim that names (as ordinarily used) have *only one* semantical role: to designate the objects (if any) with which their uses are thus connected.[1] If nothing else, senses are redundant.

The Description Theorist now launches his counteroffensive with the following argument.

(I) The truth-condition of a sentence is a function of its logical form and the semantical contributions made by its extralogical constituents. (Shared assumption)

(II) When S is any sentence, E is any extralogical constituent of S, E, is the ith occurrence of E in S, and F is an expression of the same grammatical category as E which is foreign to S: S has the same truth-condition as $S(F/E_i)$, provided that F makes the same semantical contribution to $S(F/E_i)$ as E_i does to S. (From (I))

(III) Where a and b are distinct but codesignative names and $\emptyset$ is a predicate: the sentence ⌜John believes that a is $\emptyset$⌝ has the same truth-condition as ⌜John believes that b is $\emptyset$⌝, provided that b makes the same semantical contribution to the latter sentence that a makes to the former. (Instance of (II))

(IV) But even under the stated condition (which guarantees the truth of '$a = b$') the sentence 'John believes that a is $\emptyset$' can differ in truth-value from 'John believes that b is $\emptyset$'. (Alleged fact)

Therefore,

(V) These two sentences do not have the same truth-condition. (From (IV) and obvious principles)

Therefore,

> (VI) *a* and *b*, though codesignative, do not make the same
> semantical contributions to their respective hosts – i.e.,
> there must be some semantical feature in which *a* and *b*
> differ. (From (III) and (V))

The extra and obviously *non*redundant feature mentioned in (VI) is, of
course, just what a "sense" is supposed to be. So the Description
Theorist concludes that we must posit "senses" for proper names after
all – these presumably being captured by descriptions of some sort.[2]

This is a very seductive argument. It should be noted, however, that
admitting its soundness would only force the abandonment of what I
have called the negative component of the Causal-Historical Theory of
Names. One could still consistently maintain the positive claim about
the causal mechanics of reference; but once in bed with the Description Theorist, it would be difficult to resist his further advances.

If we are to avoid such a compromising situation, at least one of (I)-
(VI) will have to go. The most plausible victim is (IV), or so it seems
at first glance. But (IV) turns out to be no easy mark: it has a rationale
so familiar and intuitive as to merit the title "The Canonical Defense".
Here is a particular instance of the Canonical Defense. Pick an
average, middle-aged American movie-goer and ask 'Is Tony Curtis an
actor?', eliciting the response, 'Yes, Tony Curtis is an actor'. Now ask,
'How about Bernie Schwartz?'. You will probably get the reply, 'Never
heard of him' and, with a little coaching, an avowal like 'I don't have
any opinion'. This is precisely the kind of dialogue we would normally
cite as evidence that our subject believes that Tony Curtis is an actor
but does *not* believe (i.e., neither believes nor disbelieves) that Bernie
Schwartz is an actor – and this in spite of the fact that *we* know that
Tony Curtis *is* Bernie Schwartz. Hence 'The subject believes that Tony
Curtis is an actor' seems true, whereas 'The subject believes that Bernie Schwartz is an actor' seems false, yet 'Tony Curtis' and 'Bernie
Schwartz' designate one and the same man. Since this little *Ge-
dankenexperiment* obviously can be extended to *any* pair of distinct but
codesignative names (given a suitable informant), what better defense
of (IV) could there be?

Kripke, however, has demonstrated (in [14]) that the Canonical
Defense is no defense at all. For he has shown that when fully articulated it rests upon principles which, though individually plausible,
jointly lead to nasty paradoxes. The Canonical Defense, like a valid ar-

gument in Naive Set Theory which covertly relies on the principle '$(Ez)(x)(x .5. z \leftrightarrow \emptyset x)$', gains its air of convincingness by concentrating on concrete details while distracting our attention from the tendentious underpinnings. This result, of course, does not show that (IV) is false, which would be tantamount to showing that belief-contexts (and, by extension, propositional attitude contexts in general) are extensional with respect to names. But it does demolish the only intuitive reason for accepting (IV) that has ever been offered. Having thus cut (IV) adrift, Kripke has revived the possibility of the extensionality of *de dicto* propositional attitude contexts with respect to names (hereafter: "N-extensionality"). In what follows, I shall take advantage of this reopening of the question and suggest a way of construing N-extensionality which makes it a very live option indeed.

2. N-EXTENSIONALITY: DE JURE OR DE FACTO?

There are two ways in which one might try to motivate the claim that propositional attitude contexts are N-extensional even on their *de dicto* readings. The first would be to proffer and defend a semantics for English which automatically validated arguments like (1):

> (1) a. John believes that Cicero was a Roman orator.
> b. Cicero = Tully.
> Therefore,
> c. John believes that Tully was a Roman orator.

No doubt such a semantics for English could be formulated, but defending it is not a very inviting task. The reason is simple. A semantics which validated (1) would presumably mark (2) as self-contradictory:

> (2) Although Cicero is Tully, John believes that Cicero was a Roman orator but fails to believe that Tully was a Roman orator.

But (2) is manifestly *not* perceived as self-contradictory by most speakers of English. Quite the contrary, (2) sounds as if it might well be true. A semantical theory, however, is part of an overall explanation of human verbal behavior and cannot simply ride roughshod over empirical facts about the ways in which people act with, and react to, bits of their language. If a semantical theory stigmatizes a sentence as analytically false, it predicts in effect that a suitably reflective and semantically competent native speaker could reason his or her way *a priori* to the conclusion that the sentence in question is false. Confronted with the data about people's reactions to sentences like (2), the semantic theorist would seemingly be forced to posit what amounts to widespread *stupidity* on the part of most English-speakers: they aren't semantically competent enough, or aren't reflective enough, or whatever. In light of the fact that (2) and its obvious analogues are short, have a low degreee of syntactic complexity, and embrace virtually *all* the propositional-attitude verbs of English, the posited onslaught of incompetence can only seem mysterious and *ad hoc*.

One possible rejoinder must be mentioned. "Surely," it might be said, "Kripke has already shown us the way to explain this lapse: people mistakenly tolerate (2) and its ilk because they have been *duped* by some version of the Canonical Defense, whose flaws are admittedly subtle and hence easily overlooked." Certainly there is some truth to this remark, but it does not immediately get the proponent of the first approach off the hook. For it overlooks the fact that, *ex hypothesi*, the native speaker ought antecedently to be equipped to reason out the falsity of (2) and its relatives *a priori*. Fooling an amateur is one thing; fooling a professional is quite another – and, in the relevant sense, we are all supposed to be "professionals". So the mystery persists: how could we be semantically armed against this deception and yet be so easily gulled? In default of some *principled* explanation of this anomaly, we have good reason to abandon the first approach and to look elsewhere.

The failure of the first approach lay in its attempt to secure N-extensionality *de jure*, by semantical decree. The alternative would be to attempt to secure N-extensionality *de facto*, by appeal to something which is *external* to semantic competence. This second approach, if feasible, would immediately result in a major benefit anent our treatment of the empirical data of speech. If, from the standpoint of the semantic theory which models semantic competence, (1) is not valid and (2) is contingent, then our theory is already in line with speakers' intuitions; there is no divergence between actual and predicted reactions to need explaining-away. The burden has been shifted elsewhere, outside the sphere of pure linguistic mastery. The question is now: Is there anything else to bear this burden? I shall approach the answer indirectly.

First of all, I want to question the traditional dogma that the validity of an inference is solely a matter of its logical form – which would rule the second approach out of court at the outset. The dogma dies hard, for it represents matters as being just the way we would *like* them to be. No one can deny the attractiveness of having a comprehensive syntax and semantics for English which would systematically assign to the components of each correct inference $S_1, \ldots, S_n/T$ syntactically plausible logical forms $\sigma_1, \ldots, \sigma_n, \tau$ such that $\{\sigma_1, \ldots, \sigma_n\} \models \tau$. The problem is just that the current state of semantic theorizing, constrained by its marriage to natural language syntax, provides no real reason for expecting that such an ideal account can be given.[3] One of the ways in which failure of the dogma manifests itself is in the eventual appeal to Carnapian "Meaning Postulates" (in one guise or another) even by those semanticists working with extremely rich logical apparatus (e.g., Montague Grammarians). The requirement that assignments of logical forms be syntactically plausible forces one to seek supplementary aids.

Meaning Postulates, of course, are not the only possible supplements. Indeed, my claim that (1) is valid in virtue of something other than its logical form must rely on some device *other* than an appeal to Meaning Postulates, lest we fall prey to the same trap which snared the first approach. For so long as Meaning Postulates are properly regarded as part of Semantics, and Semantics in turn is regarded as the theorical modelling of semantic competence in speakers, we would still be saddled with the mystery of why ordinary (presumably competent) speakers of English do not hear (2) as self-contradictory. After all, one of the principal reasons for adding Meaning Postulates in the first place is to capture "felt" implications not autonomously generated by the underlying logic, and (1) certainly does not correspond to one of these felt implications, as the standard reaction to (2) attests.

In virtue of what, then, could (1) be said to be valid? Apart from our logical theory and our empirically motivated Meaning Postulates, what else is there to do the job? Discussion of the following examples will help to motivate the sort of answer I have in mind.

> (3) Elizabeth I could have been born to Katherine of Aragon instead of to Anne Boleyn.

(4) Lincoln's death, which was caused by a head-wound, could have resulted from poisoning instead.

(5) Water might have been something other than H_2O.

(3) is true iff there is a possible situation in which Katherine of Aragon did, and Anne Boleyn did not, give birth to Elizabeth I. (4) is true iff Lincoln's death resulted from a head-wound but there is a possible situation in which it (the same event) resulted from poisoning. And (5) is true iff there is a possible situation in which water differs in kind from H_2O. These are the sorts of verdicts we would expect from a decent Semantics for English. We cannot expect, however, that Semantics will tell us *what* possible situations there are. The job of Semantics is the systematic provision of adequate truth-*conditions* for sentences of the object-language in terms of antecedently understood sentences of the semantical metalanguage. In the case of (3)–(5), these metalinguistic formulations speak respectively of certain possible relations among persons, event(-token)s, and natural (chemical) kinds. These notions – familiar ones to be sure – are primitive from the purely *semantical* standpoint. In this respect they are no different from other items of MetaEnglish such as 'table' or 'bottle', which also figure in the truth-conditions of English sentences. From the standpoint of *philosophical analysis*, however, they may well not count as primitive: indeed, it is part of the task of Analytical Metaphysics to provide us with analyses of these notions which issue in identity and individuation criteria for persons, events, and other such philosophically interesting citizens. From the purely semantical standpoint, there is nothing about (3)–(5) to suggest that they are anything other than contingent – and that is probably the way most English-speakers hear them. But from the standpoint of philosophical analysis of the notions ingredient in their truth-conditions, one might see (3)–(5) as *incapable* of truth. If various fashionable essentialistic doctrines are accepted, (3)–(5) will be branded, not just false, but metaphysically false – on the ground that the stated truth-conditions are metaphysically incoherent in some way. (E.g., a child of different parentage might look and act like the historical Elizabeth I, and even enjoy a similar career; but she would be a mere *Doppelgänger*, not the *same person* as "our" Elizabeth I.)

The distinction I am urging between Semantics and Philosophical Analysis is often obscured in practice by the fact that both arts are frequently exercised in tandem by a given practitioner, so that it is difficult to tell where one leaves off and the other begins. If, as a philosopher, one is convinced that persons are, say, "mere bundles of sensations," then one will be tempted, when speaking as a semantic theorist, to build this reductive analysis into one's semantical metalanguage, with the result that object-language sentences employing the word 'person' are given truth-conditions which talk of bundles of sensations. Such a move, however, could only be justified as *semantic* theorizing if it could be shown that the incorporation of the reductive analysis into the axioms of the semantic theory was *required* in order to get the proper distribution of truth-conditions among sentences of the object-language. In other words, it would have to be shown that a "decompositional" axiom like (6) enables us to capture some important semantical data which could not be captured by using instead the obvious "austere" axiom (7):

> (6) An object X satisfies 'person' iff X is a bundle of sensations.
>
> (7) An object X satisfies 'person' iff X is a person.[4]

Departures from Austerity can, of course, take other forms as well, as when one assigns to an object-language sentence a "logical form" which seems at odds with our intuitive understanding of the sentence's superficial structure and/or content. But, in any event, such departures must be independently motivated on syntactico-semantic grounds.

In contrast to the relatively modest aims of Semantics, the goals of Philosophical Analysis, as typified by theories about the identity and individuation criteria for persons and events, are quite ambitious. It should come as no surprise that an adequate Semantics for English should leave open certain questions about the truth-possibilities of certain English sentences, whereas a definite verdict about those possibilities might be forthcoming from the philosophical analysis of various notions which figure in the semantical metalanguage. The semantically competent native speaker, however, can only be expected to recognize semantically based truth and falsehood: the ability to detect necessary truths and falsehoods which are not "analytically" such (if indeed there are any) is no part of the speaker's linguistic patrimony.

What I want to suggest is that these considerations apply squarely to (1) and (2). (1) is not semantically valid, nor is (2) semantically false. Yet from the standpoint of a proper philosophical understanding of certain notions which would figure in an adequate semantical treatment of (1) and (2), (1c) "must" be true if (1a) and (1b) are, and (2) "cannot" be true. This is the sense in which I would claim that belief-contexts are N-extensional *de facto* but not *de jure*. The question of philosophical analysis, however, is obviously parasitic on the question of what an adequate Semantics for English would have to say about the *logical form* of belief-sentences. So I shall begin with the latter.

3. THE LOGICAL FORM OF BELIEF-SENTENCES

A plausible proposal regarding the logical form of belief-sentences (which is all that I am presently aiming at) can be obtained by modifying Donald Davidson's paratactic approach to *oratio obliqua* (Davidson $^\lceil 8 \rceil$). Let us suppose that a sentence of the form $\lceil X$ believes that $p \rceil$ is, from the standpoint of its logical form, really *two* sentences: a simple relational sentence $\lceil X$ believes that $\rceil$ (call it "the attitude sentence"), in which 'that' is a specialized demonstrative, and an accompanying sentence p (call it "the content sentence"). The attitude sentence is regarded as being asserted by the speaker, but not so the content sentence, which is to be thought of as merely tokened by the speaker to illustrate the nature of the reference being made by 'that' in the attitude sentence.[5]

More precisely – and here I depart considerably from Davidson – the demonstrative in the attitude sentence refers not to the content sentence or even to the speaker's utterance thereof but to a *class of locutionary acts* (act-*tokens*) in the Austinian sense. This class is the image, under a certain relation R, of the locutionary act performed by the speaker in his utterance of the content sentence. In other words, where '$L_u(p)$' denotes the locutionary act performed in u's (the speaker's) tokening of p, the referent of 'that' in the attitude sentence is $R"\{L_u(p)\}$, where R is a yet-to-be-specified relation on locutionary acts. It is important on this view that we think of the unasserted content sentence p as nonetheless being *used* by the speaker u in the minimal sense of being employed to perform a *sample* locutionary act $L_u(p)$, which serves as a representative of the class that it determines. We are to think of $L_u(p)$ in Austinian fashion as the deliberate production by u of a token of p *as an English sentence-token* whose context-free constituents have their customary interpretation in English and whose context-dependent constituents (e.g., indexicals and names) have their interpretations fixed by reference to the speaker u. Thus, if what the speaker says is 'Tom believes that Nixon is dishonest', we regard the unasserted content sentence 'Nixon is dishonest' as involving a reference to whatever individual is the referent of 'Nixon' *in the speaker's mouth* on that occasion. (The importance of this fact will emerge shortly.)

The role here envisaged for the demonstrative in the attitude sentence is by no means peculiar or unique. We often use demonstratives to refer to a *kind* of thing by simultaneously ostending (demonstrating) a sample member of that kind. Thus the instructor who is explaining bomb-disposal techniques may hold up or point to a particularly nasty piece of terrorist hardware and say 'The terrorists are relying on *this* to defeat us'. The case of the attitude sentence is parallel, except that the accompanying demonstration involves *creation* by the speaker of the sample item (cf. 'This is the call of the Blue-footed Booby', followed by the speaker's producing some avian noises.)

I propose to say nothing about the interpretation of the verb 'believes' beyond remarking that it is to construed in terms of some dyadic relation B between persons and classes of locutionary acts. It is the job of philosophical psychology, not Semantics, to provide the analysis of B; and for the purposes of this paper it does not much matter how we imagine B to be analysed. The minimal constraint that B should relate persons and classes of locutionary acts is compatible with all manner of sentialist, propositionalist, and behaviorist biases. And while my own bias runs towards a form of sentialism – x standing in B to $R''\{L_u(p)\}$ being (roughly) a matter of x's performing and "storing" a member of $R''\{L_u(p)\}$ in a certain way – nothing in what follows depends on thinking of B in this particular way.[6]

But what of the mysterious relation R which figures so prominently in all these doings? If my suggestions are to have any real content at this point, R must be identified, either as some relation constructed from antecedently given technical notions of semantical theory or as some "intuitive" relation belonging to the *in*formal furniture of the semantical metalanguage. I propose that R should be thought of in the latter way, as corresponding to our intuitive notion of one saying (in the locutionary sense) capturing the gist of another. It will be convenient in what follows to speak simply of $L_x(p)$ *capturing* $L_y(q)$, especially since 'capture' is a nicely neutral verb.[7] Although capturing surely has a semantical dimension, it summons nonsemantical resources as well and is a prime candidate for philosophical analysis.

So far, then, we have the following interim result. Our semantic theory tells us that an utterance of, say, 'Tom believes that Nixon is dishonest' is true (relative to a speaker u and a time t) iff Tom stands at t in the relation B to the class of locutionary acts which are captured by the sample locutionary act performed by u at t in tokening 'Nixon is dishonest'. Period. Applied to (2), our Semantics tells us that an utterance of (2) is true relative to u and t iff Cicero = Tully and John bears B to the class of locutionary acts which are captured by the locutionary act performed by u at t in tokening 'Cicero was a Roman orator' but does *not* bear B to the class of locutionary acts which are captured by the locutionary act performed by u at t in uttering 'Tully was a Roman orator'. Again, period. Whether we say that the two classes are the same or not depends upon our philosophical analysis of "capturing", to which I now turn.

4. CAPTURING: THE INTUITIVE NOTION

I have claimed that there is an intuitive, not-purely-semantic notion of "capturing" operative in our understanding of belief-sentences. The same notion also figures, I think, in the proper account of *oratio obliqua* (where Davidson speaks of "samesayers"). Consequently, a rough-and-ready test for whether $L_x(p)$ captures $L_y(q)$ is whether we would feel comfortable with the claim 'y *said that* p' produced by x in the same sort of context as was envisaged for $L_x(p)$. Of course this test cannot without circularity be taken as a *definition* of capturing, but it can prove useful in the informal motivation of various parts of that definition, the provision of which is our next task.

Since our primary concern is with the role of proper names, we can afford to be somewhat vague about the nature of the contribution to capturing made by aspects of the total locutionary act other than the production of name-tokens. Accordingly, let us briefly dispense with those necessary conditions for capturing which do not specifically involve names. There are two such conditions on $L_x(p)$'s capturing L_y (q):

Condition 1: p and q have the same logical form.

Condition 2:   With the exception of names, each extralogi-

> cal component of p (considered in the context
> of $L_x(p)$) renders the force of the corre-
> sponding extralogical component of q (con-
> sidered in the context of $L_y(q)$).

In as much as these conditions are not exactly pellucid, some commentary is called for.

Condition 1 requires that we be able to speak of sameness or difference of logical form as between sentences of any two natural languages. This tacit quantification over natural languages may be repugnant to some, on the ground that we lack "pure" identity and individuation criteria for natural languages – i.e., criteria which do not appeal to "meanings," "propositions," "intensions," or the like. (Davidson's own scruples on this matter threaten to undermine his account of *oratio obliqua*; cf. Arnaud [1].) It is, however, no part of project in this paper to rail against Intensionalism in general: in subscribing to an anti-intensionalist theory of names, I leave open the question of the proper semantical treatment of other parts of our language. As for the worries about identity and individuation criteria for natural languages, I suggest the following picture.

Suppose we begin with the purely formal specification of a class of "canonical" uninterpreted languages (e.g., the lambda-categorial languages of Cresswell [7]) and then specify their possible interpretations. Suppose further that we can make it reasonable to believe that appropriate syntactic transformations and lexicalization operations, applied to formulae of these canonical languages, would produce exactly the grammatical surface sentences of our natural languages. Then the class of possible human languages could be theoretically identified with the class of ordered triples $<L, I, T>$ in which L is a canonical idiom, I is a possible interpretation of L, and T is a set of transformation and lexicalization operations. And what makes a possible human language $<L, I, T>$ the (or an) *actual* natural language of some population P is the existence of certain *conventions* relating P to $<L, I, T>$ (see, e.g., Lewis [17] and Peacocke [21] for elaborations on this idea). Talk of sameness or difference of logical form across natural languages now has a determinate sense, since we have constructed our canonical languages to have the same categorial structure and logical apparatus, and to differ only in which categories are occupied, and by what items. If $<L, I, T>$ is the actual language of a population P, $<L', I', T'>$ the actual language of P', p a surface sentence in the output of the former and q a surface sentence in the output of the latter, then: p has the same logical form as q iff there are formulae F and F' of L and L' respectively such that p derives by T from F, q derives by T' from F', and F is structurally isomorphic to F' (i.e. they consist of symbols of the same categories arranged in the same way).

Condition 2 is the predicted fudge, since no theory of "forcerendering" has been attempted here. Roughly, however, the idea is this. At the level of logical form, corresponding atomic context-independent constituents should *translate* one another (in the sense that, say, the French noun '*pain*' translates the English noun 'bread'), and corresponding atomic context-dependent constituents – which may or may not translate one another – must *take the same value* with respect to the contexts in which their surface-forms were tokened, in the way that, e.g., 'yesterday' uttered today takes the same value as 'today' uttered yesterday. Rough as they are, these characterizations will suffice for the moment: since the crucial content sentences in (1) and (2) differ *only* with respect to the names occupying corresponding positions, we may safely assume that Conditions 1 and 2 are met in the cases that immediately concern us.

What, then, of names? What further requirements are intuitively laid upon the captured locutionary act by the tokening of a name in the capturing act? One such requirement seems to be that the occurrence of a name-token in the capturing utterance must be matched by a corresponding occurrence of a *name*-token in any captured utterance. (This is not guaranteed by Condition 1, since it is common to treat "singular term" as a single category containing names, deictic pronouns, and possibly definite descriptions as well.) Confirmation is afforded by the *oratio obliqua* test: if X points to Quine and says,

$$\left\{\begin{array}{l} \text{He} \\ \text{That man} \\ \text{The man in the grey suit} \end{array}\right\} \text{ looks very distinguished.}$$

we would be reluctant automatically to accept 'X said that *Quine* looks very distinguished' (though, of course, this reluctance may prove misguided).

Further, and more importantly, it seems to be required that the name-token in the captured utterance, whether type-identical to its correlate in the capturing utterance or not, should also be related to it in special ways. (I will assume that name-types are individuated phonologically rather than orthographically.) Where different languages are involved, we commonly allow for different "versions" or "renderings" of a given name: e.g., '*München*' and '*Munich*', though distinct names, appear to be related in the required way, for in reporting a foreign utterance we standardly replace occurrences of foreign names with their anglicized counterparts (if such there be). Sometimes a language will provide several phonologically distinct counterparts for a given foreign name: witness 'Mohammed', 'Muhammad', and 'Mahomet' as attempted renderings of the Arabic name. Another kind of permissible divergence is exemplified by names which stand respectively in a part-whole relation, as do 'John' and 'John Smith'. It seems that the capturing utterance may legitimately employ a well-formed *part* of the name employed in the captured utterance, but *not* vice versa: one may subtract from, but may not add to, the name in the captured utterance. The asymmetry is confirmed by the *oratio obliqua* test. If what X uttered was, e.g., 'Willard van Orman Quine is a philosopher', we are not bothered by the contraction in 'X said that *Quine* is a philosopher'; but if what X uttered was 'Quine is a philosopher', we are uneasy about 'X said that *Willard van Orman Quine* is a philosopher'. Then again, when a name is initially formed by "freezing" a definite description or name-*cum*-title, we commonly allow for translation of the frozen descriptive or titular elements as well. Thus 'Charlemagne', 'Karl der Grosse', and 'Carolus Magnus' are variations on a common theme.

What if the name-tokens in question *are* type-identical? Consideration of empty names in mythological contexts points toward the answer. Indeed, precisely because they lack referents, such names illustrate very clearly how capturing can be concerned with more than a name's referent. Suppose we were to encounter a "lost" tribe whose story-teller relates a saga startlingly similar to our own Santa Claus legend. He tells of a magical being called 'Zarf' who lives at the top of the world and who rewards good children every winter solstice by bringing them presents in his strange conveyance drawn by flying deer. We want to report this amazing story to our colleagues, but here we encounter a problem. When rehearsing the story in our language, may we replace 'Zarf' by 'Santa Claus' throughout, or must we instead adopt 'Zarf' for use in our report? Intuitively, the constraints on capturing seem to demand the latter approach. For, *ex hypothesi*, we are dealing with a "lost" tribe: their mythology has *no* historical connection with ours in spite of its similarity; their use of the name 'Zarf' has no connection whatever with our use of the name 'Santa Claus.' So if we are to say things which capture our native informant's utterances, we have no choice but to adopt the name 'Zarf' ourselves, in the same way we have expropriated 'Wotan', 'Vishnu', and the like. Now notice that it would have made no essential difference to our example had the lost tribe called their mythical being 'Santa Claus' instead of 'Zarf'. For the reasons given, *their* tokens of 'Santa Claus' would not be connected in the right way with *our* tokens of 'Santa Claus'; their Santa Claus is not ours. (Of course, in a derivate sense we might say that they have *a* Santa Claus, but this would not be to the point). To keep matters straight, we might use 'Santa Claus*' when transcribing their legend. In any event, type-identity of name-tokens evidently does not cancel the need for their being historically connected in certain ways when capturing is at issue.

What all these cases have in common is a certain historical pattern. In each case there is an historical ancestor-construction which has been passed down from speaker to speaker through various possible modifications – e.g., changes in pronunciation, contractions, permutations, etc. Name-tokens whose production is explained by causal chains reaching back to a common ancestor-token may be said to be *relatives* of one another – it being remembered that relatives may not physically resemble (be of the same name-type as) each other or the ancestral construction. Tokens of 'Munich' in the mouths of English-speakers and tokens of '*München*' in the mouths of German-speakers are relatives in this sense, but tokens of 'Santa Claus' in our mouths and tokens of 'Zarf' (or 'Santa Claus*') in the hypothetical natives' mouths form two classes no member of one of which is a relative of any member of the other (though, of course, the members of each class by itself are presumably mutual relatives).[8]

On the assumption, then, that name-tokens in the capturing utterance must have *relatives* at corresponding places in any captured utterance, let us ask what further constraints may be required. Examples involving a divergence of "speaker's reference" and "semantic reference" provide a convenient starting point. Fred, who was once a student of Quine's, attends a party at which he sees a man who looks exactly like Quine. Fred remarks, 'How nice, Quine is here'. In fact

Quine is not at the party; the man Fred sees is McX, the infamous metaphysician, tricked out in one of his clever disguises. How should we relay Fred's utterance? Did he say that *Quine* was at the party, or that *McX* was at the party, or both, or neither? In effect, this is the question of which (if any) of these locutions *we* can use to capture Fred's locutionary act. Such cases of misidentification can seem puzzling for a Causal-Historical Theory of Names because they seem to involve two incompatible causal routes: on the one hand, there is the antecedent chain of transmission (grounded in Quine) by which Fred acquired the name 'Quine' in the first place; on the other hand, there is the perceptual causal chain leading from McX. How are we to assess the impact of these considerations upon capturing?

To begin with, it should be noted that intuition is heavily on the side of the verdict that Fred said that *Quine* was at the party. The very concept of an ostensive *mis*identification with a name presupposes that the name (in that use) picks out its referent independently of the ostension, which supplies a different object. One way of putting this is to say that, in Fred's utterance, the *semantic* referent of 'Quine' is Quine, whereas the *speaker's* referent of 'Quine' on that occasion is McX; the misidentification consists in the fact that Quine $\neq$ McX. But why discriminate against the causal chain leading from McX? The answer hinges on the role we envisage for the perceptual event of seeing McX. In the case at hand, Fred does not *rely* on the perception to fix the reference of 'Quine'; quite the contrary, he relies on his inherited usage of the name 'Quine' to fix the reference. When his mistake is pointed out to him, Fred will presumably confess, 'I was wrong; that's not Quine', and the name 'Quine' will be functioning in Fred's mouth *just as it functioned in his previous utterance*.

Of course, the force of perceptual encounters can run in the opposite direction as well: speaker's reference can *become* semantic reference in certain circumstances. Gareth Evans (in [12]) points out that the name 'Madagascar', which we use today as the name of a certain island off the African coast, was originally used by the natives as a name for an adjacent part of the mainland. The sailors who picked up and transmitted the name back to Europe simply misunderstood the natives' speech. Yet me clearly do not want to say that when *we* use the name 'Madagascar' we are talking about part of the mainland. Corresponding native and English utterances employing tokens of 'Madagascar' fail to capture one another. Here is a plausible diagnosis of Evans' example.[9] The sailors need a name for the island. Their transactions with the natives suggest (falsely) that the island is already called 'Madagascar'. Since one name is as good as another, they adopt the name 'Madagascar' and duly inscribe it on their official charts, in the log book, in letters, etc. These entries on maps and in log books, however, have a *normative* force for them: 'Madagascar' has been adopted *as a name of that island*. There is no real *reliance* on the antecedent history of the name as a reference-fixer; the island has been ostensively dubbed 'Madagascar', and it is only the appropriateness, not the efficacy, of the dubbing which suffers from the misunderstanding. Once the new convention had become entrenched, it does no good to reveal the error. Our primary use for the name is in talking with our own kind, not with the natives, and it is simply too much bother to try to undo what has been done. The relevance of a name's history is thus selective: the origin of a name and the situations which prompt its use are both part of its history, but neither automatically takes precedence over the other in determining its referent in a given use.[10]

The foregoing examples point to the importance of the fact that only certain portions of a name-token's overall history may be operative in determining its referent (Devitt [9] puts this same point by saying that while one may have inherited the ability to refer to a number of individuals by a token of a given name-type, one may be *exercising* just one of these abilities on the occasion of a particular tokening). But things can go wrong in all sorts of ways. Most strikingly, one may rely in a given situation on *divergent* sources, the result being a referential ambiguity with important consequences for the possibility of capturing.

Imagine a family consisting of two identical twin boys, Rod and Rick, and their mother, Marsha. Together they have conspired to fool the rest of the world into thinking that Marsha has only one son, called by everyone else 'Rodrick'. Rod and Rick takes turns appearing in public, and the deception is as complete as one could wish. One of the neighbors remarks to Marsha, 'I had a nice chat with Rodrick yesterday'. In fact Marsha knows that it was Rod, not Rick, with whom the neighbor spoke. If Marsha then privately says to Rick, 'One of the neighbors said that she had a nice chat with *Rod* yesterday', we begin to feel some doubts about the accuracy of her report. These doubts stem from the fact that the name 'Rodrick', in the neighbor's mouth, is historically grounded in *both* Rod and Rick; it does not univocally refer to either but, so to speak, "partially designates" each. But 'Rod', in Marsha's mouth, univocally designates Rod.

Ambiguity of reference, it appears, can get in the way of capturing. But what if the two parties employ name-tokens which are ambiguous with respect to the *same* set of objects as referents? Even here there is room for some misgiving, inasmuch as partial designation readily lends itself to classification in terms of *degree*. A name-token n whose (operative) history leads back to dubbings of distinct objects X and Y is certainly ambiguous, but there is a clear sense in which n may be more closely involved with X than with Y, at least where n's utterer is concerned. For the routes leading from n back to X may be very numerous and/or populated by very many antecedent tokenings by the speaker of that name, whereas the routes leading from n back to Y might be correspondingly few in number and/or correspondingly under-populated.[11] In such a case, there is some intuitive force to the claim that n designates X to a greater degree than it designates Y. (Exactly how such degrees are to be measured need not concern us here.) Accordingly, we may suspect that the contribution to capturing made by an ambiguous name-token is conditioned by the degrees to which it partially designates various objects – and hence that, *ceteris paribus*, capturing would require that corresponding name-tokens n and n' should designate the same objects to the same degree. (Just how restrictive this requirement would be in practice would, of course, depend on how much precision is demanded in the measurement of degrees of designation.)

Obviously much more can and ultimately should be said about this conception of partial designation, but doing so here would take us too far afield. For present purposes, we can sum up the additional intuitive requirements for $L_x(p)$'s capturing $L_y(q)$ as follows:

> *Condition 3*: Either p and q contain no names at all, or else they contain names in corresponding positions (at the level of logical form).
>
> *Condition 4*: If $N_1, \ldots, N_k$ and $N_1', \ldots, N_k'$ are respectively the names in p and q in order of occurrence, then for each i $(1 \le i \le k)$: the name-token n_i of N_i produced in $L_x(p)$ is a relative of the name-token n_i' of N_i' produced in $L_y(q)$.

Condition 5:   For each such n_i and n_i', either both are
empty or else both designate the same objects
and to the same degree.

5. CAPTURING THEORETICALLY REVISED

As expounded above in Conditions 1–5, the intuitive understanding of
what is involved in the capturing relation R does *not* suffice to validate
the argument (1). For these could be some locutionary act captured by
the speaker's sample tokening of 'Cicero was a Roman orator' but *not*
captured by his or her subsequent sample tokening of 'Tully was a
Roman orator', with the result that distinct classes of locutionary acts
would be invoked by (1a) and (1c). The reason for this possible divergence lies, of course, in Condition 4. For the relation among name-
tokens expressed by 'is a relative of' is clearly not transitive. In
principle, a name could come into use through *conflation* of two pre-
existing but completely unrelated names: it would be a relative of *both*
antecedent constructions (and their descendants), yet neither of these
constructions would be relatives! (I take for granted the reflexivity and
symmetry of R, allowing the first member of such a chain to count as
its own ancestor.)

My response to this situation is to suggest that the intuitive notion of
capturing sketched above requires further philosophical analysis. As it
stands, R is a blend of syntactic, semantic, and pragmatic considerations not all of which are clearly determinative of the *truth*-conditions
of belief-sentences. Condition 4 is a case in point. If we were to insist
on Condition 4 as an ingredient in the truth-conditional contribution
made by R to belief-sentences, we would be forced to embrace some
highly counter-intuitive results elsewhere. In particular, we would have
to admit that an utterance of (8) could be *true* even when 'Cicero' is
employed *univocally*:

(8) John believes that Cicero was a Roman orator, and it's
not the case that John believes that Cicero was a
Roman orator.

Of course, if the utterer of (8) were not using 'Cicero' univocally, then
the possible truth of his or her utterance would be unproblematic. A
Causal-Historical Theorist would posit a reference-shift (as when, e.g.,
the utterer has a friend named 'Cicero' who lies at the end of the

causal chain behind the production of the second token of 'Cicero', whereas the first token is grounded in Cicero the Roman). The Description Theorist would simply claim that each occurrence of 'Cicero' might go proxy for a different description, so that the content of the belief ascribed to John in the first conjunct might be distinct from that of the belief withheld from John in the second conjunct. The Description Theorist is thus in a somewhat more secure position, since he or she can account for the possible truth of an utterance of (8) even when the two tokens of 'Cicero' are coreferential.

But if we insist on univocality, we rule out both avenues of escape, and Condition 4 then creates a genuine problem for theorists of either stripe. For even if the two tokens of 'Cicero' are not only coreferential but are also backed by the same description(s) (for the utterer), it still does *not* follow that they will be relatives in the utterer's mouth. The following (admittedly far-fetched) story illustrates why the entailment fails. Suppose that Cicero (the famous Roman orator), in addition to getting the name 'Cicero' from his parents, was also, much later in life and quite independently, dubbed 'Zizro' by a visiting foreigner who was unaware of Cicero's Roman name. The visiting foreigner thereupon decamps for his distant and isolated homeland, the tiny island of Utopia, where he initiates an historical chain of uses of 'Zizro' which is totally independent of the chain of uses of 'Cicero' which spreads throughout the rest of the world. Over the course of centuries, 'Zizro' gradually degenerates into an expression phonologically and orthographically identical to our 'Cicero' – let's write it '**Cicero**' just to keep matters straight. Now we can imagine that the ubiquitous Fred, who acquired 'Cicero' via the normal channels, also acquires '**Cicero**' while on a tour of Utopia. As yet Fred has no inkling that 'Cicero' and '**Cicero**' are coreferential: he may treat the orthographic and phonological identity as mere coincidence. Clearly 'Cicero' (used by Fred in speaking English) is *not* a relative of '**Cicero**' (used by Fred in speaking Utopian). Nor is there any reason why Fred should not gradually (through speaking with Utopians in their language) come to associate with his uses of '**Cicero**' in Utopian sentences the same description(s) he standardly associates with his uses of 'Cicero' in English sentences. Indeed, he may even assert '**Cicero** is none other than Cicero!'.

Now suppose that, for whatever reason one cares to imagine, Fred utters (8), referring by 'John' to his visiting cousin John and (of course) by 'Cicero' to Cicero. Suppose further, however, that Fred's first tokening of 'Cicero' is part of the standard 'Cicero'-chain but that his second tokening of 'Cicero' is part of the **'Cicero'**-chain, which he has joined by dint of his dealings with Utopians (i.e., his second tokening is a tokening of **'Cicero'**). But then $R"\{L_{Fred}(\text{'Cicero was a Roman orator')}\} = R"\{L_{Fred}(\text{'Cicero was a Roman orator')}\}$, since, given Condition 4, neither locutionary exemplar captures the other. Therefore, we would have no intrinsic ground for saying that Fred's utterance of (8) is self-contradictory! Perhaps the Description Theorist would claim that if Fred "really" associates the same description(s) with these two name-tokens then they "cannot" have the envisaged histories, but *why* this should be so is totally obscure. For if we had supposed instead that 'Zizro' retained its original form and that Fred uttered not (8) but 'John believes that Cicero was a Roman orator, and it's not the case that John believes that Zizro was a Roman orator', the contingency of his utterance would (even under the stated conditions) have been palatable. So what non-question-begging ground can there be for straining at Fred's token of (8), totally similar but for *pronunciation* of 'Zizro'?

If, as this consequence seems to dictate, we dropped Condition 4, we would still have to explain two things: first, why the argument from fictional names given earlier appears to support the need for Condition 4; and, second, why we are tempted to extend Condition 4 beyond the case of names in fiction to the use of names in ordinary, real-world discourse. The answer to the first question demands a minor concession from the Causal-Historical Theorist: the argument from fictional names *does* support the "relatives"-requirement *for the case of empty names*. Some of the sting is removed from this concession by the realization that an adequate treatment (within the Causal-Historical framework) of empty names in general, and in particular of their role in negative existentials, already forces us to pay attention to the kinds of ancestry-considerations embodied in the notion of "relatives". (Donnellan's discussion (in [11]) of "blocks" shows, in effect, that being relatives is the requisite analogue for empty names of coreference for nonempty names; see especially his discussion of why we normally take the utterer of '*Père Noël n'existe pas*' and the utterer of 'Santa Claus does not exist' to be "saying the same thing" even though the respective truth-conditions of these utterances in virtue of mentioning the histories of different expressions, are not formally equivalent.) Each empty name in the capturing utterance must, then, be matched by an empty *relative* in any captured utterance. This result, however, does not commit us to the bizarre view that belief-contexts are N-extensional with respect to nonempty names but not with respect to empty names. Such a consequence would ensue only if we allowed that equations with empty names as terms could come out *true* when those names were *not* relatives. (I assume that no one would be attracted by the desperate expedient of calling a belief-sentence false or truth-valueless merely because some name in its *content*-clause is empty!) There is certainly no difficulty in rigging our formal semantical theory to rule out this possibility anent equations: the development of Free Logic supplies all the apparatus we need. More importantly, making such an adjustment is *not* merely *ad hoc*. For the only plausible cases of literally *true* equations with empty names as terms are *precisely* those in which the ingredient names *are* relatives in the required sense! Our inclination to say, e.g., that 'Clark Kent = Superman' is true rests entirely on our connection with the *story* in which little Kal-El comes to be dubbed in different circumstances with both names. In Donnellan's jargon, the histories of the names 'Clark Kent' and 'Superman' each terminate in a "block", but the blocks themselves are related to one another by the story in which they occur. But in the case of, say, 'Zeus = Marduk' there is no appropriate linkage of underlying mythologies (cf. 'Santa Claus' and 'Zarf' in our earlier example) and no temptation whatever to call this equation true (*or* false, for that matter). By adopting something like the "story-semantics" proposed by Lambert and van Fraassen in [16], the threat to N-extensionality posed by empty names is easily and quite naturally removed.

Let us then replace Condition 4 by Condition 4*:

> *Condition 4*: If $N_1, \ldots, N_k$ and $N_1', \ldots, N_k'$ are respectively the names in p and q in order of occurrence, then for each i ($1 \leq i \leq k$): if the token n_i of N_i produced in $L_x(p)$ is empty, then the token n_i' of N_i' produced in $L_y(p)$ is a relative of n_i.

(The corresponding emptiness of n_i' is required by Condition 5.) The resulting set of Conditions 1–3, 4*, and 5 provides a theory of capturing which *does* validate (1) and make (2) self-contradictory. For the truth of the ingredient identity-statement in the speaker's mouth guarantees that his or her subsequent tokens of 'Cicero' and 'Tully' in the sample utterances meet the relevant constraints on name-tokens. [12] Thus we are assured that any locutionary act captured by that performed by the speaker in his or her utterance of 'Cicero was a Roman orator' is also captured by that performed by the speaker in his or her utterance of 'Tully was a Roman orator' (and vice versa), so that the relevant classes of locutionary acts are the *same*. (2)'s contradictoriness follows similarly.

But a theory of capturing which assures the N-extensionality of belief-contexts and explains, as we did earlier, why a semantically competent speaker of English would not *ipso facto* recognize that N-extensionality still faces the nasty problem of explaining why most English-speakers seem positively inclined to *reject* the posited N-extensionality out of hand. In the present context, this amounts to having to answer the second of our two questions: Why do native speakers of English positively expect every name-token in a capturing locution to correspond to a *relative* in every captured locution? The answer lies not in Semantics but in a mixture of syntactic and pragmatic considerations. Specifically, we must look to certain "lexical presumptions" attaching to the use of names and to certain "pragmatic presumptions" attaching to the performance of locutionary acts as exemplars.

First of all, there is associated with the use as a name of an expression token t by an utterer u the lexical presumption that some historical relative t' of t has antecedently been introduced into u's language (or an ancestor thereof) in the appropriate way. For an expression to gain the grammatical status of being a *name*, it is of course not necessary that it should actually name anything: successful dubbings, whether formal or casual, are not the only mode of introduction. Abortive naming-ceremonies surely introduce names as well. 'Vulcan', as once introduced by astronomers, counted as a name, albeit an empty one (the fact that 'Vulcan' had antecedent uses in mythology is irrelevant to this point). Pretense is another source of names: the novelist may adopt certain expressions as names for his characters, and readers of the novel may employ tokens of those expressions as genuine (albeit empty) names. The point of all this is that what *makes* an expression-token t a name is the fact that the causal chain underlying its production ultimately traces back to some relative of t which was introduced into speech in one of the aforementioned ways. It is important to note that this chain of transmission which legitimizes t's grammatical status as a name *need not* be a chain of the reference-preserving variety, even when it leads back to a successful dubbing. The name 'Santa Claus', we are told, is a relative of 'St. Nicholas', and both presumably depend for their namehood on some historical dubbing of Nicholas; but in the course of time some chains of transmission became garbled, resulting in a distorted version of the original name (whatever it was) which no longer refers to Nicholas but has found a new life in folklore.

But what has all this fuss about namehood to do with "lexical presumptions"? The answer is simple: any ostensible sentence-token s containing ostensible name-tokens $t_1, \ldots, t_n$ (i.e., expressions in singular-term position which clearly are not any other sort of singular term) lexically presumes that $t_1, \ldots, t_n$ qualify for namehood (have historical relatives introduced in the right sort of way) in the sense that s is *syntactically ill-formed* (ungrammatical) unless $t_1, \ldots, t_n$ *do* qualify as genuine names. Consider the expression-token which occurs inset below:

> Blufzegnaf eats onions.

This token is not an English sentence-token, for the token of 'Blufzegnaf' it contains is a mere mark without a role – a piece of gibberish. Yet I could easily *make* subsequent tokens of the same type into English sentence-tokens simply by bestowing a token of 'Blufzegnaf' in a successful naming-ceremony, or in the context of a story I am writing, etc. All of this is part and parcel of the oft-remarked fact that names do not "belong" to specific languages in the way that, say, common nouns and verbs do, but are instead essentially *ad hoc* devices that transcend linguistic borders. Individuals have a kind of power over namehood that they do not have (qua individuals) over nounhood and verbhood. These points can be summarized by saying that (ostensible) sentence-tokens containing ostensible name-tokens are only *relatively grammatical* – i.e., grammatical relative to satisfaction of the "presumption" in question. In fact, relative grammaticality is a fairly widespread phenomenon in natural languages and can owe to a variety of constructions other than names. For example, the relative pronouns 'who' and 'which' respectively induce presumptions of personhood and non-personhood. Roughly: 'The one *whom* I saw was large' presumes that the intended referent is granted at least provisional status as a person, whereas 'The one *which* I saw was large' presumes denial of such status. Each would be ungrammatical relative to a context in which its presumption failed. (cf. Lakoff [15])

I have deliberately spoken of lexical *presumptions* rather than *presuppositions* of the classical Strawsonian sort. Although there are obvious parallels between the two notions, there are also some crucial differences. Failure of a lexical presumption, like failure of a Strawsonian semantic presupposition, results in truth-valuelessness. But presupposition-failure does not affect the well-formedness of the original sentence, whereas failure of a lexical presumption does just that. 'The Present King of France is bald', uttered today, is perfectly grammatical and meaningful in spite of its alleged truth-valuelessness; but the token of 'Blufzegnaf eats onions' displayed earlier is also truth-valueless but neither grammatical nor meaningful as a whole. The source of the truth-valuelessness in the former case is presumably failure of reference or denotation, a semantic defect; but the source of truth-valuelessness in the latter case is of a radically different sort – viz., the fact that gibberish isn't even a candidate for possession of a truth-value! Moreover, the seriousness of such presumption-failure, un-like the effects of presupposition-failure, seems to be a matter of de-gree, corresponding to an intuitive difference between the "mildly" and the "severely" ungrammatical. The string 'This is the man which I told you about' evinces "mild" ungrammaticality: though strictly truth-valueless (since truth-conditions are assigned only to well-formed strings), the slip is so unimportant for most purposes that we would probably gloss over the mistake and treat it just as we would 'This is the man whom I told you about'. The displayed token of 'Blufzegnaf eats onions', however, is at the other end of the scale, in roughly the same position as '#$%&@¢ eats onions'. Our ability as hearers charitably to patch things up is strained to the limit: it is only the fact that 'Blufzegnaf' can be pronounced and '#$%&@¢' cannot that in-clines us provisionally to treat the token containing the former as a sentence whose subject term is an "unfamiliar" name. (For further dis-cussion see Boër and Lycan [3].)

Secondly, the practice of performing and demonstrating a locution-ary act as a representative or exemplar of a certain *kind* of locutionary act is subject to the pragmatic presumption that the exemplar is a *fair sample* or *typical member* of that kind in salient respects. This is not surprising in light of the fact that such a presumption seems to attach to all cases of referring to a kind by means of demonstrating one of its members. In general, an exemplar, apart from possessing the essential characteristics of its kind, should not display prominent accidents peculiar to it alone or to only a small subset of the relevant class. For unless the audience is antecedently acquainted with many members of that kind as such, there is the strong likelihood that any prominent but atypical features of the exemplar will be taken as "normal" for the kind, with the result that the audience will form a distorted picture of the things in question. Worse still, if the audience does not antecedently possess a theory of the basis of the kind, there is the danger that they might be misled into supposing a certain atypical feature displayed by the exemplar to be criterial for membership in that kind. When the presumption of a fair sample is violated, the speaker can still succeed in referring to the relevant kind, but will have produced a *deviant* performance in the sense that the audience will be *misled* in certain predictable ways about the makeup of the referent.

Where kinds of locutionary acts are concerned, the presumption of a fair sample requires of the demonstrated locutionary act that (*inter alia*) it should not induce any lexical presumptions regarding the locution which are not common to other locutionary acts of that kind. This may sound exotic, but it is in fact quite straightforward, as some examples will show. Suppose I utter, 'Tom believes that the one *whom* he touched weighs 200 lbs.' In so doing, I use the special demonstrative 'that' to demonstrate L_1 ('The one *whom* he touched weighs 200 lbs.') as a fair sample of the kind to which I am referring, viz. the class of locutionary acts captured by the exemplary locutionary act thus demonstrated. However, *my* use of the personal form of the relative pronoun in my exemplar induces the lexical presumption that the salient item touched by Tom is a person (or at least "counts as a person" for contextual purposes). The pragmatic presumption of a fair sample then *projects* this very lexical presumption onto the other locutionary acts captured by mine. In other words, it is suggested to the audience that those other locutionary acts *also* induce the lexical presumption that the salient item touched by Tom is a person, which of course they could do only if their respective locutions contained appropriate linguistic devices. If now we imagine one member of this class being an overt or inner locutionary act performed by Tom himself, we can see how the presumption of personhood can migrate from my mouth into Tom's: from *my* talking about that thing as a person we slide quickly into *Tom's* talking about it as a person. Yet no entailment is involved, since the transition is pragmatically, not semantically, based. Or again, if I have heard that Tom's boss has picked a successor, and I believe (whether truly or falsely) that Tom's boss is a woman, then I might say, 'Tom believes that his boss has picked *her* successor'. My use of 'her' rather than 'his' in the demonstrated locutionary act induces the lexical presumption that Tom's boss is female, and the pragmatic presumption of a fair sample then projects this onto the rest of the class of locutionary acts captured by mine. So if we once more think of one member of this class being performed by Tom overtly or *in foro interno*, then the presumption of femaleness is transferred to it as well – i.e., we will suppose that Tom talks about his boss as a female.

Of course, lexical presumptions are not the only features of the locutionary exemplar that are pragmatically projected under the presumption of a fair sample: various pragmatic characteristics carry over as well. For example, "emotive words" such as racial and ethnic slurs have a way of transferring attitudes from the speaker to the believer when they occur in the content-clause of a belief-sentence. Ralph, who hates Italians, may say 'Tom believes that his sister is dating a wop'; one who does not know Ralph or Tom will tend to hear the negative attitude towards Italians associated with the term 'wop' as tainting *Tom* as well as Ralph – though the transference in these cases is perhaps not so strongly felt as it is in the case of lexical presumptions.

Let us see now how these considerations might help us with names. Consider (1a), represented by the lights of our theory as (9):

(9) John believes *that*.─────────────────────────┐
                    └─→Cicero was a Roman orator.◄──┘

The unique feature of the lexical presumptions generated by the presence of ostensible *name*-tokens is that the presumptions involve *mention* of those very tokens: a tokening of (1a) involves a sample tokening of 'Cicero was a Roman orator', and the resulting tokening of the content sentence presumes that the ingredient token c of 'Cicero' has an historical relative which was introduced into the language (or an ancestor thereof) in one of the ways appropriate to creating a name. The presumption of a fair sample then projects *this very lexical presumption about c* onto the other locutionary acts of which the demonstrated one is an exemplar. All are now viewed as having the grammaticality of their ingredient locutions conditional upon c's having relatives with the right connections. This requires firstly that what corresponds to c in these other, captured locutions is in each case an ostensible name-token. Let t be one such ostensible name-token, and s the sentence-token containing it; then s obviously depends for its grammaticality upon t's having relatives with the right connections. How can we reconcile this with the projected dependence of s upon the status of c, a wholly distinct item from any in s? The answer is plain: when t and c are *relatives*. For then their convergent ancestry would explain how a faraway utterance in a foreign tongue could have its grammaticality conditioned by the very specific presumption which conditions the grammaticality of my English utterance here and now. In this way the pragmatic projection of the lexical presumption leads us – via inference to the best explanation – straight to the conclusion that corresponding to each name-token in the capturing utterance must be a name-token in the captured utterance which is a relative of it. But being pragmatically led to a conclusion is not the same thing as being logically required to draw it.

But if X's utterance of (1a) thus pragmatically suggests that every locutionary act captured by L_x('Cicero was a Roman orator') involves the production of a corresponding name-token which is a relative of X's 'Cicero'-token, then the same reasoning, applied now to X's utterance of (1c), shows that it will pragmatically suggest that every locutionary act captured by L_x('Tully was a Roman orator') involves production of a corresponding name-token which is a relative of X's 'Tully'-token. In light of these pragmatic suggestions, it is easy to see why there is a strong tendency on the part of speakers of English to *reject* the claim that the truth of X's utterances of (1a) and (1b) warrants the truth of X's subsequent utterance of (1c). For the satisfaction of what is pragmatically suggested by X's utterance of (1c) is *not* guaranteed by the presumed truth of X's utterances of (1a) and (1b), since we obviously cannot conclude from the truth of an utterance of $\ulcorner a = b \urcorner$ anything at all concerning whether a and b are *relatives* in the utterer's mouth. Hence X's utterance of 'Cicero is Tully' will seem curiously irrelevant to the question at hand: even when presumed true, it will fail to forge the seemingly required ancestral connection between X's token of 'Cicero' and X's token of 'Tully' needed to vouchsafe the conclusion that $R''\{L_x(\text{'Cicero was a Roman orator'})\} = R'' \{L_x(\text{'Tully was a Roman orator'})\}$. And, of course, the same considerations, applied to (2), show why an utterance of (2) would not normally be felt to be self-contradictory.

We have now completed our primary task. We have seen, on the one hand, reason for thinking that capturing, properly analysed, is a much weaker relation that it seems at first blush to be – weak enough, in fact, to validate each use of (1) and to make each use of (2) self-contradictory. And we have seen, on the other hand, why pragmatic and syntactic factors conspire to incline us to perceive capturing as a stronger requirement – and hence to suppose that (1) as invalid and (2) contingent. A Causal-Historical Theorist can indeed hold that belief-contexts (and, by straightforward extension, propositional attitude contexts in general) are N-extensional, while accomodating our manifest intuitions to the contrary. And with this result falls the last major obstacle to embracing the Causal-Historical Theory of Names.

APPENDIX:

On iterated belief-contexts

Unless some modifications are made, the foregoing account of capturing will be threatened by circularity when we attempt to apply it to iterated belief-sentences of the form (10):

(10) X believes that Y believes that p'.

Our current logical form proposal calls for treating (10) on the model of (11):

(11) X believes $\underline{that_1}$.
$\longrightarrow$ Y believes $\underline{that_2}$
$\longrightarrow p'$.

The problem is that we have thus far been silent on the nature of the contribution made by an embedded special demonstrative like 'that$_2$' to the capturing relation R. Given that 'that$_2$' refers to $R''\{L_u(p')\}$, what would it mean for a locutionary act $L_v(q)$ to capture L_u ('Y believes that$_2$')? The obvious answer is that, *inter alia*, q would have to contain (either overtly or at the level of logical form) a corresponding special demonstrative whose tokening in $L_v(q)$ is connected with the demonstration of an exemplar-token which captures $L_u(p')$! But we clearly cannot mention capturing in the analysis of capturing unless steps are taken to prevent a vicious circle.

Perhaps the simplest way out is to impose some system of stratification upon sentence-tokens (or tokenings). Let us assume that our canonical languages include a syntactically marked item 'THAT' which underlies 'that$_1$', 'that$_2$', etc., and their surface counterparts in other natural languages, i.e., a syntactically special demonstrative for this special role. Then we could group together at Level O all those sentence-tokens which contain no reflection of 'THAT'; at Level 1 those sentence-tokens containing reflections of 'THAT' in tandem only with demonstrations of sentence-tokens of Level O; and, in general, at Level k ($k \geqslant 2$) those sentence-tokens containing reflections of 'THAT' in tandem only with demonstrations of sentence-tokens of Level $i \leqslant k\text{-}1$. We could then attempt to define capturing by induction on "Level", it now being understood that capturing is a relation between locutionary acts (whose locutions are) of the *same* Level. For locutionary acts $L_x(p)$ and $L_y(q)$ of Level O, the account of R is as given in Conditions 1–3, 4*, and 5, since no reflections of 'THAT' are involved. For $L_x(p)$ and $L_y(q)$ of Level k, we repeat the same definitional material together with a new clause: Corresponding reflections of 'THAT' in p and q are such that their tokenings in $L_x(p)$ and $L_y(q)$ occur in connection with demonstrations of tokens of sentences r and s such that $L_x(r)$ and $L_y(s)$ are of Level $i<k$ and $L_x(r)$ captures $L_y(s)$.

Professor Steven E. Boër
Department of Philosophy
The Ohio State University

NOTES

I wish to thank William G. Lycan for his helpful comments on an earlier draft of this paper. The current version was submitted in 1980.

[1.] The restriction to "ordinary" uses of names is meant to rule out certain exotic cases, such as the use of names as *titles* (e.g., 'Ronald McDonald' and 'Handsome Dan' (the Yale mascot's title)), about which the Causal-Historical Theory of Names has nothing directly to say. It is arguable that these "titular names" are indeed tied to descriptions in more or less the way that Description Theorists suppose that all names are so tied.

[2.] It is of course abstractly possible that this "extra feature" should be nondescriptive and ineffable, though I doubt that very many proponents of the argument (I)-(VI) would find this view congenial, smacking as it does of mystery-mongering.

3. E.g., what syntactically plausible logical forms for 'This is red' and 'This is colored' would show that the former entails the latter? Similar remarks obviously apply to a host of other felt entailments.

4. Perhaps it is the fact that Object-English is contained in Meta-English that obscures for some people the crucial difference between metalinguistic axioms such as (6) and (7) on the one hand and, on the other hand, object-language sentences of the form 'X is a person iff p_1 & ... & p_n', where $p_1, ..., p_n$ are supposed to offer individually necessary and jointly sufficient conditions not containing the word 'person' or any synonym thereof. Insofar as one is inclined to regard the latter sort of statement as analytic and necessary, it may help to remember that axioms like (6) and (7) express purely contingent facts about the object-language.

5. Although the specific logical form proposal owes to Davidson, the spiritual father of the present approach is Wilfrid Sellars (cf., e.g., [23], [24]).

6. For elegant elaboration of one version of this way of analysing B, see W. Lycan. 'Towards a homuncular theory of believing', *Cognition and Brain Theory* 4, pp. 139–159, 1981.

7. For this reason it is better than artificial coinages like 'samesay', which connotatively prejudge such issues as, e.g., whether R is an equivalence-relation. 'Capture' also has the advantage of minimizing the unwanted connotation of temporal asymmetry inherent in such possible substitutes as 'reproduce' and 'echo'.

8. The relation of the "relatives"-requirement to the part-whole asymmetry mentioned earlier is less obvious but seems to reside in the fact that, barring specific information to the contrary, we view addition to the name tokened in the captured utterance as *increasing the risk* that the expanded name, in our mouths, will *not* be a relative of the original name in the utterer's mouth. Insofar as we know *what* name the utterer tokened, the safe course in this regard is obviously to use *it*, or a well-formed part thereof, in a spirit of pure borrowing.

9. Evans himself suggests a somewhat different analysis of this example. I leave it to the reader to decide which is more plausible.

10. What happens in the 'Madagascar'-case might seem to happen in the previous Quine/McX case as well, were the latter story to be continued in certain ways. If, e.g., we imagine Fred's exposure to McX to extend over a long stretch of time, during which many uses of 'Quine' by Fred are prompted by observations of McX, then there is some temptation to say that McX becomes a *semantic* referent of 'Quine' in Fred's mouth. However, given Fred's envisaged past, there is still some temptation to say that Quine is semantically designated in these uses. The resolution, I think, lies in the considerations anent "partial designation" adduced below. In effect, the name 'Quine' becomes semantically ambiguous in Fred's mouth, partially designating Quine but also (and to an ever-increasing degree) partially designating McX as well. I would still distinguish this sort of eventuality from the 'Madagascar'-example, inasmuch as the latter seems to involve the (informal) establishment of a new convention which nullifies the semantical relevance of the previous history of 'Madagascar' insofar as *we* are concerned.

11. It should be remembered that "routes back to X" need not terminate in some one en-

counter by someone with X but can each terminate in a different X-encounter (possibly) involving the utterer of *n*), thus "reinforcing" the connection between X and *n*.

12. Assuming, as always, that the ingredient names are used univocally throughout.

REFERENCES

Arnaud, R.: 1976, 'Sentence, utterance, and samesayer', *Nous* **10**, pp. 283–304.

Boër, S.: 1972, 'Reference and identifying descriptions', *The Philosophical Review* **81**, pp. 208–28.

Boër, S. and W. Lycan: 1976, 'The myth of semantic presupposition', in A. Zwicky (ed.) *Papers in Nonphonology* (OSU Working Papers in Linguistics, Vol. 21: May, 1976).

Burge, T.: 1978, 'Reference and proper names', *Journal of Philosophy* **70**, pp. 425–39.

Burks, A.: 1951, 'A theory of proper names', *Philosophical Studies* **2**, pp. 36–45.

Burks, A.: 1977, *Chance, Cause, Reason*, University of Chicago Press, Chicago.

Cresswell, M.: 1973, *Logics and Languages*, Methuen, London.

Davidson, D.: 1969, 'On saying that', in D. Davidson and J. Hintikka (eds.) *Words and Objections*, D. Reidel, Dordrecht.

Devitt, M.: 1974, 'Singular terms', *Journal of Philosophy* **71**, pp. 183–204.

Donnellan, K.: 1970, 'Proper names and identifying descriptions', *Synthese* **21**, pp. 335–58.

Donnellan, K.: 1974, 'Speaking of nothing', *The Philosophical Review* **83**, pp. 3–32.

Evans, G.: 1973, 'The causal theory of names', *Aristotelian Society Supplementary Volume* **47**, pp. 187–208.

Kripke, S.: 1972, 'Naming and necessity', in D. Davidson and G. Harman (eds.) *Semantics of Natural Language*, D. Reidel, Dordrecht.

Kripke, S.: 1979, 'A puzzle about belief', in A. Margalit (ed.) *Meaning and Use*, D. Reidel, Dordrecht.

Lakoff, G.: 1971, 'Presupposition and relative well-formedness', in D. Steinberg and L. Jakobovits (eds.) *Semantics*, Cambridge University Press, Cambridge.

Lambert, K. and B. van Fraassen: 1972, *Derivation and Counterexample*, Dickinson, Encino and Belmont.

Lewis, D.: 1975, 'Languages and language', in K. Gunderson (ed.) *Language, Mind, and Knowledge*, University of Minnesota Press, Minneapols.

Loar, B.: 1976, 'The semantics of singular terms', *Philosophical Studies* **30**, pp. 353–77.

McKinsey, M.: 1978, 'Names and intensionality', *The Philosophical Review* **87**, pp. 171–200.

McKinsey, M.: 1978, 'Kripke's objections to description theories of names', *Canadian Journal of Philosophy* **8**, pp. 485–497.

Peacocke, C.: 1976, 'Truth-definitions and actual languages', in G. Evans and J. McDowell (eds.) *Truth and Meaning*, The Clarendon Press, Oxford.

Schiffer, S.: 1978, 'The basis of reference', *Erkenntnis* **13**, pp. 171–206.

Sellars, W.: 1956, 'Empiricism and the philosophy of mind', in H. Feigl and M. Scriven (eds.), *The Foundations of Science and the Concepts of Psychology and the Psychoanalysis*, Minnesota Studies in the Philosophy of Science I, University of Minnesota Press, Minneapolis.

Sellars, W.: 1954, 'Some reflections on language games', *Philosophy of Science* **21**, pp. 204–228.

RICHARD LAING

MACHINES AND BEHAVIOR

1. INTRODUCTION

In this paper we consider the capacity of machines to exhibit human-like behaviors. In "Logic, Computers, and Men" Burks (1973) takes as his thesis that for each human there is a deterministic finite-state automaton behaviorally equivalent to that human. That is, he argues that machines can exhibit all natural human functions.

His argument can be made in the following seemingly roundabout (but, I think, fair enough) fashion. Suppose that neurophysiologists are at last able to identify every behaviorally relevant physical component of the human nervous system, as well as all the interconnections between these components; suppose as well that the neurophysiologists are able to specify, for each component, all of its behaviorally pertinent physical properties and transactions with every other component to which it is connected.

We can now carry out the following thought-experiment. Let us suppose that for some human for whom such precise and detailed neurophysiological information is available suffers some damage to peripheral componentry of his central nervous system (in the auditory input, for example). Then we can imagine that, knowing the precise normal relevant physical functioning of the damaged componentry, we could, in theory at least, design and construct an artificial device (with electronic and electrochemical and electromechanical constituents, perhaps) which would substitute in all behaviorally relevant respects, for the injured portion of the nervous system. It would be our expectation (and a laudable and reasonable medical goal) that the patient would honestly and correctly report that hearing had been completely restored, and that indeed, all, as far as the patient can tell, was exactly as it was before the injury.

We now continue our thought experiment and (with seeming perversity, and without regard for the ethics of our actions) persist in one-by-one replacement of the remaining biological nervous componentry of

131

Merrilee H. Salmon (ed.), *The Philosophy of Logical Mechanism*, 131–144,
© 1990 *Kluwer Academic Publishers*.

our patient, by behaviorally equivalent artificial prostheses. What will take place, subjectively and behaviorally? As biological componentry is replaced by artificial componentry would the patient report any change? Would consciousness slowly dim, and finally disappear as more and more replacement is carried out, or as the advancing margin of the artificial pushes into some ultimate citadel of the central nervous system? With the replacement of some single or a few critical neural components will consciousness suddenly dissolve?

Burks, I believe, would say that though the physical basis of implementing the behaviorally pertinent nervous system transactions had been totally transformed, that all externally observable behavior of the patient (including his reports of his subjective states) would remain unchanged by this substitution process. Moreover, Burks would go on to point out that whatever the nature of the physical componentry of the nervous system, whether the original biological "wetware" or substituted hardware, whether the pertinent physical transactions are continuous or discrete, are coded in analog or digital form, and whether they are deterministic or probabilistic in nature, that there is software, for a general-purpose computing machine which can cause the machine to simulate all the transactions of the original biological system or substituted artificial system, to whatever degree of fidelity is desired or required, to achieve behavioral indistinguishability in the computer simulation as well.

Since for every general-purpose computing machine equipped with a program there is a deterministic finite-state automaton which is the formal equivalent of the programmed physical computing machine, Burk's claim of behavioral equivalence between each human and some deterministic finite-state automaton follows. Thus, as Burks would perhaps point out, it is possible to have a continuous and probabilistic system if that is what a behaviorally adequate model of the nervous system should prove to be) embedded in a discrete and deterministic system (a general-purpose digital computing machine with program, whose computational actions are the physical embodiment of the behavior of a discrete deterministic, finite-state automaton) which computing machine itself is embedded in the actual physical universe (with whatever properties – deterministic or probabilistic or continuous – the physical universe may in fact possess or exhibit).

Burk's argument therefore assures us that an artifical machine can indeed duplicate the input/output behavior of any human being.

Supplied with the detailed information neurophysiologists might provide us on the way in which the biological human nervous system brings about this behavior, or by examining the artificial prosthetic or computer program, or idealized automaton behaviorally equivalent forms of the human system, one might hope to be able then to understand how brains work. But it seems to me unlikely that either such neurophysiologist's detailed knowledge of the human nervous system or its artificial counterpart (in prosthetic apparatus or computer program or automaton state transition diagram) is likely to be of much use in understanding how brains work, how they came to work as they do, or how machines act so as to behave like brains. It is not merely a matter of the complexity of the system to be understood that is at issue here, although this complexity is clearly considerable, but rather that, with few exceptions, we will have no notion what *significance* to assign to the physical activity so completely exposed before us. (It seems likely in fact that the neurophysiologists will not even be able to identify the behaviorally relevant physical transactions of nervous system componentry unless they *first* have some notion of the possible significance of the transactions. Burks, I think, might here agree with my objections, and suggest that we then should merely duplicate by protheses *all* physiological phenomena to whatever degree of fidelity seems necessary.)

We may be able to identify some of the neural pathways which lead to a knee being twitched, or a hand being raised, or some of the peripheral consequences of a light flashing, or a bell ringing, but despite having a detailed chart of the physical system, with this alone we shall not be able to say how and why humans behave the way they do, nor to come to understand how machines could duplicate these effects. If we seek understanding how the behavior of humans, or artificial counterparts of humans, takes place, and how it came to be in the first place, having before us a completely detailed description of the physical system may not be of much value in this quest for understanding.

To come to *understand* what is going on, another sort of approach entirely is needed, and we suggest that a successful alternative approach to understanding how human brains work and how machines might be designed and constructed to behave like brains can be found in another branch of automaton theory pioneered by Burks: his elucidation and extension of the theory of general automata proposed by John von Neumann, and the continuation of this program by Burks's colleagues and students.

2. AN ALTERNATIVE APPROACH

If machines are to act so as to exhibit human behaviors, we must first of all possess a machine system in which complex behaviors, including behaviors of the sort possible for humans, can be exhibited. Moreover, since we wish to *understand* how these behaviors may be *implemented*, we shall require a machine system which permits the direct and easy expression of the sorts of behaviors possible for humans. This means that the machine system must be able to express directly such actions as machine reproduction, machine spatial position and movement, machine encounters with other machines, and the like.

John von Neumann had hoped to develop just such a general theory of complex automata, a theory which would encompass both natural and artificial mechanisms. His successful design of an automaton system in which machine reproduction can take place is most germane to our concerns here.

The most complete explication of von Neumann's theory of machine self-reproduction is to be found in *Theory of Self Reproducing Automata* (von Neumann, 1966). This treatise on self-reproducing machines was left unfinished by von Neumann, and was later edited and completed by Burks. In this work, von Neumann describes how a "kinematic" machine, a machine which can both process information and manipulate physical components (of the same sort as those of which it is composed) can construct a duplicate of itself. Von Neumann also describes (and this is the principal subject matter of his treatise) how a "cellular" machine, a machine embedded in a universe consisting of a regular tessellation of connected rudimentary automata, can construct duplicates of itself. (The easiest, adequate introduction to these notions is to be found in Burks's paper "Von Neumann's Self-Reproducing Automata" (Burks, 1970).)

In these automaton systems (the kinematic and the cellular) there are (or can, under simple redesign, be) analogies to the spatial position and movement of physical machines, and machine componentry, and to machine creation or identification of physical componentry, and the like. Employing such systems, we want to come to understand not only how machines might exhibit human behaviors, but how humans in fact implement their behaviors. Since the machines of systems of this sort can exhibit a vast range of behaviors, we shall require some criterion by which this vast number of machine behaviors can be reduced to those most likely to be relevant to an understanding of the behavior of naturally evolved systems.

Those behaviors of evolved systems which are so ubiquitous that they are often called "natural" are likely to be those which are *adaptive*; that is, which have in the past tended to persist and spread in natural populations in an environmental context of reproductively competing individuals.

We should then therefore concentrate our attention on *machine* behaviors which are adaptive. The precise implementations of machine behaviors which are adaptive then become our candidates for explanatory mechanisms of the origin and implementation of the analogous behavior of naturally evolved organisms (including humans).

In the rest of this paper we will first list some of the basic actions which can be directly represented in any of several general machine systems. We will then combine some of these basic actions to form some complex behaviors possible for machines. Finally, in the remainder of the paper, we will form some putatively adaptive machine behaviors and discuss their relevance for understanding the behavior of evolved systems.

3. SOME BASIC ACTIONS OF MACHINES AND SOME POSSIBLE MACHINE BEHAVIORS.

Without going into technical details of the various proposed formal systems, stemming from von Neumann's original conceptions and Burks's explications, and the extensions (by Burks and his colleagues and students) we now list (in very informal fashion) a few of the possible behaviors of machines, behaviors which can be combined to produce yet more behaviors. The precisely characterized forms of behaviors listed have all been shown to be possible for some (precisely characterized) machine system or another.

1. For each algorithmic process there is a machine of the system which can carry out that process.
2. There are single machines of the system which can be programmed to simulate the actions of any of the other machines of the system.

3. A machine can simulate its own actions.
4. A machine can construct any machine for which it has a (structural) description.
5. A machine can make available to itself its own complete structural description.
6. A machine can construct a duplicate of itself.
7. A machine can inspect encountered machines and ascertain their structural descriptions.
8. A machine can inspect itself and ascertain its own structural description.
9. A machine can employ a structural description of itself or of an encountered machine, to simulate its actions or the actions of another machine.
10. A machine can duplicate itself or construct other machines on the basis of descriptions acquired by inspection.
11. A machine can dismantle encountered machines.
12. A machine can (partly) dismantle itself.
13. A machine can make machine componentry available to itself.
14. A machine can make machine componentry available to machines it may encounter.
15. A machine can permanently conjoin itself with an encountered machine.
16 A machine can break contact and withdraw from the vicinity of a machine it encounters.
17. A machine can change its behavior as a consequence of internal or external conditions (for example the number or kind of actions already taken, elapsed time, distance traversed, machines encountered, etc.) or combinations of internal or external conditions satisfied.

The above list of basic machine actions is not intended to be exhaustive but will be sufficient for our purposes. We shall now set down some examples of specific machine behaviors.

1. A machine might remain inactive.
2. A machine might do nothing but calculate.
3. A machine might do nothing but produce duplicates of itself.
4. A machine might do nothing but produce duplicates of other machines.
5. A machine might retreat from all machines it encounters.

6. A machine might attach itself to all machines it encounters.

7. A machine might move about in its environment, either according to a fixed plan, or by employing some pseudo-random process to determine the direction and distance of movement.

8. A machine might dismantle all machines it encounters.

9. A machine might submit itself to dismantling by all machines it encounters.

10. A machine might inspect all machines it encounters.

11. A machine might submit itself to inspection by all machines it encounters.

12. A machine might provide components to all machines it encounters.

13. A machine might accept components from all machines it encounters.

The simple behaviors we have so far listed might be combined in various ways to form more complex behaviors.

14. A machine might construct duplicates of any encountered machine which it is able to inspect and for which it is able to acquire a description.

15. A machine might make copies of its own description and attach these to that of any machine it encounters.

16. A machine might make copies of its own description and substitute them for the descriptions of any machine it encounters.

17. A machine might substitute the description of any machine it encounters for its own description.

18. A machine might attach a copy of the description of any machine it encounters, to its own description.

Particular machine behaviors might be exhibited only under certain conditions.

19. A machine might reproduce (construct another machine, inspect, compute, dismantle, aid, etc.) for some initial period of its existence, and not thereafter.

20. A machine might refrain from reproducing (constructing other machines, inspecting, computing, dismantling, aiding, etc.) for some initial period of its existence, and not thereafter.

21. A machine might reproduce (construct another sort of machine, inspect, compute, dismantle, aid, etc.) within a certain spatial region, and not outside.

22. A machine might reproduce (construct, inspect, compute, dismantle, aid, etc.) outside a certain spatial region, and not within.

More generally,

23. A machine might act (reproduce, construct, inspect, compute, dismantle aid, etc.) or refrain from acting until some single condition was satisfied, or until some sequence or combination of conditions was satisfied.

4. GOAL-DIRECTED AND ADAPTIVE MACHINE BEHAVIOR.

Of the list of possible machine behaviors (a list which could be expanded indefinitely) we wish to concentrate our attention on those behaviors which are most likely to be of importance in coming to understand how machines can behave like humans, how human behavior is in fact implemented, and how and why these implementations of human behavior came to be characteristic in the first place. We are of course ultimately interested in behavior of the sort which appears to be humanly relevant, goal-directed or natural. None of these terms is precise but it seems to me likely that what we tend to see as relevant, goal-directed, and natural will be behaviors which are adaptive in environments of the sort with which we are familiar.

A behavior is *adaptive* if those individuals exhibiting the behavior in a context of reproductive competition, tend to increase their numbers, relative to those individuals not exhibiting the behavioral trait (all other things being equal).

Although *adaptive* may be a more precise notion than "natural", in practice, determination of which behaviors are adaptive and which not may be rather difficult, since it will depend on the complex interactive outcome of many individuals, competing under particular environmental conditions, and a determination of ultimate reproductive success for individuals exhibiting particular behavioral traits may be ascertained only by extended modelling and computer simulation. We shall therefore content ourselves here with informal arguments and discussion of cases where the reproductive outcome or significance of the behavior seems quite clear.

For example, the capacity to *repair* oneself to some substantial degree, would seem to be an adaptive behavioral trait since if an individual machine is fully operative for a longer time, there is a likelihood that its reproductive life will be extended as well. Machines of the sort we have been considering can exhibit self-repair, at least in part. Self-repair is implementable as follows.

A machine can make available to itself a description of what its complete structure *ought* to be. It can also make available to itself a complete description of its *present actual* structure. These two descriptions can be compared, the discrepancies noted, and construction action taken to bring the actual structural state of affairs back into coincidence with the correct state of affairs. (It should be noted that such *complete* self-inspection and self-repair has been shown possible only when sufficient inspection, comparison, and construction routines themselves are not the site of structural flaws and consequent behavioral disability or malfunction.)

This machine capacity for repair can also be used to assist other machines. As we have pointed out earlier, a given machine can examine the present actual structure of an encountered machine, and so produce a description of the structure of the machine. If the given machine is provided with, or can make available to itself, a description of what the structure ought to be, then the given machine can compare the actual and the correct descriptions, note the discrepancies (if any) and, by construction actions, reconcile the differences, and so restore the encountered machine to its correct structural state of affairs.

Of especial interest is the case where a given machine might provide repair or other services *selectively* to certain other machines. Since machines can obtain descriptions of their own structures, and those of machines they may encounter, a machine can selectively aid machines of the same structure as itself (and on the other hand could dismantle all encountered machines *not* like itself).

This sort of selective assistance to machines of the same sort as the helping machine (or aggression directed toward non-kin) is of great significance since it is clearly adaptive. A type of machine which aids (and perforce may itself receive aid from) other machines of the same type is a machine type which will tend to increase its relative numbers in the population of machines and will ultimately become near-ubiquitous. Thus, viewed from the outside, such a population of machines generally would seem to be deliberately pursuing the "natural" goal of demonstrating affectionate regard and concern for kin (and indifference or antipathy toward non-kin).

In order to implement this "kin-preferring" adaptive behavior, our given machines have been assumed capable of inspecting at length any encountered machines. In a society of reproducing machines, where the dismantling of machines and the utilizing of machine components for further self-duplication can take place, a machine which assumes the passivity required for extensive inspection of its structure by another machine, is not likely to survive for long and propagate more of its kind, unless it somehow has already identified the encountered machine as "friendly." This means that, in an environmental context of machine reproduction and variation, the identification of "like" machines or "friendly" machines, will probably come to depend on partial and brief inspections, or behavioral actions or signs short of complete inspection.

With the relinquishing of complete inspection as a determiner of kinship though, far more subtle, indeed seemingly devious relationships among machines may arise. For there can be machines which possess partial structures or exhibit behavioral signs which would seem to mark them as kin, when in fact this is not the case. Thus in a population of machines which comes to behave preferentially toward kin, there can next arise a machine type which could reap the benefit of seeming kinship, and yet spare itself the effort of helping these other machines in turn. In such a situation, these "selfish" or "deceitful" machines might greatly increase their numbers, and the persistence of the originally kin-supporting machine types might depend on the adoption by them of more subtle means of identification of encountered machines. Thus something like machine acuteness or insight might become necessary to survival of type.

On the other hand it is also quite possible (though behaviorally considerably more complex) to have adaptive, mutually advantagous preferential behavior toward certain machines which are *not* kin. That is, it is even possible that in a system of reproducing machines, a form of *reciprocal altruism* will arise and become pervasive, a behavior in which machines will act in seemingly unselfish fashion toward other machines which are not kin (and need not be deceitfully posing as kin either). Evolutionary biologists, especially Trivers (1971) have argued that in situations where the reproducing entities have 1) long life spans, 2) remain in contact with others of their group, and 3) experience situations in which they are mutually dependent, reciprocal altruism may arise out of chance variation and evolutionary selection. In human terms, if helpful actions can be taken which are low risk to the giver and have a high value to the receiver (where high and low risk, etc. are relative to the impact on the reproductive potential of the individuals) and there is a likelihood that the individuals will remain in fairly close association for a long time, then any genetic predisposition to take altruistic actions will tend to spread in the population. For, in effect, it will lead to reciprocal assistance in times of need, to the greater survival (and hence increased reproductive opportunity) of those members of the population bearing this trait. A good example of this is that of an individual saving another from drowning by reaching out a branch. The risk to the giver is small, and the benefit to the receiver is great, and over a long time the benefits in terms of increased numbers of offspring are likely to be great to those members of the population genetically predisposed to behave in this reciprocally altruistic fashion.

Needless to say, opportunities for deceit and cheating in the case of hoped for altruistic reciprocity are even more numerous and complex than for kin selection strategies. In particular, each individual machine must possess the memory capacity to remember the altruistic acts and the partners in them, since the opportunity for reciprocity may not arise for some time. In addition a sort of cost benefit analysis must take place, in which the value of the act, the character of the reciprocity partner, the capacity of this partner to repay, the likely lifespan of the giver, and receiver (to decide whether there will likely be time for repayment to take place) etc. all must be calculated.

What is the relevance of all this to our general approach to discovering how machines might behave like humans, and humans arrive at and implement their behaviors? We wish to be able to generate numerous examples of machine types exhibiting adaptive behaviors, for which the implementation of the behavior is explicit. We can then employ these strategies of implementation as candidates for the mechanism (and thus explanation) of analogous such behavior in naturally evolved systems. How do we generate such implementations of machine adaptive behaviors? One approach is to continue what was done here, namely to design such systems. Such an approach requires some prior notion of what behaviors are likely to be adaptive, and has in this case been very much dependent on the insights of evolutionary biologists. See for example Hamilton (1964) and Alexander (1979). (It should be pointed out that the evolutionary biologists very often fail to propose an explicit testable mechanism for the implementation of those behaviors they claim are adaptive and so would be selected for, and it is precisely these mechanisms which we seek to uncover.)

Another approach would be systematically to introduce into our automaton system successive populations of machines possessing all combinations of the basic possible actions, starting with the shortest combinations, and then after simulating the interactions of such machine types, note which types have in fact persisted and proliferated.

It should also be noted that the machine behaviors can be ranked in terms of their complexity (the shortest sequences of basic actions, or the shortest computer program which implements a particular behavior being taken as the measure of the complexity of the behavior). We might by this means thereby by able to suggest which human or other animal behaviors are most primitive, and on this basis reconstruct the likely temporal order of their appearance in naturally evolved systems.

5. SUMMARY AND CONCLUSIONS

The conviction that all natural human behaviors can be realized in a deterministic finite-state automaton does not automatically lead to an understanding of the implementation of machines which can duplicate human behavior, or an understanding of how in fact human behavior is implemented, or how it arose in the first place. Sophisticated systems of automata, such as those suggested by von Neumann are capable of exhibiting very complex behaviors, the implementation of which behaviors can be explicit and understandable. Some of these machine behaviors are of the sort exhibited by naturally evolved organism (including humans) and many are not. In order to reduce the full set of possible machine behaviors to those most like to be relevant to an understanding of the behavior of naturally evolved systems we suggest a systematic consideration of machine behavioral types, starting from the simplest, and an evaluation of these types for their persistence and proliferation in a setting of reproductive competition; that is, we suggest concentrating attention on *adaptive* machine behaviors.

In attempting to identify the likely mechanisms of behavior of naturally evolving entities by this means, it may be important to bring the artificial evolutionary process and the historically occurring evolutionary process into much closer correspondence (by employing only those basic machine actions for which there are clear molecular or organismic counterparts, by setting up similar initial distributions of individual behavioral types, and by submitting the evolving machine population to similar *external* environmental pressures, including accidents). This would of course radically complicate the investigation. But though in this paper we contented outself with an informal analysis of some putatively relevant adaptive machine behaviors (these behaviors including self-repair, and kin selection) it seems likely that all adapted populations, whether of machines or natural entities, would evolve to include some such traits in their behavioral repertoires. There are as well many other rather general behavioral traits which are also likely to become pervasive in an evolving machine population, the precise mechanisms of which might be useful in explicating and understanding sophisticated human or animal behavior.

For example, machines can exploit or dismantle other machines over which they gain control, and this behavior may be adaptive in some circumstances, though it would not, of course, be adaptive when the dismantled machines are selectively of the same sort as the dismantling machine. It should also be pointed out that there is a point at which it probably is adaptive for a machine to calculate beforehand the consequences of certain sorts of encounters (in particular when the contemplated action is inherently dangerous, or when the expense of carrying it out is high) and such capacity for "foresight" is likely eventually to become fairly common in an evolving machine population. Thus, if two machines each asserted kinship and control over a third machine, it is not impossible that a sophisticated machine (call it Solomon) could simulate the action of dismantling the third machine and dividing it between the contending machines, and by this calculation foresee those actions of the machine disputants which would distinguish the true from the false.

Professor Richard Laing
290 E 37 Ave
Eogene, OR 97405
U.S.A.

REFERENCES

Alexander, R.: 1979, *Darwinism and Human Affairs*, University of Washington Press, Seattle.

Burks, A. W.: 1970, *Essays in Cellular Automata*, University of Illinois Press, Urbana.

Burks, A. W.: 1973, 'Von Neumann's self-reproducing automata, in Burks (1970), pp. 3–64.

Burks, A. W.: 1973, 'Logic, computers, and men', *Proceedings and Addresses of the American Philosophical Society*, **XLVI**, pp. 39–57.

Hamilton, W.: 1964, 'The genetical evolution of social behavior (I and II)', *J. Theoret. Biol.* **7**, pp. 1–16 ; 17–52.

Trivers, R.: 1971, 'The evolution of reciprocal altruism', *Q. Rev. Biol.* **46**, pp. 35–57.

Von Neumann, J.: 1966, *Theory of Self-Reproducing Automata*. Edited and completed by A. W. Burks, University of Illinois Press, Urbana.

ANDREW LUGG

FINITE AUTOMATA AND HUMAN BEINGS

Arthur Burks is best known among philosophers for his views on induction, probability and causation. He has elaborated and refined
these views over many years, and his major work, *Chance, Cause,
Reason* (1977), develops them in great detail and with much subtlety.
No less interesting, however, is Burks's application of automata theory
to such philosophical problems as the formalizability of intelligent behaviour, freedom of the will and the dispute between empiricists and
rationalists concerning abstract ideas. While automata theory plays a
role in Burks's investigations of causality, chance and inductive reasoning, it becomes especially prominent in his discussion of these other issues. Unsurprisingly, Burks, who is as much a computer scientist as a
philosopher, is exceptionally circumspect and scrupulous in bringing
computer science to the aid of philosophy.

In what follows, I consider whether Burks's strategy of reformulating
philosophical problems as problems of computer science can render
them more precise, intelligible, tractable. In agreement with Burks, I
argue that certain philosophical problems – including the formalizability of intelligent behaviour, freedom of the will and the source
of abstract ideas – can benefit from an infusion of computer science.
But I also argue that application of Burks's strategy results in a significant transformation of these problems as they are traditionally understood. Treating philosophical problems as problems of computer
science may in the course of clarifying them reveal that they are improperly formulated or that they rest on controversial assumptions.

I

A philosophical problem to which automata theory has direct application is whether intelligent behaviour can be reproduced by machines.
From a computer-theoretic standpoint, this is a problem, as Burks
points out in "Logic, Computers and Men" (1972–73), about the

145

Merrilee H. Salmon (ed.), *The Philosophy of Logical Mechanism*, 145–158,
© 1990 *Kluwer Academic Publishers*.

capability of finite automata to perform human functions. Can a finite (deterministic) automaton equipped with appropriate sensing devices behave intelligently or are human beings machines of another kind or not machines at all? This formulation of the problem is more precise than traditional formulations, being in terms of the capabilities of physical models or realizations of precisely defined logical systems (Burks, 1972–73, p. 40).

To say that a finite automaton can perform human functions is not to say that human beings are *structurally* similar to finite automata, but that finite automata equipped with appropriate sensors may be designed which are *behaviourally* equivalent to human beings.[1] What a human being and a finite automaton do may be indistinguishable, yet what makes them do it may be radically different. The idea is that we can in principle explain all human functions in the way that Charles Sanders Peirce explains the response of a frog's leg to electrical stimulation. In a passage that Burks singles out as especially significant, Peirce argues that "the connection between the afferent and efferent nerve, whatever it may be, constitutes a nervous habit, a rule of action, which is the physiological analogue of the major "premiss" of a syllogism of the form: all S arc R; this is an S; therefore this is an R.[2] (On this view, the stimulation of the afferent nerve corresponds to the minor premiss and the resulting explosion in the efferent nerve to the conclusion of the syllogism.) For Burks, the crucial point is that Peirce characterizes the phenomenon in terms of a rule of action and represents the process (at a high level of abstraction) by a logical formula. Can we, he asks, reasonably hope to find rules of action (i.e. transition rules relating the output to the input of finite automata) for intelligent behaviour? In other words, given that every finite automaton is representable by a logical formula, can intelligent behaviour be formalized?

In support of an affirmative answer, Burks argues first against those who insist that some human functions cannot be formalized. He points out that self-reproduction is not beyond the capabilities of finite automata, referring us to John von Neumann's remarkable proof that a finite automaton embedded in a particular type of cellular network can produce a copy of itself.[3] And he rejects the claim of Peirce and others that since human beings are capable of novel behaviour and of correcting their mistakes, human behaviour cannot be captured by a rule of action. Here Burks observes that Arthur Samuel (1959, 1967) has designed a checker player that is capable of learning from experience and that Richard Laing (1976) has demonstrated the possibility of self-monitoring, self-diagnosing and self-repairing automata.[4]

Secondly, Burks argues that the restriction on automata to finite, deterministic ones is not overly stringent. He provides a "quantum argument" to show that if human behaviour can be reproduced by an analog automaton it can be reproduced by a finite automaton (i.e. an automaton with a finite number of inputs, outputs and internal states) (Burks, 1972–73, pp. 56–57). And, noting that any randomness in human behaviour, whatever its source, can be reproduced by a (deterministic) source of pseudorandomness, he argues that the deterministic version of the thesis is as plausible as the probabilistic version (1972–73, pp. 55–56).

Without denying that the mind-machine thesis is clarified by reformulating it in computer-theoretic terms, one may question the arguments Burks presents in support of the thesis. As Burks emphasizes in other contexts, models of natural phenomena are always models relative to "some respect, purpose, behaviour, criterion of importance, or point of view" (1974a, p. 296). A finite automaton may be a good model of the functioning of a general purpose digital computer without being a good model of its physical composition; a cellular automaton may be a good model of heart fibrillation without being a good model of heart failure caused by congestion in the veins.[5] Thus, to evaluate what Burks calls the "man-machine" thesis, we need to know what purposes or interests are being assumed. What level of abstraction is being considered, and why this level and not a higher or lower one?

This is not an insignificant matter, since it is only when we adopt a particular point of view that we can set aside as unimportant certain aspects of human behaviour and proceed as Burks does at a high level of abstraction. Moreover, most interesting points of view impose stringent constraints on what counts as a good model of human behaviour. For instance, constraints pertaining to the speed of operation of subsystems or the manner in which these are interrelated may force us to conclude that human behaviour can be modelled, if at all, only by nondeterministic or by analog automata. (Compare the quantum mechanical "impossibility theorems".)

Finally, it is important to realize that when we take purposes and interests into account, a radical shift in perspective results. What Burks has in effect proposed is a transformation and not merely a clarification of the traditional problem. For once the problem is seen to be about the possibility of constructing models of human behaviour, it becomes clear that different kinds of models, corresponding to different purposes and interests, may be appropriate, and that only in some contexts will it be reasonable to assert the "man-machine" thesis. Resorting to computer theory reveals that whether human beings are machines is not a question which admits of a single answer.

II

The problem of freedom of the will can also benefit from a study of computer models. Burks argues, quite rightly I believe, that the problem is clarified when seen in terms of the possibility of embedding choice systems in wider deterministic systems. What he does not observe is that his reformulation of the problem shows it to have been improperly posed.

A person's choice system is almost always indeterministic, since the states of such systems do not normally have unique "next states". Yet we do not suppose that this shows that our choices are free. The indeterminism of choice systems may reflect nothing more than the way that we conceptualize or model them; there may be hidden variables that determine the choices we make. Thus, we are led to the view that the free-will problem is a problem about whether a "voluntary choice process can be embedded in a deterministic system of events governed by causal laws" (Burks, 1979, pp. 413–414; see also 1977, pp. 578–580).

It would thus appear that the free-will problem may be clarified by reformulating it as a problem of computer science concerning the possibility of embedding one system in another. But, contrary to what is sometimes suggested,[6] stating the problem this way is not entirely free of difficulty. Suppose we are able to simulate the system comprising an individual's choices and that part of the environment with which he or she interacts using the Monte Carlo method.[7] Then, if freedom of the will depends on the possibility of embedding choice processes in underlying deterministic systems, the freedom of the individual's choices depends on the source of randomness used by the Monte Carlo process. The choices are free if the source is natural, and they are unfree if it is a (deterministic) pseudorandom-number generator. But this result is implausible: whether an individual's choices are free or not should not depend on so trivial and fortuitous a consideration. The situation is analogous to that which arises when the fairness of a lottery is considered. Just as it would be a mistake to think that a lottery employing a pseudo-random number generator must be unfair, so it would be wrong to think that all determined choices are necessarily unfree. The absence of chance is antithetical neither to fairness nor to free choice.

One response to this argument is to reject the idea that freedom of the will concerns the possibility of embedding an individual's choice system in a wider, deterministic system. Another, more promising, response retains this intuition and points out that the embedding relation is triadic, since it interrelates embedded and embedding systems with respect to a concern, purpose, interest or point of view. Since an embedded subsystem is a model of the system in which it is embedded, the fact, stressed above, that models are always models relative to a concern or interest applies equally to embedded subsystems.[8] This means that from a systems-theoretic standpoint, the free-will question is about the possibility of embedding one system in another in conformity with a specified concern or interest. Here, as elsewhere, we cannot avoid considering the concerns and interests that underlie our inquiries.

When considering whether a choice is free or not, our basic concern is with responsibility. It is thus reasonable to insist that the relevant embedding relation be defined in terms of legal and practical criteria for being free. Individuals are free just in the event that their choice processes and those parts of the environment singled out by the legal and practical criteria do not constitute a deterministic system. It is irrelevant whether their choice processes and the relevant parts of the environment can be embedded in underlying deterministic systems, since these embedding systems are not related to choice processes in an appropriate manner. Put otherwise the point is the familiar one that determinism does not militate against freedom, since freedom requires not that our actions be undetermined but rather that they not be determined in certain particular ways. This, I believe, is a feature of the legal and practical conditions for freedom that now prevail, and it is, besides, what a careful examination of the lottery analogy mentioned earlier would lead us to conclude.[9]

The systems-theoretic point of view enables us to state the compatibilist's position in an elegant way, but it does not establish this view as it is usually stated. If embeddings are taken to be relative to interests, concerns and purposes, there will be as many questions about

free will as there are interests, purposes and concerns that motivate
our investigations. In different contexts, different interests will be
prominent, and we should not expect to obtain the same result
however we proceed. This point, moreover, holds even in cases where
our main concern is with responsibility: considerations appropriate to a
court of law differ from those appropriate to day-to-day attributions of
blame no less than considerations having to do with responsibility differ from those appropriate to an examination of the physical character
of systems of choice.

III

A third traditional philosophical issue that Burks advocates analyzing
in system-theoretic terms is the contrast between a priori and empirical
concepts and the related debate between rationalists and empiricists.
By resorting to computer theory and the thesis that human beings are
finite automata, we can, according to Burks, provide a more precise
characterization of the a priori-empirical contrast; and we can transform and clarify the debate about the acquisition of abstract concepts
by viewing it as a debate about learning theory. Burks's view is that
these matters are better seen as problems of computer science than of
philosophy.[10]

To clarify the traditional distinction between a priori and empirical
concepts, i.e. concepts obtained by reflection or intuition and concepts
obtained by experience, Burks adopts a computer-theoretic standpoint
and views human beings as information-processing devices. He characterizes a priori concepts as concepts contributed by the individual's innate capacity to process information: a concept is a priori, he suggests,
if it is "a basic aspect or feature of the structure-programme complex
constituting a person's innate processing capacities, in terms of which
this complex processes information and organizes experience"; and a
concept is empirical if it is "abstracted" from "directly experience(d)
instances" of the concept itself by means of a structure-programme
complex.[11]

This proposal leaves obscure the idea of directly experiencing an instance of a concept and the idea of being a feature of a structure
programme complex. Nevertheless, it is an improvement upon older
accounts, since it provides a clearer account of the character of the
contributions of the mind and the world. A fuller account would clarify

the obscurities just noted as well as the status of concepts the instances of which are neither directly perceived nor basic features of an individual's structure-programme.

Burks's proposal about a priori and empirical concepts has an interesting bearing on the contrast between reflective and experiential knowledge. It is true that Burks holds that individuals "become aware of ... an a priori concept by having experiences involving the concept and seeing on reflection that the concept is involved in these experiences", and that his discussion of the nature of causal necessity focuses on whether this concept can be acquired by reflecting on one's experience (1977, pp. 613, 614). Nevertheless, as Burks characterizes a priori and empirical concepts, we need not be able to tell when an innate structure-programme is being used to abstract a concept from experience and when it is not. Sophisticated experimentation and high-level theorizing may be required to ascertain whether a concept is a priori in Burks's sense. For it may not be at all easy to determine whether an intuited concept is an aspect of an individual's innate as opposed to his or her acquired structure-programme. (Burks introduces the idea of an "acquired structure-programme complex" in recognition of the fact that an individual's capacity to process information is in part acquired through interaction with the environment (1977, p. 612).) Indeed, I suspect that, the more realistic our treatment of concept formation, the more we shall depart from the spirit of the traditional distinction between a priori and empirical concepts.

Burks regards his discussion of a priori and empirical concepts as a contribution to the philosophical debate concerning their character, but one might reasonably argue that what he actually shows is that the debate rests on a misunderstanding, that it is a debate without substance. Burks himself hints at this when he observes that the information capacities of human beings have evolved, but he does not elaborate the point (1977, p. 613 ftn.). We may press it home by noting that human beings can learn to learn, i.e. that our innate structure-programmes may be revised as well as supplemented. In Burks's terminology, this means that concepts acquired "a priori" may at a later date be acquired by "the direct experience of instances" and conversely. But then, if the contributions of the mind and the world interact with and modify one another, the value and even the possibility of isolating them becomes moot. In other words, if an individual's structure-programme is subject to revision, it is difficult to envisage any useful philosophical role for the distinction between a priori and empirical concepts as traditionally characterized. By clarifying the distinction, I want to suggest, Burks has in fact revealed that we would do well to view it in the way we view the distinction between nature and spirit – as having outlived whatever useful function it may once have had.[12]

Turning to Burks's discussion of the debate between rationalists and empiricists, we discover an interesting tension between his approach to learning and his attempt to clarify traditional ideas. Burks's intriguing suggestion concerning this debate is that we view it as a debate about what must be innate in our innate structure-programmes if we are to acquire the concepts, ideas and knowledge that we do. The key question for Burks is whether innate structure-programmes are "organized in terms of such specific forms and categories as space-time, substance-property, and cause-effect, (employing) such presuppositions as the uniformity of nature" or whether such categories and presuppositions are acquired by experience.[13]

When evaluating this way of reconceptualizing the philosophical issue underlying the traditional debate, it is important to bear in mind that historically the question of innateness was approached in a variety of ways. In particular, as Roy Edgley has observed, philosophers fluctuated between seeing the issue as one about innate ideas, beliefs or propositions and seeing it as one about innate tendencies, dispositions or habits (Edgley, 1970, pp. 3–16). Of these two aspects of the debate, Burks seems to have the former in mind. He notes, for instance, that although the empiricist's approach is often alleged to be more modest than the rationalist's, it actually assumes a "more specific and structured" basic learning programme. Here the idea appears to be that the empiricists, we discover an interesting tension between his approach to (Burks, 1979, p. 407).

In light of the computer analogy, however, this very way of posing the problem seems misconceived. The advantage of thinking about learning in computer-theoretic terms is that it obviates the need for minds containing ideas. We can avoid both the view that abstract ideas can be compounded from experience and the view that individuals possess complex, abstract ideas prior to experience. There need be nothing that an individual is born with other than an innate structure-programme complex. If Burks is right, and it seems to me that he is, an individual's experiences are structured by his or her structure-programme complex, and we have no reason to invoke innate concepts, ideas or principles to explain how this happens.

These remarks suggest that the traditional debate, to the extent that it is a debate about ideas, rests on a false premiss, and that we should opt for the more modest view that there are no innate ideas, only innate dispositions. Not only is this suggested by the findings of computer science, it tallies with Peirce's observations about habits and rules of action, which Burks finds so congenial. From this standpoint, the question becomes an empirical one concerning the character of the structure-programmes of human beings, and in particular how specific they must be if individuals are to acquire information as rapidly as they do. In this regard, Burks is surely right to observe that the investigation of machine learning can illuminate the structure and specificity of human learning.

It might be argued that historically the issue between rationalists and empiricists related for the most part to epistemology rather than psychology, and that empiricists might agree with Burks about human learning yet maintain that nothing other than the deliverances of experience are well-founded. I would not deny either of these points. However, it should be borne in mind that Burks himself does not attend to the epistemological strand of the debate. And, more importantly given the concerns of this paper, once we accept the computer-theoretic approach to learning and insist on the inseparability of the contributions made by the mind and the world, new conceptions of evidence and confirmation come to the fore so that the epistemological problem takes on a quite different character.

In short, neither empiricists nor rationalists can derive much comfort from Burks's reformulation of the traditional debate.

IV

The strategy of recasting philosophical problems as problems of computer science throws light on social processes as well as on cognitive ones. This is not a point that Burks has pursued,[14] but it supports his claim that philosophy can benefit from computer science. Here I shall confine myself to a brief examination of the character of group decision making and the justification of coercive institutions.

The problem of group decision-making concerns the way individuals who hold different views can best achieve consensus. What we wish to know is how individuals rationally and objectively come to a common view, which may of course simply be to suspend judgement pending further inquiry. From the computer-theoretic standpoint, a group of decision makers may be viewed as an information processing device, and the problem is to specify an appropriate structure-programme for this device. In other words, the task is to specify a transition function for the group as a whole which updates the information that each of its members share. (Compare Burks's treatment of the dispute between empiricists and rationalists as a problem of specifying the learning programme of an individual.)

Once again the traditional problem is transformed as well as clarified by viewing it in computer-theoretic terms. Traditionally, the problem has been to spell out rules of choice, i.e. a fixed decision procedure. The computer version of the problem on the other hand, leaves open the possibility that the structure-programme of the group evolves, and hence that the constraints on its choice change as the group learns. Just as individuals may learn to learn, so too may groups as a whole. With the computer analogy in mind, we can thus avoid the twin sins of absolutism, which assumes the existence of fixed rules, and relativism, which assumes that methodological frameworks cannot be compared. Furthermore, once we adopt the proposed reformulation of the traditional problem, we shall be less inclined to think that there must be rules of choice to which all decision makers ideally adhere. We are obliged to hold neither that the master transition function is embodied in the practice of each individual nor that the structure-programmes of members of the group are functionally equivalent. On the contrary, given the computer analogy, the natural assumption is that there is a complex, hierarchical network of sub-structure-programmes, each with its own transition function. It is more plausible to think that complicated decision-making bodies incorporate a large variety of interrelated centres of authority than that they comprise groups of individuals each using a common method. Finally, we shall be more likely to consider models of decision making that include the possibility of decision makers' having limited knowledge, their being misinformed and their having different levels and different kinds of expertise. Such considerations, an adequate examination of which is unlikely to be accomplished without the aid of a computer, almost certainly play a crucial role in explaining how highly differentiated social communities are able to achieve agreement concerning issues of fundamental importance and why it often takes so long for this to happen.[15]

Similar observations can be made about the problem of justifying infringements on individual rights and liberties by the state. As the problem of coercive institutions is usually posed, to solve it one has to show that individuals with certain characteristics – rationality, self-interestedness, etc. – would agree that such institutions are advantageous to them. The problem is basically a decision problem, but in this case the question is whether and to what degree individuals will agree to do something, specifically transfer authority to a central governing body. To what extent will rational, self-interested individuals agree to the state's being empowered to interfere with their lives? When framed in computer-theoretic terms, this question concerns a set of individuals (each with his or her own structure-programme) interrelated by a master structure-programme. The problem is whether and to what degree a master structure-programme which imposes minimal constraints on the individuals in the system will develop into a structure-programme that imposes more rigorous constraints, given that the individual structure-programmes conform to the traditionally assumed characteristics of rationality and self-interestedness. Put otherwise, the reformulated problem concerns the kinds of master structure-programme that can be obtained given the minimally structured structure-programmes that the traditional approach envisages.

This formulation of the problem is more general than some that have been canvassed. It is notably more general than approaches which state the problem in terms of a representative individual's deciding what powers to yield to an external authority and approaches which state it in terms of individuals' bargaining in the absence of any external authority. These approaches should be viewed as special cases involving stringent and controversial idealizing assumptions; they are obviously not to be ignored, but neither should they be thought of as exhausting the possibilities. With the mathematical sophistication provided by computers we can envisage and investigate much more complicated configurations.

When we adopt the standpoint of computer theory the idea that individuals have fixed psychologies appears particularly problematic. I emphasized in the last section that the structure-programmes of individuals develop, so that I need add here only that such development depends in part on the character of the social structure (i.e. the master structure-programme), which is itself evolving. On this view, we must reject the relatively common assumption that the transfer of power to an external authority occurs in a single step by a decision to leave the state of nature and enter society. Far more plausible, 1 suggest, is the view that there is a continual interaction between the master structure-programme, the individual sub-structure-programmes and the environment.

Insofar as this interaction occurs, the allegation that coercive institutions should be justified with reference to the preferences of self-interested individuals loses much of its initial plausibility. The key point is that the adoption of the computer analogy prevents our separating the personal from the social realms.[16] For what is social at one time (i.e. a part of the master structure-programme) may be individual at another (i.e. a part of the sub-structure-programmes). What we can do – and what computer simulation may considerably aid us in doing – is demonstrate that certain configurations of individuals and social constraints are stable, that certain configurations can be improved by modifying the prevailing social constraints, and that certain modifications would be counterproductive. It is difficult to see what more one could reasonably want.

I have been suggesting that the emphasis on structure encouraged by the systems-theoretic approach can lead to a revision in our thinking about traditional philosophical problems. I have not claimed that such revisions can only be achieved by resorting to computer science, even less that all philosophical problems are appropriately treated this way. Some of the points made here were emphasized by philosophers before the advent of the modern electronic computer; and although computer science can indicate a variety of models, whether a model is realistic or not has to be determined by other means. Nevertheless, the present observations support Burks's contention that computer science provides us with a tool for analyzing and clarifying traditional philosophical issues: there can be little doubt that philosophy in consort with computer science will produce more than philosophy alone – indeed, it already has.

Professor Andrew Lugg
Department of Philosophy
University of Ottawa

NOTES

1. As Burks notes, his thesis is more general than the traditional thesis about human beings and machines and that "the actual mechanism used to perform a given function is not at issue, either at the level of parts . . . or at the level of programs and organization" (1972-73), pp. 39–40.

2. Quoted in Burks (1975), p. 304.

3. Cf. Burks (1979), pp. 402–403. What von Neumann actually produced was a specification of an infinite cellular network of finite automata, a finite number of which are in an excited state (and so constitute a finite automaton) such that when let to run, quiescent states in the empty part of the network are activated, ultimately producing a set of states similar to those initially activated.

4. These papers are discussed in Burks (1979), p. 400 and p. 403.

5. Computer simulation of the heart is discussed by Burks in (1974b). Unfortunately, this is not an appropriate occasion for me to review the important work discussed in this paper.

6. This view is intimated in Burks (1977) and Burks (1979), and it appears somewhat more explicitly in Burks's unpublished note "A New Approach to the Traditional Free-Will Controversy."

7. For a brief discussion of the Monte Carlo method, see Burks (1977), p. 598.

8. Cf. Burks (1974a), p. 306, and note that models are always isomorphic to embedded subsystems of the systems they model.

9. Reflection on the lottery analogy can perhaps lessen the appeal of the position of those, like Isaiah Berlin (1969, pp. x-xxv), who think that compatibilism is a subterfuge and an evasion.

10. This is the central theme of Burks (1979).

11. Burks (1977), pp. 612–613 and p. 612. Concepts "constructed" from a priori concepts are also said to be a priori, and concepts "constructed" from empirical (and possibly a priori) concepts are said to be empirical.

12. In response to the present argument, one might maintain that there is – perhaps common to all members of the species – a fixed structure-programme within which each individual's (sub) structure-programme develops. This is essentially an empirical issue, but we may be forgiven for wondering whether the proposed division between kinds of structure-programmes yields the division between concepts that Burks requires. Why, in particular, is it reasonable to think that the concept of causal necessity is a feature of our fixed rather than our developing structure-programmes?

13. Burks, (1979), p. 407. Burks, it should be noted, emphasizes that the computer version of the problem is still speculative. "If we have a computer program which could [convert unorganized sensory input data into scientific knowledge]," he observes, "it might be difficult to tell whether it employed the concept of causal necessity or the uniformity of nature principle" (*ibid.*).

14. Burks has however commented on the possibility of simulating evolutionary processes and on the formalizability of language, which may be viewed as a subsystem of the social system, broadly understood.

15. The considerations of this paragraph also bear on the problem, which stems from the work of Paul Feyerabend and Thomas Kuhn, of whether science is furthered more by persistent critical examination of its foundations or by dogmatic adherence to them.

ANDREW LUGG

16. Compare the discussion of a priori and empirical concepts in section III.

REFERENCES

Berlin, I.: 1969, *Four Essays on Liberty*, Oxford University Press, Oxford.

Burks, A. W.: 1972–73, 'Logic, computers and men', *Proceedings and Addresses of the American Philosophical Association* **46**, pp. 39–57.

Burks, A. W. 1974a, 'Models of deterministic systems', *Mathematical Systems Theory*, **8**, pp. 295–308.

Burks, A. W.: 1974b, 'Cellular automata and natural systems', in W.D. Keidel. W. Handler and M. Spreng (eds), *Cybernetics and Bionics*, R. Oldenbourg, Munich, pp. 190–204.

Burks, A. W.: 1975, 'Logic, biology and automata', *International Journal of Man-Machine Studies*, **7**, pp. 297–312.

Burks, A. W.: 1977, *Chance, Cause, Reason*, University of Chicago Press, Chicago.

Burks, A. W.: 1979, 'Computer science and philosophy', *Current Research in Philosophy of Science* in Asquith and H. Kyberg (eds.) Philosophy of Science Association, East Lansing, pp. 399–420.

Edgley, R.: 1970, 'Innate ideas', in G.N.A. Vesey (ed.) *Knowledge and necessity*, Macmillan, London, pp. 1–33

Laing, R.: 1976, 'Automaton introspection', *Journal of Computer and System Sciences*, **13**, pp. 172–183.

Samuel, A.: 1959, 'Some studies in machine learning using the game of checkers', *IBM Journal of Research and Development*, **3**, pp. 211–232.

Samuel, A.: 1967, 'Some studies in machine learning using the game of checkers II — Recent progress', *IBM Journal of Research and Development*, **11**, pp. 601–617.

Vesey, G. N. A. (ed.): 1970, *Knowledge and Necessity*, Macmillan, New York.

R.J. NELSON

ON GUIDING RULES

Professor Arthur Burks, in his Presidential Address to the Western Division of the American Philosophical Association in 1973 developed the thesis that a man is a finite, possibly probabilistic, automaton (Burks, 1972–73). This mechanist or computationalist philosophy of man is of course a version of the fairly widespread view that the mind is an information processing system of some kind. Burks' typically and admirably clear address both developed this picture of man far more definitely and precisely than is usual in most philosophical discussions and added an ingenious argument in support of the thesis based on the psychological principle of *just noticeable difference*.

I am in almost total agreement with what he said; beyond that I am indebted to him as are many others for the work he has done to bring the logic of computers and automata to a position of substantial philosophical relevance. I am bothered, however, by a certain question as to what his thesis really is meant to maintain. At the opening of his remarks, Professor Burks asked, "Am I really saying that man is a machine?" He answered, "Yes, in a sense, you and I *are* machines {italics mine}." But elsewhere in the address he quite consistently maintained that a finite automaton, i.e. a stored program digital computer, can *simulate* or perform all natural human functions. I say "simulate" purposely, for a computer might perform the functions of a thing without being at all like that thing, even in abstract logical structure. To be a machine, really, is not the same as to be simulated by one. A related point, harking back to the heyday of logical empiricist and instrumentalist theories, is that a man could be truly *explained* in mechanical terms without being a machine, that is to say, without any commitments to realism whatever. This issue is a reflection of the familiar controversy between instrumentalist and realist philosophies of science. I am not certain, from what he said, which side Burks is on.

In this paper I wish to argue for a realist version of Arthur Burks's thesis, which I share (although I prefer to think of the mind, rather

159

Merrilee H. Salmon (ed.), *The Philosophy of Logical Mechanism*, 159–177,
© 1990 *Kluwer Academic Publishers*.

than of all natural functions, as a system of recursive rules), and will attempt to do so in essentially automata-theoretic terms.

According to mentalist philosophy use of language depends on knowledge of the rules of grammar. These rules are "internalized," it is said, at the time we learn our native tongue: They guide speech and guide understanding when we heed the talk of others.[1] A broader philosophy ascribes rules of the same general kind to other mental processes including perception, thought and belief, although it is not always claimed that we have tacit *knowledge* of such rules.[2] This doctrine of tacit knowledge of rules or of rules that we unconsciously follow in our mental workings distresses some empiricists who balk at inner things other than dispositions to act in certain ways, qualitative spacing of ideas, and habits. There are habits, even recursive habits of mind, but no underlying rules that guide mind any more than there are rules that guide falling bodies. Other empiricists, including myself, find no difficulty in the idea of a rule-following mental life that goes on without attendant consciousness, but would like to clarify and naturalize the idea, transferring it to a nonmentalist and nonrationalist setting (Nelson, 1978).

The rules I am concerned about come in bundles or systems, each system being equivalent to some Turing machine. The family includes phrase structure grammars, transformational grammars, phonological rules, computer programs both at the machine language and compiler levels, automata, finite transducers, Markov algorithms, Post productions, the rules of inference of logics (excepting infinite rules), computer semantics, and possibly natural language semantics. All of these rules might be a part of or stored in, a finite automaton, as Professor Burks would certainly agree. Some functionalists and all mechanists, (the second and fourth groups just alluded to) regard the having or internalizing of such rules as a necessary condition for having a mind. I will suppose that such rules are easily differentiated from habits, customs, rules of conduct, commands, dispositions, and so forth. If there are problems in thus identifying them, let us agree to set them aside for another time.

The job is to distinguish rules of this kind that *guide* behavior from those of the same kind that descriptively *fit* behavior.[3] Jones speaks English, and Professor X's transformational grammar (supposing it to have been completely developed) fits her speech. The sentences she produces when she talks or writes up to her full competence thus satisfy a certain recursive definition – a system of recursive rules, as I shall term it. However, the rules that *guide* Jones, or indeed that guide her grammarian in his psycholinguistic deliberations, need not be the same as the rules of the theory, even when they are equivalent to that theory. Both systems are perhaps satisfied and are minimal; they fit, but only one guides. Jones' linguistic behavior and that of her community underdetermines the theoretical grammars that fit. Is the "correct" one (or many, if individuals do not necessarily learn precisely the *same* grammars) right because it *really* guides or simply because it is the simplest and most economical grammar-cum-hypothesis according to the usual conservative empirical criteria?

I will suggest an answer to this question, hoping to show that: (I) guiding rules can be adequately distinguished from fitting rules, *even in the* tacit case;[4] (II) such rules are *real* – there *are* guiding rules, and selection of a right grammar by the linguist is not only a matter of the pragmatics of hypothesis formation but of the truth.

(I) Michael Root (1975) has written on a similar theme, but with a different accent. According to him, a behavior-guiding rule is one involving "nesting" and "hierarchies," a pair of concepts which he sets out in analogy to certain phrase structure rules in the special case of linguistic behavior. The stock examples are sentences having syntactically different readings for one and the same sequence of morphemes, such as Chomsky's familiar

> They are flying planes

or Root's own delightful

> Intriguing women can be a drag.

Each of these is a part of some fragment of generative English that produces two derivation trees in the familiar way. Root's main point is that there are dependencies among the terminal symbols that transcend the temporal sequence (when uttered or read), and that these dependencies are traceable to properties of the underlying generative rules. In a way this is nothing new. However, Root goes on to generalize and extend the concept of nonadjacent dependencies to behavior beyond language and to suggest that behaviors showing this type of sequence are the ones that are rule-guided.

I like this idea, and would only add that the concept of recursive rule already captures everything he needs, so far as is known, for it covers all kinds of embedments, hierarchies, nestings and the like which are of relevant interest. As Chomsky has argued for years, *some* such rules (finite state grammars, for example) are not rich enough to be interesting; and there is, perhaps, some call for a sharper line to be drawn between those rules that would account for complex behavior and the ones provided by Root's definition. But I think this question can be safely left up to psychologists and psycholinguists. More should and perhaps can be said about the intentional aspects of rule guidance (which is surely guidance to an end and which in mental activity includes semantic elements, even in the dumb). However, we already have enough to keep us busy for one paper.

Root's remarks on the categorial nature of guiding rules do not tell us, unfortunately, which guiding rules guide and which guiding rules merely fit, behavior. For instance, the reader can surely write down a context-free grammar for a language including "They are flying planes" in all of its readings. But from this minor feat you can conclude neither that there really are such rules in the mind, nor, if there are any at all, that they are the same (in some appropriate sense of "same") as the ones you just invented, that is to say, that your rules are the ones that guide and not just fit the production of the readings. This calls for a different treatment of our common theme.[5] In the following I will revert to the early terminology calling guiding rules in Root's sense "recursive rules." The problem so disambiguated then becomes to distinguish adequately between recursive rules that guide and those that fit, including those heeded tacitly.

Fitting versus guiding is related to behavior or "function" versus structure, and it is this latter distinction which I will use to explicate the former. The discussion requires some concepts from model theory applied to languages that formalize systems of recursive rules. I will repress the details as much as possible without compromising, I hope, the adequacy of the argument.

Jones is a computer systems engineer who specializes in logic design. One Sunday she finds a black box on her technician's bench, and she wonders if it is the answer to her call for a working hardware "model" of a certain device she is developing. The box has input and output ports, which Jones readily identifies owing to her knowledge of common layout practices of inputs and outputs of similar black boxes. Using the proper instrumentation (a power supply and an oscilloscope) she makes a number of experimental tests and finds that the box responds to input pulses of +v volts and 0v volts at a fixed clock rate. Behaviorally, the box is a synchronous, binary, digital device. Her tests show that each arbitrary stream of 0v and +v pulses input to the box generates a response of equal length having the property that alternate +v pulses of the input are deleted, i.e., replaced with 0v pulses in the response stream. Thus, if 1 stands for +v and 0 for 0v an input stream 011011101 yields output 010010100. There being no exceptions, Jones generalizes to a law-like statement of the behavior of the box: if x is input then y is output where y is the result of deleting even occurrences of 1 in x, for every x and y. She is pleased with this discovery as it shows her technician succeeded in putting together a circuit that performs according to the desired design specifications.

Her interest now turns to the inner workings of the box. Did the technician literally follow her logic diagrams or did he follow his own nose for good electronic gadgetry? From our philosophical point of view as interested by-standers she has a hypotheses – her own design – that accounts for the box behavior; is that design logic the real *guiding* system of rules? Or contrariwise does the behavior really follow another logic? Being a functionalist (like all systems engineers) our heroine doesn't give two hoots about the physical properties of the box, to begin with, but seeks to find just the logic scheme. Her own scheme is expressible in the system of rules (1)–(4):

(1) $q_0 s_0 \rightarrow s_0 q_0$
(2) $q_0 s_1 \rightarrow s_1 q_1$
(3) $q_1 s_0 \rightarrow s_0 q_1$
(4) $q_1 s_1 \rightarrow s_0 q_0$

These inscriptions are four *rewriting* or *production* rules of the familiar kind due to Post. They are ways of writing Turing machine tables that I (not Jones) have chosen simply to emphasize that all such can be expressed in a manner analogous to that of phrase structure grammars. The emphasis is not idle: I will later want to *plausibly* shift from easy models, such as Jones' box, to the mind where we do not have any independent evidence that the forthcoming technicalities apply. These rules can be formalized in a logic with identity in a manner analogous to that of other systems of abstract relations such as groups. We insist on this step as we wish to talk in a precise way about concrete models or *realizations* of the abstract scheme (1)–(4), thus pinning down endemic functionalist talk about automata being realized in this or that medium. The natural formalization is in a many-sorted language consisting of two sorts of individual constants $s_0, s_1, s_2, \ldots$ and $q_0, q_1, q_2, \ldots$; functors f and g, the relation $=$, and the usual notation and formational definitions of an underlying first order logic. In this language (1)–(4) can be written in functorial notation as eight special axioms:

(i) $f(q_0, s_0) = q_0$
(ii) $f(q_0, s_1) = q_1$
(iii) $f(q_1, s_0) = q_1$
(iv) $f(q_1, s_1) = q_0$
(v) $f(q_0, s_0) = s_0$
(vi) $f(q_0, s_1) = s_1$
(vii) $f(q_1, s_0) = s_0$
(viii) $f(q_1, s_1) = s_0$

Here (i), (v) correspond to (1) in the rewriting system, (ii), (vi) to (2), (iii), (vii) to (3) and (iv), (viii) to (4). These axioms constitute the defining relations of an abstract, finite transducer (finite automaton with output) where f is the transition functor and g the output functor; q_0 and q_1 are internal state symbols and s_0 and s_1 are input and output symbols. There are other proper axioms of the theory and logical axioms which we need not write down explicitly.

Now, going back to (1)–(4), if one were to take any string of symbols made up of s_0's and s_1's, place q_0 at the head of it, and apply rules, he would generate a new string with a q symbol at the extreme right. For example, take $s_1 s_1$, annex q_0, getting $q_0 s_1 s_1$, apply rule (2) and then (4) and you obtain $s_1 s_0 q_0$. The set of all pairs of input-output strings such as the pair $\langle s_1 s_1, s_1 s_0 \rangle$ of the example is the behavior of the rule system (1)–(4).

What we need to do next is grasp this concept within the theorems of the formalized system with axioms (i)-(viii). I will call the behavior of this latter system H: H is the set of ordered pairs consisting of input strings x and corresponding output strings y. Behavior theorems in this system will be identity formulas of the kind $H(x) = y$.[6] A syntactical *metatheorem* of this system, which is easily proved by induction, says in effect that H is the infinite set of ordered pairs of equal length strings x, y on the symbols s_0, s_1 having the deletion property that Jones generalized from her observation of the black box. Thus the actual behavior in terms of +v and 0v matches the theoretical behavior; the finite transducer provides a formalized "high level" theory that accounts for the behavior of the box. Indeed, rules (1)–(4) as formalized in (i)-(viii) *fit* that behavior. Whether the specific structure of (i)-(viii) matches that of the box is something else that must be considered.

So far the fit of the hypothesis is established by noting that +v (=1) and 0v (=0) correspond to s_1 and s_0. There is an interpretation map from the symbols of sort s to the physical entities such that *streams* of 1's and 0's correspond to *strings* of the s's in the formal alphabet. If we let B be the 2-ary box predicate "*x* is the input to output y," *x*, *y* ranging over streams, and if B is the interpretation of H, then to say that *H fits* is just to say that H is true in the interpretation (omitting some quite dispensable detail). So far so good.

However, the physical organization of the black box, the logic system, might be considerably different than the abstract transducer structure (i)-(viii) hypothesized by Jones. In other words it is quite possible that the real system of the box be other than an interpretation of those rules in the formalized version. For consider the two tables below.

Transition		Output			Transition		Output	
0	1	0	1		0	1	0	1
A A	B	0	1		A' C'	B'	0	1
B B	A	0	0		B' B'	A'	0	0
(a)					C' A'	B'	0	1
					(b)			

In both tables A, B, C are physical states (electronic flip-flop settings – (b) requiring two flip-flops to encode the states and (a) just one) internal to the box, and 0, 1 are as before short for 0v and +v. The table represents automaton transition and output functions in the usual manner.

Now we already have that 0, 1 are associated to s_0, s_1 in an interpretation of a simple system of concatenation (strings) that establishes the fit. Next, looking at table (a) we assign A to q_0 and B to q_1 of the formal transducer. Writing the rules (1)–(4) over once again in tabular form, (c), it is clear that we can naturally make functor f correspond to the transition function of the box and g to the output function. It is now easy to see by inspection that the structure (a) is a model or *realizes* the set of formulas (i)-(viii)[7].

	f		g	
	s_0	s_1	s_0	s_1
q_0	q_0	q_1	s_0	s_1
q_1	q_1	q_0	s_0	s_0

(c)

Let us suppose that the black box of our running account is indeed (a), and therefore that it realizes (c).

Turn to (b). Notice first that (b) has exactly the same behavior as (a). Technically, for any stream-type x on the basic symbols 0, 1, the output stream y starting in state A in (a) is typally the same as the output starting in state A' in (b), and conversely. From this fact, which can be shown using the definitions of note 6, it follows directly that (b) too is a deleter and that 0, 1 of the device is an interpretation of s_0, s_1 of the abstract automaton (c) with respect to input-output string pairs. Thus (c) fits the behavior of (b) as well as of (a).

However, (b) does not *realize* the system (c); there is no mapping of symbols s_0, s_1, q_0, q_1 to 0, 1, A', B', C' and of the functors f, g, to transition and output functions such that certain sequences on the input-output and state entities of (b) satisfy (c). This follows immediately if we require that the interpretation must be *onto* the domain. It takes some elementary but quite fussy work to show there is *no* interpretation even if we loosen the requirement to 1–1 *into* maps; but the negative result still holds. (b) does not realize (c); in other words, (c) and (b) do not *follow* the same state-to-state sequences. [8]

In the light of these examples, it seems very natural to say that (c), or better the rules (1)–(4) *guide* (a) but not (b), as (a) but not (b) *realizes* (i.e., satisfies) (c). Yet (1)–(4) *fits* both: they constitute a set of guiding rules in Root's sense that fit one case but both guide and fit the other. [9]

Before attempting to generalize away from hardware boxes to organisms it will be worthwhile to consider a much stronger notion of guidance. To present this idea I shall rely on the well-known correspondence between abstract automata and idealized neural networks. In order to keep Jones on the scene I shall suppose that electronic switching circuits will suffice to represent such networks, which they do except that switching diagrams are not iconic depictions of neural networks, and the latency intervals from input to output are different; even in the real circuit, gates take virtually no operate time, while neurons take an appreciable delay before firing.

That brains instantiate complex automata or programming-like systems is of course a conjecture underlying all mechanist or computationalist versions of functionalism. And whether and how actual neuronal networks embody such automata structures in any systems remotely like idealized networks is still more conjectural. But that there is an organized network system instantiated in some way seems to me to be beyond serious debate, given the philosophical framework. And the idealization is good enough to motivate some further distinctions without needing to be a close approximation to the actual anatomy.

Figures (d) and (e) depict two of an infinite number of possible embodiments of (a) in switching circuits. In these diagrams, $x(x')$ are inputs and $z(z')$ are outputs. The two states A and B of (a) are represented by the outputs 1 and 0 of the delay elements δ, which are labelled y and y' respectively. It is easy to veryfy that both (d) and (e) embody the deletion behavior of the box, and somewhat harder to show that there is an algorithm for recovering the table (a) from either (d) or (c).

Please observe that (d) and (e) are quite unlike, one consisting of six logic elements and the other of seven. Moreover, the logical schemes of the two circuits are not the same. This may be verified by chasing 0's and 1's through them which, incidentally, will convince you that the behaviors are exactly alike. So here again is same behavior and different structure, but in a very different sense of "structure" than before. *Both* (d) and (e) are *embodiments* of one and the same set of rules ((1)–(4) or (i)-(viii) or table (c)) in the *realization* (a) of (c).[10] For our purposes these differences may be captured in the expressions of a formal first order language which *encodes* table (c). Encodement is again an algorithmic translation. In this language, which is a form of recursive arithmetic,[11] (d) is expressed by two equations, one for the next state at a given time t, $y(t)$, and the other for the output at t, $z(t)$. The input at t is $x(t)$. Assuming each cell has a latency time of 1 before firing, we obtain for (d)

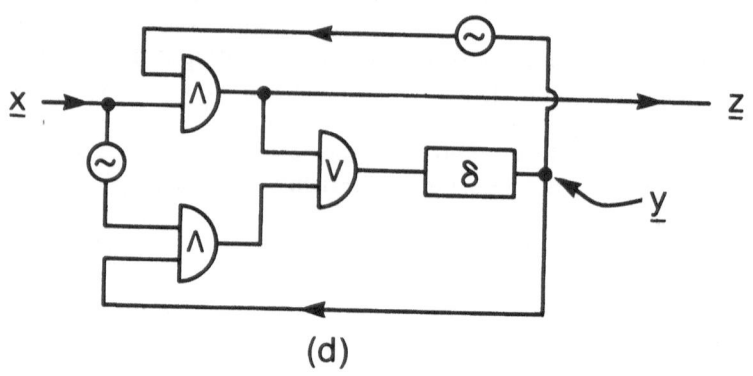

(d)

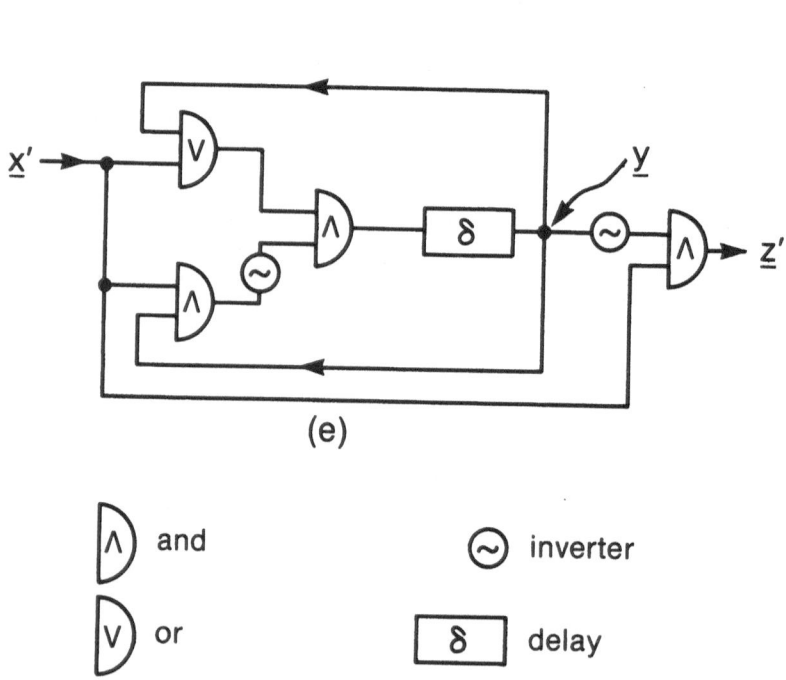

(e)

$$y(0) \equiv 0$$
$$y(t+1) \equiv (x(t) \wedge \sim y(t)) \vee (\sim x(t) \wedge y\,(t))\} \qquad \text{(d')}$$
$$z(t) \equiv x\,(t) \wedge \sim y(t) \qquad\qquad\qquad\qquad\qquad \text{(d'')}$$

Likewise, (e) is expressed

$$y'(0) \equiv 0$$
$$y; (t+1 \equiv (x'(t) \vee y'(t)) \wedge \sim(x'(t) \wedge y'(t))\} \qquad \text{(e')}$$
$$z(t) \equiv x'\,(t) \wedge \sim y'(t) \qquad\qquad\qquad\qquad\qquad \text{(e'')}$$

The output conditions (d'') and (e'') are the same for each net, and since the next sate conditions are truth functionally equivalent (truth functions on the values of the propositional functions of time), the same behavior is expressed by both as required.

The translation from nets to linguistic expressions is direct. Switching elements correspond 1–1 to well-formed parts. In figure (d), the *and* and *or* elements feeding the delay, labelled "δ," correspond exactly (this can be made precise) to the right side of the equivalence expression "$y(t+1) = \ldots$" in (d'); the inverters likewise correspond to the tildes "$\sim$" that operate on "$y(t)$" and "$x(t)$" in (d') and (d'')' and the top leftmost *and* corresponds to the right side of the equivalence "$z(t)$"

. . . Similarly for diagram (e) and expressions (e') and (e''). Accordingly, I will say that expressions *individuate* nets. Furthermore, expressions of this type of language are *identical* if and only if they match token for token of the same type from right to left up to within commutativity of $\wedge$ and $\vee$. Thus, if x, x' etc., are the same type, (d'') = (e'') but (d') $\neq$ (e'). If we were to assume that we had written the recursive definition of "sentence" for our language, then we would see at once that formulas are identical (except for commutativity) when they are generated by the same sequence of applications of the clauses of the definition.

This fact inspires another concept of guidance as follows. We say that (d', d'') *strongly* guide (d) while (e', e'') *strongly* guide (e). Both pairs of expressions describe or fit the two network embodiments of the transducer (a) as well.

Now let us attempt to extend these observations from simple automata to the general case, keeping organisms in mind. Under our blanket hypothesis that the guiding rules of mind we are worried about are recursive rules more or less like (1)–(4) on p. 163 and that they are expressible in formal languages more or less like (i)-(viii), I suggest the following definitions. *A formalized system of rules fits the behavior of an organism or organism part under some description of stimulus and responses if and only if that behavior is a true interpretation of the behavior (in our example, the H function) of the rule system. A formal rule system weakly guides the behavior of the organism or organism part if and only if the organism realizes the system.* Here I am tacitly assuming that the appropriate interpretation maps exist and that the standard notion of satisfaction of model theory is used. Finally, at the level of embodiment (nerve networks, genetic structures, muscle configurations or whatever the mechanisms be), a *set of arithmetic formulas strongly guides* the organism or part if and only if the formulas individuate the specific embodiment. For a system to be strongly guided it must imply, in a sense to be explained in the next paragraph, a system of weakly guiding rules.

You will have noticed (the fact was hinted at above) that if an organism is strongly guided by network rules it is also weakly guided by a unique set of recursive rules. In well-understood cases in automata theory, this holds true because there is a decoding procedure (an algorithm) from an arithmetic language (exemplified by (d'), (d") or (e'), (e")) back to tables, like (c), that produces a unique table up to within an isomorphism. So I guess that in other cases of substantial complexity there is at least a semi-algorithm, i.e. a deduction. However, the opposite is not true. There is no *unique* system of strongly guiding rules, given a suitable language for expressing them, that can be obtained from knowledge of weakly guiding rules. Recursive rule-type explanations, to use methodological rather than ontological talk, are underdetermined by behavior data at two levels; A hypothesis that fits need not guide even in the weak sense; and one that does guide in the weak sense may not uniquely point to a system of expressions in a formal network or embodiment-type language that individuates or strongly guides the behavior of the organism.

This completes my attempt to explicate the concept of guiding rules that truly guide. A quick and dirty summary could be put in a somewhat simpler way. The same information processing tasks can be performed by different computer programs, and a program which is not actually followed by a computer still, in a sense, describes or fits. I do not think, however, that the program paradigm is quite as convenient for philosophical purposes as Turing machines or the rewriting rule version of automata, mainly because humans are machines with respect to theoretically relevant psychological processes, and are not programs, although it is an interesting possibility that humans do follow programs that are superimposed on an enormously complex finite transducer – somewhat in analogy to computers. At any rate, the program kind of explication of guiding does have the merit of highlighting "*tacit following*" as a perfectly reasonable way of talking. The transducer model, however, enables somewhat more precise ways of analysis.

(II) If mechanism be true, there really are guiding rules that guide, not just fit, as Chomsky has argued for many years. Talk of a mechanical or commutational mind is not metaphorical, and some automaton theories of mind that turn out to fit could well be *real*, depending on whether they are true of external behavior only or also of underlying process.

Jones can, of course, check the reality of her theory (i)-(viii) by opening the box, or better yet by looking at the technician's design diagrams. As the technician is not likely to know much about abstract automata theory, Jones's actual way of proceeding would be to determine which of (d) or (e) is right, and then via the decoding algorithm alluded to earlier to compute out the automaton table. No such way is open to the psychologist, who is faced with a roughly analogous problem, short of assorted probings of cats heads. Even the neuroscientist scarcely knows *what* to look for as yet – even if he likes the mechanistic, recursive rule framework –, although he has a pretty good idea *where* to look for some features such as visual perception or speech. So we as philosophers are hard put to justify the realistic stance except on some kind of *a priori* grounds. This doesn't mean we are bad off. We don't know what the real, guiding rules of mind are short of knowing all about mind, in which case there is no issue, philosophical or otherwise. But I believe we can know *that* there are such rules, assuming mechanism.

In its weakest form that doctrine says that mental features involve informational processes of some kind, which by Church's thesis means rules of the kind we have been discussing. This way of putting the matter makes no commitments whatever about fitting and/or guiding, as that distinction as just explicated depends essentially on theory formalization. The real is relative to theory – literally a realization – and I can not claim the reality of a definite rule system without the expressing language. The idea of a strongly guiding rule which individuates a physical embodiment, makes this quite evident. Now the possibility of discursive knowledge is an article of faith that is supported by a considerable mass of inductive evidence – the accumulated results of science itself. However, for empiricists, all granting this guarantees is that fitting not necessarily guiding theories are possible. What is needed beyond that is a small remark that we borrow from mathematical logic. As we have seen, automata theory including the transition and output functions of an arbitrary Turing machine can be formalized in a first order language with identity with added nonlogical axioms. Moreover, the logic formulas that result from encoding an automaton are all formulas of a certain recursive arithmetic. In sum, the specific embodiments of such automata are expressible in formal languages.

I maintain in analogy to our simple running model of the pulse divider that any set of recursive rules (by the converse of Church's thesis, any principles underlying effective processes) can be so expressed. And this is more than a bare analogy, as it is supported by our basic intuitions about recursive relations as ones that are computable by algorithms. All algorithms we know of are expressible linguistically (if, indeed, the proposition is not merely a tautology). It seems to me that this establishes the expressibility of rules in a formal language at the level of nets or other organs and *a fortiori* in abstract form, analogous to finite or Turing automata, quite well. Hence, there are real, guiding rules. Obviously no such conclusion holds for phenomena not subject to algorithms; it probably does not hold for physics. If mechanism is true, then in a sense we perhaps can know more, in principle, about laws of mind than of other parts of nature!

Realism of this type is not a theory of truth, although it might appear to have a certain similarity to Peirce, as what is real depends on inquiry. Peirce said *the* truth is what is fated to be agreed on by the scientific community, and the real is its object (CP, 5.407). But my theory does not say that any one will ever hit on the right effective language (we are not "fated"), and even if some one does, it does not follow that true (or highly confirmed) sentences having the appropriate properties will ever be invented. What I claim to be true is that there are appropriate expressing languages, not that the true theories formulable in them will ever be discovered. That is up to science as such, not philosophy.

It seems to me that this is a truly empirical position, although it does take sides on what has traditionally been a metaphysical issue. If there are confirmed recursive rule type theories, say of speech, where this appears to be increasingly likely, then it will follow by the realist principle that there are behaviorally equivalent guiding rules in the mind. Presumably these could eventually be isolated when more is known about brain systems and nerve networks; and this knowledge could *falsify* the theory. It might turn out, for instance, that there are infinite mental states not representable by automata in the exact sense of the word.

I must add that I like the idea of a recursive rule of mind that tacitly *guides* better than that of an acquired recursive *habit* (Quine, 1972, p. 443). It does take a lot of hassle to separate guiding from fitting rules, and the idea might turn out to be hardly adequate – it certainly does not deliver a neat, overt, behavioral criterion, admittedly – but it doesn't suffer from the vague, slightly contradictory air of "recursive habit". A habit is based, I suppose, on associations of some kind (association of a response with reinforcing conditions or of ideas with ideas, etc.) whereas recursive rules do *not* associate observable entities with entities. They associate *pairs* of states and stimuli with other entities, which makes all the difference in the world. If the behaviorist never meant to omit such in his construal of "behavioral" and of "habit," let him join up.

The existence of such guiding rules explains mental life without recourse to unwholesome powers or virtues or excessively fuzzy things such as dispositions to behavior or innate knowledge. The theory of real rules is essentially an extension of Chomsky's views (if I do not completely misunderstand him) of descriptive adequacy stretched to include all mental features along with language (Chomsky, 1965, p. 24). I depart from him, however, in denying that these rules are "known" in any sense of the term having epistemic significance of any kind, instead using the relatively precise idea of guidance, which in turn goes back to elementary logical concepts. Of course, they are known to the psycholinguist who has a true theory, or an approximately true theory if each individual's internalization of rules differs from most others. For the issues I am intimating here concerning linguistic competence and the rationalist way of understanding it see my (1978).

A couple of paragraphs back I suggested that recursive rules are not like habits, especially if the acquisition and strengthening of habits be putatively explainable in behaviorist terms. This is not an issue any more if one simply enlarges behaviorism to include the richer machinery of computation. However, if the rules that guide the mind include only the recursive, how account for intention and meaning? Recent history teaches us that that family is perhaps adequate for the syntactical component of language, including transformations (which are perhaps more like rules for recursive *functionals*, as they operate on trees, not strings); but what about the semantical and phonological aspects of language, and what about belief, desire, want, thought, mental representations, and other mental features? What about meaning? What about control? I don't have pat answers, and would just ask the reader to be content with the information that this class of questions is exactly what keeps functionalists, computationalists, cognitivists, and intelligence artificers busy. To explicate *guiding* for the underlying rule category – to be sure using a very idealized model – is one thing, and to show how these primitives can be used to analyze mental attributes is quite another. It seems to me that the reality of these relations is of sufficient philosophical interest in itself.

As it stands this kind of realism bears a slight resemblance to Kantian philosophy, although some functionalists would give this more emphasis. But I myself see little point in pressing historical analogies very far unless one side of it genuinely illuminates the other or both. In theory of knowledge, Kant's aim was to demonstrate the possibility of universal and necessary knowledge, which he claimed mathematics and the basic principles of physics already manifested. So far as I can see this has little in common with an empiricist, naturalist philosophy that accepts no *a prioristic* epistemology, critical or not, although if one tries hard enough he might historically trace the idea of rules qua informing mental processes (including belief and knowledge) back to the categories and transcendental schemata. But there's a lot that could go wrong here. Conceptually a form of judgment, for one, is a far cry from a recursive rule. And while knowledge does depend on antecedent mental conditions in both, the principles involved are quite opposite. For Kant, scientific theory (i.e. knowledge) depends on *a priori* constitutive transcendental categories or forms of judgment, while in the other case the rule of mind as *real* entity is related to a theory that is true, if it is true, on empirical grounds. Of course, if psychology bears mechanism out, then the *latter* theory also depends on real guiding modes of thought. The scientists's thoughts are not exempt from the rules of mind he attributes to humans in general. There is no paradox here, but only an inevitable consequence of a naturalist epistemology that boot-straps itself out of science. Whatever be the rules that guide science they will be known only to science itself, not deduced *a priori*.[12]

There is a much more illuminating resemblance to moderate Aristotelian realism, perhaps closest to John Duns Scotus (and therefore in some respects to Peirce's *thirdness*). I have more to say about this in another place, and here just want to say enough to complete the sketch of what a guiding rule is like philosophically (Nelson, 1982). A realized automaton is a form: it does not subsist apart from material embodiment, but is *formally distinct* from the material in the sense that it is individuated by formulas that really define it. And it is *real* in that it *guides* and not merely fits by some exterior description. Man's internalized collection of guiding rules constitutes his essence, or in Chomskian terms his mental competence; it defines him as a natural kind. On the other hand, as there is no reason to suppose that your logic is mine exactly, there are only material *individuals* but individuals realizing sufficiently similar forms to found a mental science of them as a unified species (Nelson, 1978). If this be nominalism, then there is, nevertheless, a transcendental invariant "core" which is not merely "conceptual" or "subjective" although it is most certainly theory-relative. The theory is realistic, but still naturalistic and empirical. I find I am inevitably *led to* it from considerations of contemporary logic, computers, and cognitive science; and therefore *led back* to a stance in some important respects closer to classical thought than to modern.

R. J. Nelson
Department of Philosophy
Case Western Reserve
University

NOTES

*This material is based on work supported by the National Science Foundation, Grant No. SES-8012173.

1. For a fairly recent statement of his position see Chomsky (1969), printed in Gunderson (1975).

2. Ned Block gives a long list of thinkers in this camp or near it. See (Block, 1978).

3. The locutions are Quine's in his (1972) p. 442. Quine limits "guiding" to behavior wherein "the behaver knows the rule and can state it," and it is of course just this limit I want to lift.

4. This is precisely the "sticking point ... this Chomksian mid-point between rules as merely fitting, on the one hand, and rules as real and overt guides on the other; Chomsky's intermediate notion of rules as heeded inarticulately" (Quine, 1972, p. 444).

5. Fodor's idea that knowing rules is knowing how without being able to explain how coupled with a computer simulable sequence of operations seems to be essentially the same. A simulating program, too, accounts for nonadjacent string dependencies. (See (Fodor, 1968)).

6. In a little detail, we want the output functor g to be extended to arbitrary strings of symbols. Letting G be the new functor, we also extend f to F and recursively define

$$F(q_0, \Lambda) = q_0$$
$$F(q_0, xs) \; f \; (F \; (q_0, x), s)$$

and

$$G(q_0, \Lambda) = \Lambda$$
$$G(q_0, xs) = g \; (F \; (q_0, x), s)$$

where x ranges over strings of s_0, s_1, Λ is the null string, s is either s_0 or s_1, and q_0 is the initial state. Now if the behavior of the string is $<x, H \; (x) = y>$, where $x = s_{i_0} s_{i_1} \ldots s_{i_n}$, s_{i_j} being equal to either s_0 or s_1 for all j, then $y = (g \; (q_0, s_{i_0}) \; G \; (q_0, s_{i_0} s_{i_1} \ldots G(q_0, s_{i_0} \ldots s_{i_n})$. This latter formula is a theorem; and the metatheorem we asserted above says y has the property of being x with alternate occurrences of s_1 replaced by s_0.

7. Since we, following Jones, are interested in understanding the actual design of the box we must require that the interpretation of the symbols be at the *very least* a 1–1 map, and preferably *onto*; we want to understand the whole thing, redundancies, duplications, if any, and all.

8. It is somewhat counterintuitive, perhaps, that there is no 1–1 interpretation of (c) into the domain of (b) even though it can be shown that there is homomorphism of (b) onto (a). But this fact will be so only for trivial automata (complete finite transducers). At any rate "same structure" does not quite capture "follows the same" or "guides the same." which is the reason we had to work with the technicalities of interpretations rather than with the conventional mathematical structures.

9. One might complain that the defining rules for this box are too simple to be interesting. In fact there is a finite state grammar that generates a "language" consisting of precisely the input-output pairs x, y. However, the analysis that we are in the middle of goes the same way for Turing machine control boxes in general, although a more complex example would tend to blur the line of argument.

10. At the level of embodiment there are still various *material* structures that are possible; the nets (d) and (e) could be electronic, neuronal, etc. In all, there are three methodologically meaningful senses of structure, at least, within the recursive rule paradigm: the rule or Turing machine structure (c), the embodiment (d) or (e) and the material structure – electronics, nerves, muscles or what have you.

11. Cf. Church (1957).

12. Classical realism in this empirical version is related somewhat to Putnam's internal realism, although he is pushing a theory of knowledge – realism explains why language users and scientists succeed – while I am after real rules of mind. In both the

real is relative to a theory formulated in a language. However, his "real" is the same as "corresponding to true" while my notion of real as guiding (which is meant to capturer the classical antinominalist, not just the epistemological, concept) has no counterpart in his theory. Though I am far from neutral on the question of the meaning of "truth" the issues here are other ones. See Putnam (1976, 1977).

REFERENCES

Block, N.: 1978, 'Troubles with functionalism', in Savage (ed.), (1978), pp. 261–325.

Burks, A. W.: 1972–73, 'Logic, computers and men', *Proceedings and Addresses of the American Philosophical Association*, XLVI, pp. 39–57.

Chomsky, N.: 1965, *Aspects of the Theory of Syntax*, M.I.T. Press, Cambridge, Mass.

Chomsky, N.: 1969, 'Knowledge of language', first John Locke Lecture, Oxford University; printed in Gunderson (ed.) (1975), pp. 229–320.

Church, A.: 1957, 'Application of recursive arithmetic to the problem of circuit synthesis', Summer Institute for Symbolic Logic, Cornell University.

Davidson, D. and G. Harman (eds.): 1972, *Semantics of Natural Language*, Reidel, Dordrecht.

Fodor, J.: 1968, 'The appeal to tacit knowledge in psychological explanation', *Journal of Philosophy* LXV, pp. 636–638.

Gunderson, K.: 1975, Language, Mind and Knowledge, Minnesota Studies in the Philosophy of Science VII, University of Minnesota Press, Minneapolis.

Nelson, R. J.: 1978, 'The competence-performance distinction in mental philosophy', *Synthese* 39, pp. 337–381.

Nelson, R. J.: 1982, *The Logic of Mind*, Reidel, Dordrecht.

Peirce, C. S.: 1931–35, *Collected Papers of Charles Sanders Peirce*, I-V, C. Hartshorne and P. Weiss, eds. Harvard University Press, Cambridge, Mass.

Putnam, H.: 1976, 'What is realism?', *Proceedings of the Aristotelian Society*, pp. 177–194.

Putnam, H.: 1977, 'Realism and reason', *Proceedings and Addresses of the American Philosophical Association*, 50, pp. 483–497.

Quine, W. V.: 1972, 'Methodological reflections on current linguistic theory', in Davidson and Harman (eds.), pp. 442–454.

Root, M.: 1975, 'Language, rules, and complex behavior', in Gunderson (ed.), pp. 321–343.

Savage, C. W. (ed.): 1978, *Perception and Cognition*, Minnesota Studies in the Philosophy of Science IX, University of Minnesota Press, Minneapols.

EDWARD C. MOORE

ACTUALITY AND POTENTIALITY

Twentieth century philosophy has been a philosophy of the actual. The notion of potentiality does not receive the attention it once did. In 1890 *The Century Dictionary of the English Language* devoted two columns to definitions of potential and potentiality – many of them written by Charles Peirce. In 1967 *The Encyclopedia of Philosophy* entry under "Potentiality" says only "See Possibility."

In Chapter 10 of *Chance, Cause, Reason* Arthur Burks (1977, p. 645) argues for a revolutionary epistemological structure of statements by describing, in addition to the traditional categories of empirical statements and logical statements, a third category of "Inductive statements" intended to cover causal uniformity and, particularly, "inductive probability statements and theorems." Such an epistemological category re-opens the question of the metaphysics of potentiality. For while Burks's "Empirical" category (corresponding to Hume's matter-of-fact) requires only the realm of the actual, and his "Logical" category (Hume's relations-of-ideas) requires only the realm of logic, the inductive category, it seems to me, requires a realm of the potential to make its causal statements (laws of science) meaningful and to make probability statements meaningful. Since *Chance, Cause, Reason* is not a book in metaphysics, Burks discusses the metaphysical significance of the epistemological categories only peripherally. However, the problem of the ontology of potentiality has been a continuing concern of his (Burks, 1951, 1955). In 1947 when Professor Burks accepted me as his first graduate student, and we discussed a thesis topic, he suggested to me the question of whether Charles Peirce's metaphysics was compatible with his pragmatism. (At that time at least one member of the Michigan department questioned whether Peirce was of sufficient significance to merit a dissertation devoted solely to him.) I agreed to the topic with enthusiasm and in my dissertation concluded that Peirce's metaphysics was not compatible with his pragmatism because the former required the notion of potentiality and the

179

Merrilee H. Salmon (ed.), *The Philosophy of Logical Mechanism*, 179–190,
© 1990 *Kluwer Academic Publishers*.

latter could not define it. I have always suspected that Professor Burks anticipated that dissertation topic would lead me into the thickets of the metaphysics of potentiality. (At that time I was studying the philosophy of science with Burks; I was also taking a seminar from him in Kant, and Kant doth make metaphysicians of us all.)

When I concluded that Peirce's metaphysics was not compatible with his pragmatic theory of meaning, I decided that he would have to give up his metaphysics. I now believe the conclusion is that we are forced to keep the metaphysics and change the empirical theory of meaning[1] I think that with his proposal for an *a priori* basis for induction, Burks may be moving in the same direction. In the preface to *Chance, Cause, Reason*, he makes the cryptic comment (p. xvi): "I do not think pragmatism should be applied to metaphysics." However, he then adds, "My dispositional theory of empirical probability arose from my analysis of causal dispositions, with a hint from Peirce's suggestion that a probability is a 'would be'." In a recent conversation with Burks, I asked him about the statement that pragmatism should not be applied to metaphysics. He said he meant in the strict sense, but that there might be a less strict sense of the pragmatic theory of meaning in which it could apply to metaphysics. In the light of the possibility that it is time to review the empirical criterion of meaning, it may be useful to examine some of the changes it has seen. I would like to suggest that such an examination reveals a basic problem which empirical meaning theories have been trying to solve, namely, the problem of the meaning of potentiality and the consequent difficulty of allowing propositions involving potentiality to be meaningful.

I

The modern empirical theory of meaning goes back at least as far as the nominalism of William of Ockham. In its classical formulation by David Hume in the eighteenth century, and by Auguste Comte and John Stuart Mill in the nineteenth century, it rested on an empiricist epistemology that held that empirical knowledge was possible only of the realm of observable sense-experiences. Therefore, if we use the term empirical knowledge to apply only to those ideas which have a referent in sense-experience, we can have knowledge (by definition) only of observables. If we claim to have an empirical concept that does not denote either an observable particular or, at best, a series of such observables, we are referring to a fiction. Elaboration of this doctrine took place in the twentieth century in the theories of meaning developed under such names as pragmatism, operationalism and positivism.

The writings of Comte had as a basic motivation the elimination of trans-empirical metaphysics and began with a strict adherence to the empiricist meaning criterion. His original formulation advised us that a proposition was empirically meaningful only when it could actually be verified in sense-experience. Propositions for which the actual verification – or disverification – was not possible were meaningless. From this position Comte attained his notorious assertion that propositions about the chemical constituents of the stars were meaningless because no technique had been developed for verifying such propositions.

But it soon became evident that strict adherence to this meaning criterion would not do. It ruled out too many propositions that obviously had a meaning, and it did not suffice even for the purposes of science. Unless an observer knew the meaning of a proposition before he was actually able to verify it, if he did develop a technique for dealing with the proposition, he would have no way of recognizing the verification when it occurred as referring to the proposition with which he was concerned. From this it followed that a proposition must have some meaning even before it could actually be verified.

Because of difficulties of this order, the logical positivists of the Vienna Circle, in the 1930's, relaxed the criterion to what came to be known as the less rigorous formulation: A proposition is meaningful if it can, in principle, be verified in sense-experience. If in theory we know how to verify a proposition, that is, if we know what experiences would, if actualized, prove or disprove it, then it is meaningful. This formulation may be seen, for example in R. Carnap's articles (1936, 1937) on "Testability and Meaning." In substituting confirmability for verifiability Carnap says, "When we call a sentence S confirmable, we do not mean that it is possible to arrive at a confirmation of S under circumstances as they actually exist. We rather intend this possibility under some *possible circumstances* whether they be real or not. (1936, p. 457)". This relaxation of the meaning criterion is important as marking the thin edge of the wedge. It is no longer necessary for a proposition to be actually verifiable to be meaningful; it is sufficient if it would be possible to verify it. But the departure is not so drastic as might appear, since the possible experience must be reducible to an actually observable experience. "A *sentence* S is called confirmable ... if the confirmation of S is reducible ... to that of a class of observable predicates (1936, p. 456)". The possible experience is like a paper bank note; the bank note has value only so far as it is based upon a piece of precious metal. So a potentiality of experience has no reality *qua* potentiality; its only reality is that of the actual experience which could be produced under specified conditions. Carnap reiterates this view even when dealing with dispositional properties where he says, "Those predicates of the thing-language which are not observable, e.g., disposition terms, are reducible to observable predicates and hence confirmable (1936, p. 466)." The interesting question here is in the reference to a non-observable predicate. Does this mean that there are properties that can not be actualized in experience, although they can somehow be "reduced" to actual experiences?

The British empiricist A. J. Ayer in 1947 in the preface to the second edition of his *Language, Truth and Logic* talks in a very similar fashion. In re-formulating the verifiability criterion from the version he had used in the first edition, Ayer talks of "indirect verification"; he says that "a statement is indirectly verifiable if it satisfies the following conditions: first, that in conjunction with certain other premises it entails one or more directly verifiable statements which are not deducible from these other premises alone; and secondly, that these other premises do not include any statement that is not either analytic, or directly verifiable, or capable of being indirectly verifiable (1947, p. 13)." Ayer criticizes even the less rigorous formulation of the empiricist meaning criterion developed after Comte's time as "too harsh ... for it would seem to imply that it was illegitimate to introduce any term that did not itself designate something observable (1947, p. 14)." From this brief historical review it is possible to see that the empiricist meaning criterion underwent an increasing shift away from what was actually observable. The successive stages of the position seem to be: (1) a term is meaningful only if its referent is actually observable, (2) a term is meaningful if its referent is theoretically observable, (3) a term is meaningful if its referent is actually and theoretically unobservable, provided its referent has consequences which are actually or theoretically observable.

The same difficulty that led to these shifts in the empiricist meaning criterion in the twentieth century was present for Charles Peirce in developing his pragmatic criterion at the end of the nineteenth century.

In the 1878 article on "The Fixation of Belief", in speaking of the diamond whose hardness was not tested, Peirce said that such statements concern "much more the arrangement of our language than they do the meaning of our ideas" (CP 5.409). Which is to say that to talk about an unobservable hardness in the diamond is a linguistic issue, not a factual one. Peirce eventually recognized that this analysis was a mistake that limited scientific discourse to actual observables and did not allow for non-observables. A few years before his death, he wrote that the statement about the hardness of the diamond in that early paper was in error; he said, "I must show that the *will be's*, the actually *is's* and the *have beens* are not the sum of the reals. They only cover actuality. There are besides *would be's* and *can be's* that are reals" (CP 8.216).

Peirce never solved the problem of how to define non-observables. He might have done so if he had adopted a meaning criterion proposed by Chauncey Wright in 1877, just the year before the pragmatic criterion was announced. Wright says, "Thus, while ideal or transcendental elements are admitted into scientific researches, though in themselves insusceptible of simple verification, they must still show credentials from the senses, either by affording from themselves consequences capable of sensuous verification, or by yielding such consequences in conjunction with ideas which by themselves are verifiable (1877, p. 47)." This seems to be the same meaning criterion that Carnap and Ayer came to seventy years later and is, I suspect, what Peirce would have come to if he had worked the matter out. In what follows, I refer to this as "Wright's principle" in deference to Wright's priority in stating it.

II

Although Peirce would probably have accepted Wright's principle, it would not have solved all the matters at issue. This principle is ambiguous. It may simply be interpreted, as Carnap seemed to prefer, as a form of reductionism. That is, the unobservable predicate may not be asserted to exist as such but simply to be a class name for a set of observables. Thus the above quotation from Carnap tells us "Those predicates of the thing language which are not observable, e.g., disposition terms, are reducible to observable predicates and hence confirmable."

For Charles Peirce, reductionism would have been a form of nominalism. The next step necessary to solve this problem is to move away from nominalism and to follow Peirce into a metaphysical realism. Peirce's metaphysical realism is not well understood because it differs from the ordinary interpretation of the realist-nominalist issue. I have discussed this at length elsewhere (Moore, 1952), and will state it only briefly here. Peirce rejects the ordinary interpretation which sees this controversy as having to do with Platonic entities. He says, "The notion that the controversy between realism and nominalism had anything to do with Platonic ideas is a mere product of the imagination which the slightest examination of the books would suffice to disprove (Peirce, 1871, p. 454)." The issue as Peirce saw it had to do with concepts (or in his terminology "generals"), and the question was whether there was anything *in the external world* (not in a Platonic realm of ideas) which corresponded to a given concept. If so the concept was real, if not it was a fiction. Thus Peirce says, "The question ... is whether *man, horse*, and other names of natural classes, correspond with anything ... independent of our thought (1871, p. 454)." Peirce applied this metaphysical realism to scientific laws and in effect asked the question: What is there in nature that corresponds to the concept of a law in science? Let us examine this question. A basic purpose of science is to know the future – to make predictions. What a scientist seeks to know is not past actualities – they are history – nor even present actualities, for the immediate present is beyond control. He seeks to know the future in order to control it so far as possible. His business is to assert meaningful propositions (causal laws) about events which are not yet actualized. In some cases, as a result of his knowledge of as yet unactualized events, he seeks to find means of preventing those events from ever becoming actualities.

Consider a particular situation: Suppose that I am about to release an ordinary pencil and I ask a scientist whether it will fall. He knows the law of gravitation and therefore asserts that it will fall. When he knows the proposition "The pencil, if released, will fall", what is it that he knows? Certainly he does not literally know the future actual event of the pencil falling. He can not peer into the future and see something which has not yet occurred. We might say that he only anticipates the future event. But to anticipate correctly is to know now something which will be actualized only in the future. The event is not actualized now and hence can not itself be an object of knowledge. What is known now is the potentiality of this event occurring in the future; this potentiality is real now, and it is the potentiality – not the future actuality – which is the object of knowledge.

A future actuality is never known *qua* actuality; it is always known only as a potentiality. What a scientist knows whenever he knows something about the future – and whenever he knows a law he does know something about the future – is a potentiality and not an actuality.

Of course, a scientist's knowledge of the future is not coercive. A botanist who knows Mendel's laws does not know whether there will be any sweet peas in the year 2000, but he does know that if there are any sweet peas in 2000 that are the descendants of any flower alive today, the descendants will exhibit certain characteristics which are potentially present in the contemporary flower. He does not now know these future characteristics as actualities; he only knows them as potentialities. A definition of a scientific law cannot be reduced to a statement about actualities, as reductionism would suggest. Every scientific law must contain a reference to potentialities which are not now and which, if, for example, I do not release the pencil, may never become actualities.

There is, however, a difference between the potentiality involved in the pencil falling at some future time and the potentiality of a dispositional property. The potentiality of the pencil falling is what may be called an actualizable; that is, it may under appropriate conditions, become actualized. There is a one-to-one correlation between the potentiality of the pencil falling and the actuality of the pencil falling with reference to all the properties of each except for their different physical states – one is potential and the other is actual. This is not the case, however, with a dispositional property. A dispositional property, as such, is never actualizable. What are actualizable are certain consequences of the property but not the property itself. A dispositional property is a potentiality for acting in a certain manner – what Peirce described as a "would be" or as analogous to a human habit (CP 2.664). The dispositional property itself is not actualizable any more than a human habit is actualizable. What are actualizable are certain consequences of the property but not the property itself.

To say that a lump of sugar has solubility is not to say that it is dissolving now, but to say that it has now the potentiality of dissolving at some future time. A piece of sugar that is actually dissolving is an instance of dissolution, not of solubility. But to say that solubility is real is to say that there is now a real potentiality in the external world corresponding to solubility and not reducible to a class of actualities – in short, to accept Peirce's metaphysical realism.

An adequate philosophy of science must admit both real laws and real dispositional properties. Both of these involve potentiality. Any philosophical approach that renders potentiality meaningless must be inadequate as a base for science. The attempt to identify all meaning with actualities does make potentiality meaningless. The meaning of potentiality in a dispositional property or in a general law of science cannot be expressed in terms referring to the property or law as actualized in future experience but must explain its nature *now* when it is unactualized. And the terms used to describe the unactualized entity are only meaningful on the reductionist view if stated in terms of actualities. But what actualities can describe an "unactuality?"

I thus reach the conclusion that what I have called "Wright's principle" above, requires elaboration. If we allow "actuality" to denote what enters into human sense experience, that is what is observed, we need to introduce terminology for referring to properties which are not actualities. A set of definitions such as the following seems to be necessary:

(1) x is an actuality $=df\, x$ is observed
(2) x is an actualizable $=df\, x$ is not now an actuality but may in theory become an actuality
(3 x is a potentiality $=df\, x$ is not an actuality but is (a) an actualizable or (b) is a dispositional property which is not itself actualizable but has consequences which are either actualities or actualizables
(4) x is real $=df\, x$ is either an actuality or a potentiality
(5) x is a fiction $=df\, x$ is neither an actuality nor a potentiality.

(1), the first part of (4) and the first part of (5) are clearly within the framework of the reductionist interpretation of Wright's principle. What I am arguing is that the principle must be interpreted so as to state that (2) and (3) are not identical to any class of (1) and that (4) must include potentialities as well as actualities. In short, the potential can not be identified either logically or empirically with the actual, and the potential is as real as the actual.

This is the metaphysical realism which seems to me necessary to underpin Burks's third category of "Inductive statements" since this category includes causal statements and inductive probability statements. Without real potentiality in the external world, causal statements and probability statements would not be applicable to real life problems. Now our "innate structure-program complex", as Burks (1977, p. 612) calls our apparatus for interpreting the external world, is *a priori* on his view and requires us to interpret the external world as being so structured as to justify a belief in causal uniformity and inductive probability. If the external world lacks potentiality, it would not be so structured as to justify those beliefs.

If our interpretive organs require us to interpret the world as having real potentiality and if it does not have real potentiality, it would not seem possible that the human organism could have survived over millions of years and have become the most successful of all animals with so basic a mismatch between its interpretive apparatus and the world it was interpreting. I am thus prepared to go one step beyond Kant – and probably also beyond Burks – and say we have *a priori* categories of the sort Burks describes, that they include potentiality, and, based on the fact of human survival, we are justified in saying we know that the external world corresponds to them at least to the extent of including real potentiality.

III

A definition of potentiality is difficult because, given our receptor organs, we do not experience the potentialities in things directly. We only experience them indirectly when they have become actualities or have produced actualities. Thus any definition of potentiality must be reached by an inference from actualities. Although the survival advantages of being able to experience potentialities directly would appear to be very great, it seems that no organism has ever fully developed them. However, in the struggle for existence, the vital importance of a strong sensitivity to actualities may easily have been so overwhelming as to prevent organic life, in its earlier forms, from getting beyond a keen sense of actualities accompanied by no more than the dim sense of potentiality which most life forms exhibit.

It is possible to interpret the success of the human organism as being due to its having finally evolved a quasisensitivity to potentiality by means of its extended reasoning capacity. Inductive reasoning and heuristic reasoning (or what Charles Peirce called abductive reasoning) as established in the procedures of science may constitute the formalizing of these sensitivities into a productive disciplinary mode. If there is any merit to this view of things, then the strict empiricist view of knowledge, which would limit knowledge simply to observables, would be a denial of one of the most significant capacities of the human organism.

The failure to develop a meaningful concept of potentiality has led to serious philosophical difficulties in many areas of human activity. Science, of course, is an obvious one, since we cannot give any justification for scientific predictions without potentialities. Another important class of propositions are counterfactual propositions. Regret and blame are based upon contrary-to-fact beliefs. A man regrets having acted in a certain manner. He regrets it because he believes that if he had acted in some other way, a way in which as a matter of fact he did not act, then some other situation which he finds more desirable would have occurred, i.e., he believes the situation had another potentiality that was not realized. Similarly, in a court of law a man is blamed for a result produced by his actions because the court believes the counterfactual proposition that if he had acted in a way in which he did not act, some different result would have ensued. That is, the court believes that there were real potentialities present in the situation which were not realized. (And one function of the whole legal apparatus is to define such potentialities of action. Where a court believes no such potentialities exist for an accused – for example, if he is insane – the law will not find him guilty.) Also in ethics, judgments of good and bad actions are based on counterfactuals. We believe an action is bad or good if we believe there are other potential actions a man might have done, although he did not, which would have less desirable or more desirable results.

To hold to the reductionist epistemology of strict sense-experience observables is to hold to a theory on which the basic propositions of science, of law and of ethics can not be asserted to be meaningful. Under such circumstances it seems philosophically essential to hold that Wright's principle needs to be interpreted so as to assert the reality of potentiality.

Borrowing a phrase from Charles Peirce that "The real is composed of the potential and the actual together".[2] I would elaborate the notion (definition 4 above) that the potential consists of actualizable properties and dispositional properties, as follows: An actualizable property is a property in a state of being capable of full transition into an actuality; while it is, however, in the state of mere being without actuality, it correlates one-to-one with its actuality in all its other characteristics except actuality. Thus, a tool-maker is a potential tool-maker when away from his shop, whereas when working in his shop he is an actual tool-maker. His potentialities as a tool-maker are fully transformable into actualities when he returns to work. A dispositional property, on the other hand, is a physical capability such that a specifiable appropriate occasion will trigger a specific actuality, but the actuality will have different physical characteristics from the potential property. Thus, the apple already lies potentially in the blossom and, given the appropriate environmental conditions, an apple will result from the blossom and the chain of development may be traced from the blossom to the apple. That is, the potentiality and the actuality have different physical characteristics, as in the difference between the genetic material and the organism produced, or in the difference between a human habit structured into the nervous system and the response produced by the habit when a stimulus is present.

In the light of these definitions I conclude with a modification of Wright's principle: Concepts of potentiality are admitted as meaningful and real. Though in themselves insusceptible of simple verification, in order to be admitted as real they must still show credentials from the senses, either by affording from themselves, as actualizables, a set of actualized consequences capable of sense-verification, or, as dispositional properties, by yielding in conjunction with other terms a set of actualized consequences which are not deducible from these other terms alone.

Under these rules, verification that a potentiality was a reality, not a fiction, would then authorize the deduction from it of causal statements, of probability statements and of contrary-to-fact conditionals. The validity of all of these statements would, of course, be subject to continued inquiry concerning the nature of the potentiality by the usual experimental procedures of science as embodied in the community of investigators.

Edward C. Moore
University of Alabama

NOTES

[1.] See (Moore, 1978). The present paper continues the discussion in that paper.
[2.] From a letter to William James in January of 1903. Quoted in Perry (1935, p. 246).

REFERENCES

Ayer, A. J.: 1947, *Language, Truth, and Logic*, Gollancz, London.

Burks, A. W.: 1951, 'Introduction to Peirce selections', in Fisch (ed.) (1951) pp. 41–53.

Burks, A. W.: 1955, 'Dispositional statements', *Philosophy of Science*, **22**, pp. 175–193.

Burks, A. W.: 1977, *Chance, Cause, Reason*, University of Chicago Press, Chicago.

Carnap, R.: 1936, 1937, 'Testability and meaning', *Philosophy of Science*, **III**, pp. 419–471, **IV**, pp. 1–40.

Fisch, Max H. (ed.): 1951, *Classic American Philosophers*, Appleton-Century-Crofts, New York.

Ketner, K. L., *et al.* (eds.): 1981, *Proceedings of the C. S. Peirce Bicentennial International Congress*, Texas Tech Press, Lubbock.

Moore, E. C.: 1952, 'The scholastic realism of C. S. Peirce', *Philosophy and Phenomenological Research* **XII**, pp. 406–417.

Moore, E. C.: 1981, 'On an alleged incompatibility between Peirce's metaphysics and his pragmatism', in Ketner, *et. al.* (1981), pp. 169–178.

Peirce, C. S.: 1871, 'Review of Fraser's edition of Berkeley', *North American Review*, **CXIII**, p.454.

Peirce, C. S.: 1931–1935, *Collected Papers of Charles Sanders Peirce*, **I-VII**, edited by C. Hartshorne and P. Weiss, Harvard University Press, Cambridge.

Perry, R. B.: 1935, *The Thought and Character of Wlliam James*, **II**, Little, Brown and Company.

Wright, C.: 1877, *Philosophical Discussions*, Holt, New York.

SOSHICHI UCHII

BURKS'S LOGIC OF CONDITIONALS

1. INTRODUCTION

Burks's logic of causal statements and his analysis of causal connections by means of it is one of the earliest attempts to apply modal logic to the analysis of a conditional statement (conditional, for short) 'If A then B.' Of course, the problem of conditionals was one of the major motives for C. I. Lewis when he began modal logic by means of symbolic technique; for, what he called 'strict implication' is a kind of conditional, and it was intended to capture the sense of 'A logically implies B.' And it is also true that Lewis himself was aware of the notion of causal necessity (he called it 'real connection,' though; see [9], bk. ii, ch. viii). But his systems of strict implication were meant for logical entailment, and he did not apply his symbolic technique to the analysis of causal or 'real' connections.

Research on conditionals after Lewis seems to fall into four groups, depending on how one interprets the meaning of 'if . . . then.' (i) One is the study of logical entailment; i.e. the central meaning of 'if. . .then' in this case is 'A logically entails B.' In Lewis's systems of strict implication, an impossible antecedent strictly implies anything, and a necessary consequent is strictly implied by anything (paradoxes of strict implication). Dissatisfied with these results, W. Ackermann, A. R. Anderson, N. Belnap, and A. A. Zinov'ev pursued a more adequate formalization of logical entailment. On this approach, 'A entails B' is roughly analyzed as: $A \supset B$ is logically true and there is some connection (relevance) of content between A and B (see [1], [3], [18], and my review [16] of [18]).

(ii) The second approach construes the conditional 'If A then B' as follows: Together with some implicit condition, A necessarily leads to B. This is the usual modal approach to conditionals, and to this group Burks belongs. If one interprets 'necessity' in terms of logical derivability or provability (asserted in metalanguage), this analysis

191

Merrilee H. Salmon (ed.), *The Philosophy of Logical Mechanism*, 191–207.
© 1990 *Kluwer Academic Publishers.*

turns into the well-known metalinguistic analysis of conditionals. However, since most metalinguistic analysts are also extensionalists who wish to dispense with modal notions, the metalinguistic analysis is usually opposed to a modal approach such as Burks's. But on our present classification, the modal and the metalinguistic analysis of conditionals share many important features so that we may regard them as belonging to the same group. The newer form of modal analysis is formulated in terms of Kripke's possible world semantics. Robert C. Stalnaker and David Lewis are best known for this newer approach, but their analyses have much in common with Burks's older approach. I will argue for this point later, as well as their major differences.

(iii) On the third approach, the central question is the evidential connection between the antecedent and the consequent of the conditional. Thus 'If A then B' is analyzed as 'A is a reasonable ground for asserting B.' This approach is pursued by Ernest Adams and C. L. Stevenson (see [2] and [12]). If one interprets 'reasonable ground' in terms of probabilistic connections, this approach turns into the following (iv).

(iv) The fourth and last approach stresses the probabilistic aspect of 'if . . . then,' and hence it may be called the probabilistic approach. On this analysis, 'If A then B' roughly means that, together with some implicit conditions, A probably leads to B. Usually, this approach is associated with the third approach, 'evidential connection' being interpreted as 'probabilistic connection.' Patrick Suppes's Probabilistic theory of causality [13] is akin to this approach, as far as conditional statements are concerned.

It is not my purpose to review all of these approaches. I will concentrate only on the second approach (ii). I will briefly survey Burks's analysis of causal conditionals, Stalnaker's and Lewis's newer theories of conditionals, and point out their similarities as well as differences; and from these comparisons, I will suggest my own theory of conditionals, trying to exploit the merits of Burks's and newer theories.

2. CLASSICAL ANALYSIS OF CONDITIONALS

Let us consider the following series of statements (adapted from Burks's [6], 7.3):

(1) This ring is gold (G).

(2) If this ring is gold and should be placed in aqua regia
    (A), it would dissolve (D).
(3) If this ring should be placed in aqua regia, it would dis-
    solve.
(4) If this gold ring had been placed in aqua regia, it would
    have dissolved.
(5) If this ring is gold and should be placed in aqua regia, it
    might (still) be undissolved.

I understand that (4) is equivalent with a conjunction of (1), (2), and
$-A-D$, and (5) is equivalent with the negation of (2). The conditional
(2) is construed as asserting some sort of necessary connection, not
merely a truth-function of the antecedent and the consequent. (4) is a
counterfactual form which makes explicit that some condition (G) in
fact holds, and it implies that both antecedent and consequent are in
fact false. Finally, (5) is a 'might' conditional which contradicts 'would'
conditional (2).

    First, we can analyze these conditionals in terms of strict (logical)
implication. Strict implication can be defined as a logically necessary
material implication:

$$(6) \quad A \rightarrow B =_{\text{def.}} \Box \, (A \supset B) \quad (\Box \text{ is logical necessity}).$$

Then, (2) can be analyzed as: there is some law X and some fact Y
such that $XYGA$ strictly implies D. The law X and the factual statement Y are assumed to be true. In symbols,

$$(7) \quad (EX) \, (EY) \, (X \text{ is a law \& } Y \text{ \& } (XYGA \rightarrow D)).$$

This is the first pattern of classical analysis. I will not attribute this
analysis to any particular author, since it seems to be accepted with
minor variations, by many philosophers. Although the strict implication in (7) is logically true, if true, the whole statement is empirically
true, if true, since the existence of such a law and fact is an empirical
question. If we replace the strict implication by a metalinguistic assertion of logical derivability and make necessary changes in other
parts, (7) turns into a form of the well-known metalinguistic analysis of
subjunctive or counterfactual conditionals. The crucial problem of this
analysis is how to characterize laws of nature.

    One may consider the notion of the law of nature to be too complex
and the notion of causal or physical necessity more fundamental. Then

he will replace the first and the third parts of (7) by a statement of such necessity. This is essentially Burks's analysis. Burks introduces the notion of causal necessity ($\square^c$). and defines causal implication ($\rightarrow$) as causally necessary material implication ($\square^c (A \supset B)$). Then, according to Burks, (2) should be analyzed as: there is some fact Y such that GA together with Y causally implies D. However, there are two reservations to be made for this analysis. First, he wishes to exclude the case where GA together with Y *strictly* implies D; the connection between the antecedent and the consequent should be empirical, rather than logical. Secondly, the causal implication should not hold vacuously by virtue of the antecedent alone or the consequent alone; thus the antecedent should be causally possible, and the consequent should not be causally necessary by itself. This can be seen from the fact that the following two are theorems of Burks's logic of causal statements:

$$(8) \; -\diamondsuit {}^cA \supset (A \underset{c}{\rightarrow} B)$$
$$(9) \; \square {}^cB \supset (A \underset{c}{\rightarrow} B).$$

These may be called 'the paradoxes of causal implication.' Thus, what Burks calls 'the nonparadoxical causal implication' (*npc*, for short) is defined as follows:

$$(10) \; A \; npc \; B =_{\text{def.}} -(A{\rightarrow}B) \; (A \underset{c}{\rightarrow} B) \diamondsuit {}^cA - \square {}^cB.$$

This is his original definition in 'Dispositional Statements' (1955). In *Chance, Cause, Reason* (1977), he revised the definition by changing the first condition into that the antecedent and the consequent are logically independent. In my opinion, the original is better than this, because, even if B strictly implies A, it is perfectly all right that A is nevertheless a sufficient causal condition for bringing about B, i.e. A *npc* B. Therefore I will continue to use the original definition throughout this paper.

The *npc* is adequate for formalizing those conditionals in which the antecedent states a causally sufficient condition for the consequent. Thus if we may assume that being made of gold and placed in aqua regia is sufficient for this ring to dissolve, the conditional (2) may be translated as:

$$(11) \; GA \; npc \; D.$$

However, (3) cannot be analyzed in this way, because its antecedent does not state a causally sufficient condition for the consequent. In general, most conditionals specify only a part of causally sufficient conditions in the antecedent and in this sense they are *elliptical*. This feature can be captured by using an existential quantifier for statements; thus (3) can be formalized as:

$$(12) \quad (EY) \; (Y \; \& \; (YA \; npc \; D)).$$

These translations in terms of *npc* constitute the second pattern of classical analysis of conditionals. Compare (11) and (12) with (7).

With the same idea as (12), Burks's *elliptical causal implication* (*ec*, for short) can be defined:

$$(13) \quad A \; ec \; B =_{\text{def.}} (EX) \; (X \; \& \; (XA \; npc \; B)).$$

Then, using *ec*, (3) becomes:

$$(14) \quad A \; ec \; D.$$

The elliptical causal implication is weaker than *npc* (i.e. $A \; npc \; B$ implies $A \; ec \; B$), and stronger than the material implication (i.e. $A \; ec \; B$ implies $A \supset B$). We will see further properties of *npc* and *ec* later.

As we have seen, the two patterns of classical analysis of conditional are quite similar. The major difference is that the first appeals to the notion of the law of nature, whereas the second makes use of the concept of causal necessity. And it must be remembered that these formal translations are supplemented by informal analyses and considerations of the character of tacit assumptions which are expressed by an existentially quantified statement. I will come back to this point in the next section.

3. NEW ANALYSIS OF CONDITIONALS

The first attempt to apply the so-called possible-world semantics for modal logic to the analysis of conditionals was made by R. C. Stalnaker in 'A Theory of Conditionals' (1968). Stalnaker's basic idea is this: In order to see whether or not a conditional 'If A then B,' which is symbolized as '$A > B$', is true, consider a possible world in which A is true and which otherwise *minimally* differs from the actual world; then the conditional is true if and only if B is also true in that possible world. In order to formulate this truth-condition within Kripke-type semantics, he introduces a function $f(A, w)$ which selects, for each pair of statement A and possible world w, the world which makes A true and in

other respects differs *minimally* from the world w (in other words,
which *maximally resembles* the world w). If A is impossible in world w
then $f(A, w)$ is empty, and vice versa. Notice that $f(A, w)$ is a single-
valued function which selects at most *one* possible world.

The maximal similarity to world w, of course, cannot be determined
by logic alone; it will depend on many factors such as the supposed
antecedent, the world w itself, the context of the conditional, the intention of the speaker, and so on. But there are certainly *formal* conditions which f must satisfy, in order that f selects a maximally similar
world, whatever that similarity may be. Thus, according to Stalnaker,
(i) if A is in fact true in world w, then w itself is the world which
makes A true and maximally resembles w, i.e. $f(A, w) = w$. And (ii) if
B is true in $f(A, w)$ and A is true in $f(B, w)$, $f(A, w)$ must be identical
with $f(B, w)$; in other words, if $A > B$ and $B > A$ are both true in w,
then the maximal similarity is the same for the two.

Let us apply this analysis to our example (2) and (3). According to
Stalnaker, (2) should be translated as $GA > D$, and (3) as $A > D$. Thus
(2) is true in the actual world w if and only if there is a unique world
$f(GA, w) = w'$ and D is true in w'. The possible world w' is of course
possible relative to w (in symbols, wRw'); i.e. any sentence true in w'
is possible in w. And G, A, and D are all true in w'. Now, there seems
to be something strange in this analysis. For, suppose there is another
possible world w'' such that w'' is possible relative to w, GA is true in
w'', w'' is sufficiently (but not maximally) similar to w in that the aqua
regia is in liquid state, etc., but D is false in w''. Then we would think
that $GA > D$ is false in w, even if GAD were true in w' which is maximally similar to w. Thus, the consideration of a single possible world
w' does not seem sufficient for determining the truth condition for 'If
GA then D.' To be more specific, there are two difficulties in
Stalnaker's analysis: First, is there any logical ground for supposing
that there is *at most one* possible world which is maximally similar to
the actual world? Second, why should we assume the *maximal*
similarity to the actual world when we consider or assert a conditional
statement like 'If GA then D'?

With respect to the first, David Lewis has made an improvement.
According to his improvement, given a possible world w, we should
consider a *centered system of spheres around w* (see [10], p. 14). A
sphere around w is a set of possible worlds which resemble w to at
least a certain degree. Thus there can be many different spheres

around w, since there are various degrees of similarity. And since spheres are meant to carry information about the comparative overall similarity, there are certain formal constraints similar to Stalnaker's conditions for $f(A, w)$; but details of such conditions are not very important here. In any case, all possible worlds are ordered around w according to their degree of similarity to w by such a system of spheres. Then,

(1) A conditional $A > B$ (Lewis uses symbol '$\square \rightarrow$' instead of '$>$') is true in w if and only if either
    (i) there is no sphere (around w) which contains at least one possible world in which A is true, or
    (ii) there is a sphere (around w) which contains at least one world in which A is true, and in every such world in that sphere B is also true.

Of course, (ii) is the principal case. According to this condition, $A > B$ is true if and only if A is possible and $A \supset B$ is true in each world belonging to the *narrowest* sphere containing a world which makes A true (in other words, in each world which resembles the actual world as much as the supposition A permits it to). Thus, there is no assumption that there is at most one world which is maximally similar to the actual world. However, David Lewis still maintains, like Stalnaker, the requirement of maximal similarity.

Now, quite frankly, these analyses of conditionals are, on the whole, worse than the classical analyses, in particular Burks's analysis, although, of course, they do contain ingenious ideas of how we can apply the Kripke-type semantics to the analysis of conditionals. For, intuitively, we consider neither a single possible world nor a set of maximally similar worlds when we entertain a conditonal 'If A then B' or a counterfactual 'If it were the case that A, then it would be the case that B.' Rather, we consider a set of possible worlds in which A is true and which, *in the relevant respects*, resemble the actual world. For example, we regard the conditional 'If this ring should be placed in aqua regia, it would dissolve' ($A > D$) as true, because we think the consequent true in every possible world in which the antecedent is true and which, in other relevant respects, resembles the actual world. The relevant respects are, for instance, that the ring is gold (or silver, platinum, etc.), that the aqua regia is in liquid state, that the usual physical or chemical laws hold, and so on. If we may assume sufficient similarity in these respects, it is simply redundant and misleading to require any stronger similarity. Who cares, as regards the truth condition for $A >$ D, that today is Monday or that there are seven pencils in the laboratory, even if these are in fact true in the actual world? – Only Stalnaker, David Lewis, and their followers.

Burks's analysis of causal conditionals, including causal dispositions and cause-effect relations, has made it amply clear that the notion of *the relevant similarity* is necessary and sufficient for considering the truth condition for a conditional, although he did not explicitly state this point in these terms. The relevant passages can be abundantly found in 7.3 and 7.4 of *Chance, Cause, Reason*. For example, in 7.4.2, he puts the point this way:

> When asserting a causal subjunctive, one is, or should be, dealing with a relatively iso-lated, self-sufficient system and a relatively definite theory or set of laws governing this system. Correspondingly, when modeling a causal subjunctive, one should include enough of its context to obtain a portion of discourse that is complete and self-sufficient with respect to the relevant forces and laws, and one should model this portion of dis-course as a unit. We call this general principle the *rule of completeness for causal sub-junctives*.

Burks's example to illustrate this point is a dictator who is about to drink a glass of poisoned wine. Suppose the wine contains several grains of strychinine. (i) If a normal adult takes one or more grains of strychinine without receiving an antidote, he will die. But (ii) the dic-tator is normal and received no antidote. Then the following con-ditional seems to be true:

> (2) If the dictator had taken the strychinine (S), he would
> have died (D).

In this counterfactual, the relevant similarity of the supposed possible worlds to the actual world is assumed to be that (i) and (ii) hold in each of them. However, suppose further that (iii) the dictator is afraid of being poisoned and has arranged for a physician always to be present with a complete stock of antidotes. Then, of course, the relevant similarity must include this information too, and hence (2) seems to be false. The point of this example is clear: We must consider all possible worlds which are similar to the actual world in all the relevant respects.

The requirement of maximal similarity can accomplish this automati-cally, but it does more. Continuing the same example, suppose the dictator likes Wagner, dislikes Jews, has a blonde mistress, and all other trivial things which are in fact true. Burks's rule of completeness does not require their inclusion; or rather, it requires to exclude these factors. For, the rule is meant for a good modeling of a causal subjunctive, and it is certainly not a good modeling if we are forced to say that such conditionals as the following are also true: 'If the dictator should take the strychinine, he would like Wagner,' 'If he should take the strychinine, he would have a blonde mistress,' etc.

Stalnaker-Lewis's requirement of maximal similarity has further undesirable consequences. Suppose the antecedent of $A>B$ is true in the actual world. Then $A>B$ reduces to $A \supset B$. This is so because in Stalnaker's system, there is at most one world which makes A true and maximally resembles the actual world; since A is true in the actual world, there are no other worlds more similar to it than itself. As for Lewis's system, we must mention that the unit set of the actual world is a sphere around the actual world (it is the narrowest nonempty sphere). Thus, according to his truth condition (1), $A>B$ with true antecedent is true if and only if B is true: i.e. $A>B$ reduces to $A \supset B$. Therefore, according to their theories, for any pair of true statements A and B, $A>B$ is true. In symbols,

(3) $AB \supset (A>B)$

is valid. Of course, this (replacing $>$ with *npc* or *ec*) is invalid in Burks's system. Thus, with our example, 'If the dictator should like Wagner, he would have a blonde mistress,' 'If he should have a blonde mistress, he would dislike Jews,' etc., are all true. David Lewis has argued that this feature is odd but innocuous [10], pp. 26–8). But it is not. We aim at a good analysis and a good modeling of conditionals. Then, why is it innocuous that their analysis makes such conditionals true, which are false according to our intuitive understanding?

It must be added, to be fair to David Lewis, that he does suggest a way to make (3) invalid, keeping all other merits of his system ([10], pp. 29–30). But this move merely revises the interpretation of maximal similarity. What I am arguing is that the notion of maximal similarity is inappropriate, and it should be replaced by the notion of relevant similarity. In considering such examples as Burks's dictator, it is simply irrelevant whether or not a possible world is more similar than another to the actual world. The question is whether it resembles the actual world in all the relevant respects.

4. COMPARISONS OF BURKS'S AND NEW ANALYSIS OF
CONDITIONALS

In order to be fair, I will compare Burks's analysis of conditionals with
Stalnaker's and Lewis's in more detail.

To begin with, let us notice basic similarities between the two. First,
Burks's *ec* and Stalnaker-Lewis's conditional connective (>, which is
called 'the corner' by Stalnaker) are both modal and elliptical: they assume, in addition to the antecedent of a conditional, implicit conditions which hold in the actual world; and some modal connection,
together with these implicit conditions, leads to the consequent.
Second, as a consequence of this elliptical character, both *ec* and > are
nontransitive, and contraposition becomes invalid for them. That is,
both

(1) (If A then B) (If B then C) $\supset$ (If A then C),
(2) (If A then B) $\supset$ (If $-B$ then $-A$)

are invalid with *ec* or > replacing 'If...then.' The reason is clear: The
tacit conditions for A and B (or $-B$) may be different.

However, there are several differences between Burks's *ec* and Stalnaker-Lewis's >. First of all, Burks's analysis excludes, whereas Stalnaker-Lewis's does not exclude, the paradoxical case of implication
where the antecedent is impossible or the consequent is necessary. Although many people may consider this a great difference, I think this
more or less a matter of technical convenience. For, if Stalnaker and
Lewis wish to exclude the paradoxical case, they can simply add the
corresponding conditions. Lewis's reason for not excluding the
paradoxical case is presumably that the relation between 'would' and
'might,' such as (2) and (5) of Section 2, becomes simpler: 'might' conditional can be expressed by a negation ($-(A>B)$) of 'would' conditional ($A>B$). Incidentally, the major difference of Stalnaker's system
from Lewis's is that

(3) $(A>B) \vee (A>-B)$

is valid in the former, but invalid in the latter. Accordingly, the distinction between 'would' and 'might' collapses in Stalnaker's system. As a
result, maybe it is improper to say that Stalnaker's theory is a kind of
necessity view, although it can be obtained as a limit of necessity view
such as Lewis's. I think it is one of the merits of Lewis's analysis that

he noticed the relationship between 'would' counterfactual and 'might' counterfactual.

Secondly, a more important difference is that Burks's analysis is *local* in that it covers only causal conditionals, whereas Stalnaker-Lewis's analysis can cover a wider range of conditionals. For, the interpretation of the concept of necessity (or possibility) is widely left open: it may be logical, causal, historical, or any other sort of necessity, provided that it is alethic, i.e. such that necessity implies actuality. And Lewis extends his analysis to deontic modalities too. (Incidentally, I have applied the same idea as Burks's to the analysis of conditional obligations; see [14] and [15].)

Thirdly, speaking of more technical points, the law of exportation holds for *ec*, but it fails for $>$. That is to say,

$$(4) \quad \text{(a)} \quad (AB \; npc \; C) \supset (A \supset (B \; ec \; C))$$
$$\text{(b)} \quad (AB \; ec \; C) \supset (A \supset (B \; ec \; C))$$

are valid (see [6], p. 446), but

$$(5) \quad (AB > C) \supset (A \supset (B > C))$$

is invalid. For suppose (5) is valid; then

$$(6) \quad (AB > A) \supset (A \supset (B > A))$$

is valid, and since $AB > A$ is valid ($>$ does not exclude logical implication),

$$(7) \quad A \supset (B > A)$$

is valid. (7) shows a paradoxical case of $>$, and this is of course invalid. Further, substitute $-A$ for B in (7); then

$$(8) \quad A \supset (-A > A)$$

is valid. But $-A > A$ is true if and only if $-A$ is impossible or A is necessary ($\square A$); hence (8) means $A \supset \square A$, and this is certainly invalid. Thus, although $>$ is essentially an elliptical implication, it cannot have a nice property of *ec*.

Fourthly, for $>$

$$(9) \quad (A > B)(A > C) \supset (A > BC)$$

is valid, but for *ec*

(10) $(A \ ec \ B) \ (A \ ec \ C) \ \supset \ (A \ ec \ BC)$

is invalid. The reason is that, even if A is causally possible, $A \ ec \ B$ and $A \ ec \ -B$ are compatible, although $A \ ec \ B-B$ is of course invalid (see [6], pp. 446–7). This shows that the truth condition of ec is too local, so that, formally, we cannot combine the consequents of two ec's with the same antecedent. This is another reason why Burks emphasizes the importance of the rule of completeness for causal subjunctives (see his discussion of Lewis Carroll's barbershop paradox, [6], pp. 460–3). According to our informal understanding, if A is possible then 'If A then B' and 'If A then $-B$' are incompatible, at least within the same context. The ec does not capture this, and hence we must be careful to include enough context in order to translate causal subjunctives by means of ec. The underlying idea of ec is that an elliptical conditional presupposes the notion of the relevant similarity; but the formal definition of ec uses only an existential quantifier over statements, and hence it cannot capture the idea adequately. Stalnaker-Lewis's analysis satisfies this requirement automatically, because their semantical conditions for $>$ include the requirement of maximal similarity. But this condition went too far, as we have seen in the previous section.

These comparisons are summarized in the table in the next section.

5. LOGIC OF CONDITIONALS: MY OWN THEORY

Burks's analysis has made it clear that the relevant similarity is assumed in a conditional, although he has not included this idea into his truth condition for npc or ec. And Stalnaker-Lewis's analysis has made it possible to apply Kripke-type semantics to conditionals. In the remainder of this paper, I will combine these two ideas and develop my own theory of conditionals (actually, this theory was first published in Japanese in 1972 ; see [14]).

The general framework of my theory is about the same as Stalnaker's. Assume a standard language for modal propositional logic. I will take necessity ($\Box$, where the intended interpretation is left open: it may be logical, causal, or other sort of necessity) and conditional connective ($>$) as primitive. Then the conditional equivalence is defined as follows:

(1) $A \geqslant B =_{\text{def.}} (A > B)(B > A).$

And the interpretation of > is given, informally, as follows:

> (2) $A > B$ is true if and only if B is true in every possible world (a) which makes A true, and (b) which resembles the actual world in the relevant respects.

In order to select all possible worlds which satisfy the conditions (a) and (b), the selection function $f(A, w)$ is introduced. For each antecedent A and each possible world w, $f(A, w)$ is the set of all possible worlds which make A true and resemble w in the relevant respects. Then $f(A, w)$ must satisfy the following five conditions:

(i)   If w' belongs to $f(A, w)$ then wRw'; i.e. any world belonging to $f(A, w)$ must be possible relative to w (R is the relation of relative possibility in Kripke-type semantics).

(ii) If w' belongs to $f(A, w)$ then A is true in w'.

(iii) $f(A, w)$ is empty if and only if there is no possible world which satisfies both (i) and (ii), i.e. A is impossible in w.

(iv) If A is true in w, then w belongs to $f(A, w)$; for w is of course relevantly similar to itself.

(v) $f(A, w) = f(B, w)$ if and only if $A \geq B$ is true in w; i.e. if A and B are equivalent in terms of >, then the relevant similarity is the same for them, and vice versa. In other words, the relevant similarity does not depend on how we describe the antecedent.

All of these conditions seem reasonable in view of our informal notion of the relevant similarity, except, perhaps, that it seems a little artificial to consider the relevant similarity as a function of the antecedent A and the possible world w. I will come back to this point later.

Given these conditions for the selection function f, we can define a model for our logic of conditionals. As a background modal system, we will assume von Wright-Feys's T (of course it can be strengthened to S4 or S5, etc.). Then an ordered quintuple $<V, w, W, R, f>$ is a *TC-model* if and only if w is a member of nonempty W, R is a reflexive relation over W, and f satisfies the preceding five conditions; V is a usual valuation function (which assigns a truth-value to each sentence in each possible world) which satisfies our interpretation of > as follows:

> (3) For each $w \, \varepsilon \, W$, $V(A > B, w) = T$ if and only if for each w' belonging to $f(A, w)$, $V(B, w') = T$.

Any sentence A is true in $<V, w, W, R, f>$ if and only if $V(A, w) = T$ (A is true, according to V, in the actual world w). And A is said to be *TC-valid* if and only if there is no TC-model in which A is false. The properties of $>$ are determined by TC-valid formulas containing $>$. For example, the following formulas are all TC-valid:

(A1) $\Box (A \supset B) \supset (A > B)$
(A2) $(-A > A) \supset \Box A$
(A3) $(A > B) \supset (A \supset B)$
(A4) $\Diamond A \supset ((A > B) \supset -(A > -B))$
(A5) $(A > B)(A > C) \supset (A > BC)$
(A6) $(A > B) \Box (B \supset C) \supset (A > C)$
(A7) $(A \gtreqless B) \supset ((A > C) \supset (B > C))$.

And it can be proved that the system T with (A1)-(A7) as additional axioms is complete with respect to my semantics; i.e. all TC-valid formulas are theorems of it and vice versa (see [17], pp. 128–137).

Aside from the preceding, the conditionals characterized by TC-validity have the following properties: our $>$ is, like *ec* and Stalnaker-Lewis's $>$, nontransitive and contraposition is invalid for it; further, even if A is true, $A > B$ is not equivalent with $A \supset B$; and of course

$$(4) \quad AB \supset (A > B)$$

is invalid.

Finally, unlike *ec*, exportation fails for $>$; i.e.

$$(5) \quad (AB > C) \supset (A \supset (B > C))$$

is invalid. These properties, together with the corresponding properties of *ec*, Stalnaker's and Lewis's $>$, are summarized in the following table.

As is clear from the table, my $>$ improves all defects of Burks's *ec* (see the first three rows); it also improves all defects of Stalnaker's or Lewis's $>$ (see fourth and fifth rows). The only nice property of *ec* which my $>$ cannot capture is the law of exportation (see the last row). So let us consider this law in more detail: on what condition it is valid, and whether or not it can be justified within my theory of conditionals.

As we have already examined (Section 4, (5)–(8)), exportation is invalid for $>$ mainly because $>$ does not exclude the logical implication and the paradoxical case of implication. Namely, since $AB > A$ is true (which is logical implication), exportation

TABLE
Characteristic Features of Four Kinds of Elliptical Implication

	>			
	Burks's *ec*	Stalnaker	Lewis	Uchii
Assumed Similarity	Local	Maximal	Maximal	Relevant
(If *A* then *B*) (If *A* then *C*) ⊃ (If *A* thfn *AC*)	Invalid	Valid	Valid	Valid
(If *A* then *B*) (If *A* then −*B*) where *A* is possible	Compatible	Incom-patible	Incom-patible	Incom-patible
(If *A* then *B*) v (If *A* then −*B*)	Invalid	Valid	Invalid	Invalid
AB ⊃ (If *A* then *B*))	Invalid	Valid	Valid	Invalid
(If *AB* then *C*) (*A* ⊃ (If *B* then *C*))	Valid	Invalid	Invalid	Invalid

$$(6) \quad (AB>A) \supset (A \supset (B>A))$$

leads to $A \supset (B > A)$ which is undesirable. And further, since $A–A > A$ is also true (which is a paradoxical case of $>$), exportation

$$(7) \quad (A–A>A) \supset (A \supset (–A>A))$$

leads to $A \supset \Box A$ which is disastrous. From these facts, it is evident that the law of exportation can be valid only within some limits. However, it is also clear that with such examples as the following, exportation is certainly valid and desirable:

> (8) If this ring is gold (*G*) and should be placed in aqua regia (*A*), it would dissolve (*D*); but this ring is gold; hence if this ring should be placed in aqua regia, it would dissolve.

For both *G* and *A* are relevant to the consequent *D*; and if *G* is in fact true, this should be counted as a relevant respect in which the supposed possible worlds resemble the actual world, when we consider 'If *A* then *D*.' In other words, in the context of (8), all possible worlds which make *A* true and resemble the actual world in the relevant respects should also make *G* true, when we consider the conditional 'If *A* then *D*.' These considerations involve the whole context of (8), not merely *GA* or *A* taken independently.

But the formal translation of (8) in terms of $>$ is invalid, as the following countermodel shows:

Actual world w_1: G–A–D $f(GA, w_1) = \{w_2\}$
           w_2: GAD $f(A, w_1) = \{w_3\}$
           w_3: $-GA$–D

Let $W = \{w_1, w_2, w_3\}$ and $R = W^2$. Then in this TC-model, both $GA > D$ and G are true, but $A > D$ is false. The selection function f is determined, in my formal semantics (as well as in Stalnaker-Lewis's), only relative to an antecedent and a possible world. Therefore, formally, $f(GA, w_1)$ and $f(A, w_1)$ can be determined independently, without considering the whole context of (8); even if G is true in w_1, it need not be included in the relevant similarity for $f(A, w_1)$. This shows that, although the selection function f can cover the relevant overall similarity for a single, isolated conditional, it is still local when we have to consider a set of interrelated conditionals like 'If GA then D' and 'If A then D' in (8). And in view of the informal nature of such contextual interdependencies of conditionals, I do not think there can be any formal method to capture them. Thus, as regards the limits or conditions for the validity of exportation, we must appeal to such rules of application of our semantics as Burks's rule of completeness for causal subjunctives.

It should be clear that the same is true of the laws of contraposition, transitivity, and other principles of conditionals such as

(9) If A then C; and if B then C; hence if A or B then C.

Sometimes such principles are valid and sometimes not; the validity depends on what the relevant similarities are. The question should be settled within concrete cases of application of our semantical apparatus, and it is simply misleading to try to answer it on formal grounds alone. Thus I think it is a great virtue of Burks's analysis that he has been perceptive both to the need for formal tools of analysis and to the need for informal rules of applications.

Soshichi Uchii
Osaka City University

REFERENCES

[1] Ackermann, W.: 1956, 'Begründung einer strengen Implikation', *Journal of Symbolic Logic*, **21**, pp. 113–128.
[2] Adams, E. W.: 1966, 'Probability and the logic of conditionals', in *Aspects of Inductive Logic*, J. Hintikka and P. Suppes (eds.), North-Holland Publishing Company, Amsterdam.
[3] Anderson, A. R. and N. D. Belnap: 1962, 'The Pure Calculus of Entailment', *Journal of Symbolic Logic* **27**, pp. 1952.
[4] Burks, A. W.: 1951, 'The logic of causal propositions', *Mind* **60**, pp. 363-382.
[5] Burks, A. W.: 1955, 'Dispositional statements', *Philosophy of Science* **22**, pp. 175-193.
[6] Burks, A. W.: 1977, *Chance, Cause, Reason*, The University of Chicago Press, Chicago.
[7] Hughes, G. E. and M. J. Cresswell: 1968, *An Introduction to Modal Logic*, Methuen, London.
[8] Kripke, S. A.: 1963, 'Semantical analysis of modal logic I: Normal modal propositional calculi', *Zeitschrift fur mathematische Logik und Grundlagen der Mathematik* **9**, pp. 67-96.
[9] Lewis, C. I.: 1946, *An Analysis of Knowledge and Valuation*, Open Court, La Salle.
[10] Lewis, D.: 1973, *Counterfactuals*, Basil Balckwell, Oxford.
[11] Stalnaker, R. C.: 1968, 'A Theory of Conditionals', in *Studies in Logical Theory* (*American Philosophical Quarterly* Monograph Series No. 2), N. Rescher, (ed.).
[12] Stevenson, C. L.: 1970, 'If-ficulties', *Philosophy of Science* **37**, pp. 27-49.
[13] Suppes, P.: 1970, *A Probabilistic Theory of Causality*, *Acta Philosophica Fennica* **24**.
[14] Uchii, S.: 1972. 'Modal logics and conditional statements', (In Japanese) *Zinbun Gakuho* **35**. pp. 21-66.
[15] Uchii, S.: 1974, '"Ought" and conditionals', *Logique et Analyse* **17**, pp. 143-164.
[16] Uchii, S.: 1974, 'Critical notice: *Foundations of the Logical Theory of Scientific Knowledge (Complex Logic)* by A. A. Zinov'ev', *Philosophia* **4**, pp. 583-599.
[17] Uchii, S. and K. Kamino: 1976, *Ronrigaku: Model Riron to Rekishiteki Haikei* [Logic: Model Theory and Historical Background] (In Japanese), Minerva, Kyoto.
[18] Zinov'ev, A. A.: 1973, *Foundations of the Logical Theory of Scientific Knowledge (Complex Logic)*, Reidel, Dordrecht.

F. JOHN CLENDINNEN

PRESUPPOSITIONS AND THE NORMATIVE
CONTENT OF PROBABILITY STATEMENTS

INTRODUCTION AND SYNOPSIS

Arthur W. Burks has argued with care and in detail for a theory of ampliative inference with a number of important and unique features. The outcome of his work in this area is contained in *Chance, Cause, Reason* (1977). Central to his system is his concept of inductive probability. His analysis of this concept has been strongly influenced by C. S. Peirce's pragmatic theory of meaning. According to this theory we should equate the meaning of a statement with "the set of those practical conditionals that are logically implied by the statement" (p. 167). This can not be applied directly to probability statements; since the practical consequences of scientific statements must so often be expressed in terms of probability. So as to develop a satisfactory pragmatic theory of probability, Burks relates inductive probability on the one hand to a Calculus of Choice and on the other to empirical probability and causation.

I believe that Burks' insistence on the practical relevance of a system of inductive probability and the consequential pragmatic components in the meaning of probability statements is correct and that if generally accepted would help to avoid some quite widespread confusions. On the other hand, I will argue, his account of the nature of the pragmatic component is unsatisfactory. It provides no adequate basis for a normative interpretation of this pragmatic component. Burks says that there is normative content in probability statements but he fails to give an adequate account of what this comes down to; and he explicitly denies that there are reasons for accepting, as a guide, inductively based probability statements rather than those of some competing system. I will argue that the normative question which must arise about any system for accepting statements with pragmatic content, can only

209

Merrilee H. Salmon (ed.), *The Philosophy of Logical Mechanism*, 209–232,
© 1990 *Kluwer Academic Publishers*.

be answered by identifying some feature of that system which marks it off as superior to alternatives.

Burks' acceptance of Hume's doctrine that no justification of induction is possible is hardly idiosyncratic; however I will argue that his and other's acceptance of this conclusion is based on an improperly limited notion of what would constitute a justification. Indeed his own presupposition theory of induction can be taken as the starting point of a justifying argument. Burks holds that the widespread acceptance of inductive reasoning can be explained by presuppositions expressing certain concepts which are a priori in the sense that we are programmed to think according to them.

It is necessary to examine carefully Burks' notion of a priori concepts. It is not at all clear that this notion can do anything towards establishing that probability statements are normative. It is also profitable to examine the notion of presupposition; for there is an ambiguity here that needs resolving. In one sense of the word we may mean by "this action (or policy) presupposes P" that a person so acting could only be doing so rationally if he accepts P as true. But there is a weaker sense of "presuppose" which might be less misleadingly expressed by "acting on the assumption that". Someone doing an action which he knows will succeed if and only if X is the case, acts on the assumption that X. It is clear that a person can in some cases act rationally on the assumption that X when he has no way of knowing whether or not X is the case. (This will be so when there is no other way of acting which improves the chance of success.)

The "presuppositions" of induction should be understood in the second sense. It is then possible to consider acting on some other assumptions. When we do so we see difficulties which render any such course of action irrational. In this way the policy of acting on Burks' "presuppositions" when wishing to predict can be justified as the only rational policy. Such a justification does not prove that induction will generally succeed – it does establish that induction is the only rational way of trying to anticipate the future.

1. BURKS' PRAGMATIC THESIS

The thought basic to Arthur W. Burks' work on induction and probability is that the beliefs we form are intimately related to the actions we do; in particular, expectations we form about the future play a crucial role in every decision we take. Since we differ in the degree of confidence with which we hold expectations from case to case we need some quantitative notion to express this variable. Burks accepts probability as playing this role. (He allows that according to a generally accepted usage this term can also play a different role in expressing statistical laws. So he distinguishes two concepts calling the former "inductive probability" and the latter "empirical probability". In the sequel, the unqualified term will mean "inductive probability" unless otherwise indicated.)

Burks is broadly in agreement with the tradition of developing precise formal systems to "explicate" the less precise concepts with which we normally function. Thus he develops an axiomatic system which determines all true atomic inductive probability statements, that is, statements of the form $P(h,e) = x$. He takes such a statement as asserting a degree of confidence, x, in hypothesis h given e as the total relevant evidence. If these probability statements are synthetic then so are the axioms of the system; and the question of their truth or falsity must arise. Burks accepts this consequence and one of the major issues dealt with in his book (1977)* concerns the grounding of inductive probability statements.

There is a danger of misinterpreting what is done in the axiomatic treatment of a set of one or more concepts. If the explication is seen as clarifying meanings and the axiomatization is seen as fulfilling this task, then it may seem that the axioms, and therefore the statements which follow from them, are analytic. Carnap seemed to adopt this position and imply that all true probability statements were analytic. He has often been criticised on this point; for if probability statements are simply true as a consequence of propositions which function as definitions, it is hard to see how the acceptance of one of them could have any practical consequences. It was, however, clear to Burks that any meaningful concept will play a role in practical activity; and, in addition, that "probability" is meaningful in this sense. In a discussion of Carnap's concept of probability he suggested that Carnap should allow that probability statements do have pragmatic content and are consequently not analytic (1963, pp. 742–748). In response Carnap largely agreed with Burks (1963, p. 981). Carnap in fact held that each of his c*-statements is analytic, but that the corresponding probability statement implies not only the c*-statement but also the claim that the c*-function is a rational way of forming expectations. This latter component of any probability statement means that it has practical and normative content and is not made true simply by the axioms which define a particular c-function.

According to the Pragmatic Thesis probability statements express degrees of confidence or partial belief in hypotheses; and these partial beliefs are made manifest in actions, like betting, which result in the promise of some quantifiable gain if the hypothesis should be true. Thus this thesis asserts that in using probability statements we express a commitment to certain practical activities; and what these commitments are determines the meaning of the probability statement. On this view the axioms which entail probability statements are not analytic but carry implications as to what actions are correct. So supporters of this thesis must consider the reasons for accepting any such axiom set.

2. ANALYTIC CONNECTIONS AND RATIONALITY

It is a confusion typical of the more formally oriented approach to philosophy to think of probability statements as made a priori true by the adoption of an axiomatic system. There is a closely related error which is commonly accepted in the ordinary language approach to philosophy. Here it is often held that there is an analytic connection between the claim that it is reasonable to accept a certain proposition, and the fact that that proposition is related to certain evidence according to a certain rule. In a classic statement of this point of view, which has gained wide currency, P. F. Strawson says that placing reliance on inductive procedures "is what 'being reasonable' means" in the appropriate kind of context (1952, p. 257). The Pragmatic Thesis tells just as forcefully against this as against the a priorist theory of Probability, Statements about what it is reasonable to believe have a pragmatic component. In asserting that the belief in some statement is rational we express a commitment to act in a certain way. However the claim that a statement is inductively based simply means that its relationship to data conforms to a certain rule; and there is no analytic implication between this and any commitment.

It is, of course, true that some words gain their meaning in contexts where the propriety of a certain policy is taken for granted. It may be that the applicability of a certain word presupposes a commitment to a certain policy; but in this case we must settle first whether we have the commitment to the policy in question before we can decide whether that word should properly be used. What is not possible is that the correctness of a certain description could be settled by considerations which took no account of practical commitments already accepted and that then such commitments be derived analytically from the description. If we can determine, without any consideration of what it is proper to believe, that a certain hypothesis is derived by certain rules from given facts then it is not possible to deduce from this that it is reasonable to believe the hypothesis and consequently that we are committed to a certain line of practical action. Any such contention would involve claiming that two logically distinct sets of criteria were both definitionally linked to a certain phrase, one constituting a definitionally sufficient condition and the other a definitionally necessary condition.

Some words in their normal usage have a certain ambiguity between a normative and a descriptive meaning, and to try and embrace both at once can lead us into the kind of logical error referred to above. The word "inductive" has something of this character. Another more obvious example is the word "to lie". If it is held that it is a contradiction to say "on some occasions one should lie" then it must always remain possible to ask on any occasion "would my uttering this falsehood constitute a lie?" We cannot establish a moral truth by appealing to two propositions both of which are held to be analytically true; namely: that knowingly to utter a falsehood is to lie, and secondly, that to lie is always wrong. It may be that each of these propositions taken separately has some claim to being considered analytic, but without logically cheating we cannot accept them both as analytic at once. Likewise we cannot appeal to two analytic propositions involving the word "induction" in order to establish the propriety of a certain policy of forming expectations. The simple point is that decisions which are purely concerned with linguistic policy cannot involve commitments in a quite different kind of activity, and if the criteria for using a certain word includes commitment to a certain policy then the acceptability of that commitment must be settled before the appropriateness of using that word can be settled.

That there is a logical gap between descriptions about what is the case and statements which express a commitment to an action or policy, is fairly widely accepted, but recently an ingenious case against such an unbridgeable gap has been advanced. This has been put forward in the area of moral philosophy. But if it stands, it might well be adapted so as to rebut the argument of the previous paragraph. It has been argued by John Searle that certain rules of procedure are constitutive of certain kinds of institutional actions and that consequently these rules necessarily apply to some situations. Thus, he argues, facts that determine that a person is in an institutional situation of a specified kind may entail the applicability of rules and hence what that person ought to do. If this is valid Strawson's argument might be elaborated along these lines by holding that inductive predicting is an institutional activity. Consequently, we should examine Searle's thesis with care.

Searle presents a putative argument from an initial premiss, which is purely descriptive, via four intermediate stages to a final conclusion which is normative. He argues that the step from each line of the argument to its successor is deductive and that any extra premisses are "in no case moral or evaluative in nature. They consist of empirical assumptions, tautologies and descriptions of word usage". (1970, p. 181). This he holds to be possible because the facts involved are institutional facts and because certain rules are constitutive of institutions like promising.

It is undoubtedly true that there are rules which are constitutive of certain kinds of human activity in the sense that persons could not properly be described as participating in the activity in question unless they conformed to these rules. Games with fixed rules are an obvious example. It is therefore analytically true that if a person is indulging in the activity in question that they will be abiding by the rules of that activity. Let us also concede, although the point may well be contested, that the question whether or not a certain person at a certain time is participating in a certain institutional activity can be settled by purely behavioural facts. Is it then possible to combine the analytical and empirical propositions to derive a normative one? If we allow that a conditional ought-statement is normative then the answer will be that we can in certain cases. Consider a statement of the form "if you would act so as to be properly described as doing x, then you should do y", where doing y is definitionally constitutive of doing x, i.e. a necessary criterion of a person being properly described as doing x is that they are doing y. In this case the foregoing statement is analytically true; and is therefore entailed by any sentence at all.

Let us consider what can be derived from such a sentence. We can know that a person has been doing x. However, to use the preceding analytic proposition to conclude that they ought to do y we need not a description of past activity but a specification of intention. And such a statement cannot follow from a behavioural description. That someone has been involved in a certain enterprise up to a certain instant is quite compatible with their ceasing to be so involved thereafter. People can and do change their minds and it is clearly false to hold that changing the nature of one's enterprise is always irrational.

In the face of the foregoing it may be urged that the only kinds of imperatives we ever need are conditional ones. In practical decisions what we need to know is what we should do if we would achieve such and such a goal. This answer will, of course, not be satisfactory to the ethical objectivist but this is not our concern here. From our point of view we must agree that questions about what it is rational to do are typically relative to some explicit or implicit goal. We choose our goals. We then need a way of deciding what is the rational way of pursuing these goals. Thus it may seem that all we need to derive are conditional imperatives, and we have seen that conditional imperatives can be derived from the analytic connection between institutional activities and the rules which are constitutive of them. However, this line of argument fails to note that in all conditional imperatives of this kind the antecedent specifies an aim to be acting under a certain institutional description. For it is only in such cases that we can conclude analytically that conformity to a certain rule is a necessary condition for the achievement of the goal.

Thus it is clear enough that if we would make predictions according to, say, the Carnapean system then we must, with a force of deductive certainty, proceed in a certain way. However, this will be of no help whatsoever to someone whose sole concern in predicting is to get it right. And this is the case that we are interested in. "Coming up with the correct prediction" is certainly a description of a goal of human activity, and it carries no implication whatsoever about the way the person proceeds in making that prediction. For an action to be properly so described all that is necessary is that the person makes an utterance which is properly interpreted as making a prediction and that this statement is in fact true. Thus, when this is our goal there will be no rules of procedure which are analytically constitutive of the activity of pursuing that goal. No conditional hypothetical of the form "If you would make a correct prediction then you ought to proceed in such and such a way" can ever be derived from rules which are analytically constitutive of any institutional activity.

3. HUME'S THESIS

I have argued that Burks' Pragmatic Thesis is not only of value in disposing of the a priori theory of inductive probability but also in bringing out the error in the ordinary language argument that the use of induction is analytically rational. This latter doctrine especially has been used to support the claim that the problem of justifying induction can be dissolved rather than resolved. If this approach is to be rejected, it would seem natural enough to move on towards providing a justification for the inductive method; and I will argue that this is the direction in which we should proceed. However, Burks holds that no justification for induction is possible or necessary; and on the latter point he comes quite close to the position of the ordinary language philosophers, (pp. 136–137). I will argue that his position concerning the justification of induction is unsatisfactory and that it is particularly incongruous in the light of his Pragmatic Thesis. However, let us first consider the reasons he advances for believing that a justification is impossible; for they are substantial and cannot be lightly put aside.

The basic structure of Burks' argument is as follows. Firstly, there are other systems of inductive probability which we have every reason to believe are logically consistent. He describes two alternative systems. He calls these "random logic" and "inverse logic" respectively. He shows that these systems are logically possible for a simplified model of reality; and he cites good grounds for thinking that this model adequately represents a possible reality. It follows, according to Burks, that we could only choose between induction and these competing systems if we knew which would most probably be successful most often. However to know this we would need to know the nature of that part of the universe of which we as yet have no direct knowledge. The only way that we could form any view about the nature of this part of the universe would be to employ some inductive logic. But to use an inductive logic in the argument intended to establish it as the proper one would be circular and so invalid. Thus, he concludes, there can be no non-circular justification of induction. This he calls Hume's thesis (although he believes the way that David Hume originally formulated his argument was not entirely satisfactory) (p. 134).

This point of view is persuasive – the regress of justification must surely end somewhere. However it is not decisive. Consider deduction. Here it can be argued even more forcefully that a justification in terms of another more basic system is impossible. And it would certainly be circular to use deductive argument to justify deduction. But the schematisation of valid reasoning offered by deductive logic is clearly valuable. Because we can not employ this schematisation, or any other, in its own justification, we can not hope for any once-and-for-all decisive proof that deductive logic is valid. But we will, naturally, employ whatever intellectual tools which are to hand in evaluating this system. In fact we are all convinced that all deductive arguments possess a property which justifies our trust in them – they are truth-preserving. This conviction is supported by various considerations, such as the use of truth-tables and appeals to intuitions of meaning, and these each carry some weight. But in the end the most important point is that no-one has ever produced a deductive argument which does not preserve truth.

The distinctive problem about the justification of induction arises just because we know that inductive arguments do not always preserve truth. But it is a mistake to think that this is the only property which makes a method of inferring reasonable to employ. (For instance it may be possible to show that induction is the only method which lacks a rational defect.) And if we can identify such a property we may employ deduction in proving its presence.

Thus Burks over-estimates the case for Hume's thesis. I will argue that a justification of induction is possible; but first we should consider the relationship between Hume's thesis and the Pragmatic Thesis; for, on the face of it, the latter seems to have a normative force which is incompatible with the former.

4. NORMS AND PRACTICE

Is it possible to interpret the Pragmatic Thesis so as to avoid any normative content? I will argue that it is not. Burks himself seems to equivocate on this point. His contention, challenged by Carnap, that there is a non-cognitive component of a probability statement, which expresses a commitment to act on that prediction, (1963, p. 755) may suggest a non-normative interpretation. However elsewhere he explicitly commits himself to the claim that they are normative. For instance in his early support for Peirce's contention that logic (including inductive logic) is a normative science (1943, pp. 189–192), elsewhere in the article cited above (1963, p. 743) and repeatedly in his book (e.g. pp. 306–652). Indeed in summing up on his Pragmatic Theory he says "probability values are and *should be* assigned according to the rules of the calculus of probability, and standard inductive logic" (1977, p. 319; my emphasis).

I believe that there can be no doubt that the kind of pragmatic content which Burks has identified in inductive probability statements is normative. Phrases which we naturally use in paraphrasing probability assertions, such as "rational degree of belief", on any ordinary interpretation, have normative force. But what is more important is the nature and context of practical decisions. Most of our deliberate actions are directed at goals. We know that they will influence what happens and we are concerned, often profoundly, about what will happen to us. The predictions we make guide our future-directed actions. We must seek to make correct predictions for only then can they aid us in the achievement of our goals. So a question must arise as to whether any prediction is the best one we can make about the outcome of the given situation. The whole point of inductive logic is to provide a system of answering such questions. And the whole enterprise would be pointless if there were no reason to believe that the predictions so generated were any better than those made in any other way.

The paradoxical nature of Burks' position is apparent when we recall that he accepts the following three theses:

I   Making a prediction has practical consequences.
II   We should use induction in making predictions.
III No justification, of any kind, is possible for predicting by induction.

I and II imply that induction is a better way of predicting than alternative methods. But III implies that it is not possible to know that this is so; for any way of establishing that it was would, thereby, be a justification of induction. Even if it is logically possible both that induction is the best method and that it is not possible to know this, no one could consistently assert both. In any case the pragmatic thesis surely implies not only that induction is a norm but that we can know that it is.

Burks holds that induction needs no justification and that this is shown by our way of thinking – "... standard inductive logic needs no justification. Indeed our faith in it is so strong that no argument would cause us to abandon it" (p. 136). If induction were a self-contained activity which everyone did because of a strong inclination so to act then, in lieu of reasons against so acting, there would be no need for any further justification. If such were the case the correct statement would be that the activity was justified by people's inclination and the lack of counter considerations. But the unanimity of opinion would be relevant to the pointlessness of seeking any discursive justification. However, induction is not self-contained; it is directed to a specific end. People do not induce just because of an inclination so to do; they make predictions in the hope of correctly anticipating the future. And it is not possible to avoid asking whether induction is the best way of pursuing this goal. Widespread agreement may suggest that there is an answer in the affirmative but it cannot render the question otiose.

5. THE RATIONALE OF INDUCTIVE PROBABILITY

In his presupposition theory Burks develops the idea that it would be unthinkable to abandon induction. Briefly, he believes that presuppositions, which express a priori concepts, underlie our use of induction and that this explains the universal acceptance of this method of reasoning. There is also the suggestion that the existence and nature of these presuppositions establishes that there is no need for a justification of induction. Let us then examine Burks' presupposition theory and how it relates to his system as a whole.

The system of inductive probability developed by Burks may be seen as consisting in a number of layers. At the base there is deductive logic. Erected on this is probability calculus, which can be supported by an explication of inductive probability. Both these systems can be justified by identifying a property of each system which establishes it as an appropriate means, given the purpose for which it is employed. We have said something about the justification of deductive logic. Burks' treatment of probability calculus is similar to that of the personalist school of Ramsey, De Finetti and Savage (p. 253). He refers to the Dutch Book argument but develops his own more thoroughgoing justification which involves the development of a calculus of choice which aims at capturing the basic rationale of decision-making.

Thus the normative aspect of the conjunction of these two systems is a straightforward matter. However this system only establishes relationships between certain probability statements and provides no basis for determining what Burks calls atomic inductive probability statements, that is statements of the form "P(h,e) = x" (except in the atypical cases where e entails h or not-h). To be in a position to derive such statements it is necessary to add a third layer to the system: namely inductive logic. Given probability calculus, Bayes' theorem can be used to determine the probability of any hypothesis relative to any body of evidence provided the prior probabilities are known (See Appendix A). Thus a system of assigning prior probabilities determines a system of inductive logic. In particular standard inductive logic can be specified in this way.

It is sometimes thought that Bayes' theorem on its own has substantial inductive content. The suggestion is that given only a non-zero probability for hypothesis h, the accumulation of positive instances in the evidence supporting h must lead to a continual increase in the posterior probability of h. And this, it is held, is all that induction amounts to. However this won't do. We must be assured not only that the probability of h continually increases but also that it reaches a sufficiently large value (it would be of little help to know that the probability of all hypotheses increases but only to an upper bound of .01!). An equally important and related point is that if we are to act on an hypothesis, we must be assured that its probability is significantly higher than that of any competing hypothesis.

These minimal requirements for a viable system of induction can only be met by constraints on the assignments of prior probabilities to hypotheses. This is made clear when we consider a set of mutually exclusive hypotheses, all of which deductively explain a body of evidence. If there are n hypotheses and all have the same prior probability then each will have the same posterior probability and the upper bound of this value would be $\frac{1}{n}$ (Appendix B). Any hope that n, that is the number of hypotheses which deductively explain the data, may be small in at least some cases is dispelled by Goodman's "New Riddle of Induction" (1965, pp. 72 et seq.). He in effect shows that there must be an infinity of mutually exclusive hypotheses all of which are compatible with any given set of data. Such hypotheses are generated by assuming that regular patterns which are universal in one space-time region are replaced by different regularities in other regions. Thus all data collected prior to T which are deductively explained by "All emeralds are green" are also deductively explained by "Any emerald at t is green if t is prior to T but blue if t is not prior to T". Since we have a large number of non-green colours to play with and an unlimited number of space-time boundaries over which patterns may mutate, there must be an infinity of hypotheses which all account for the observed colour regularity; and likewise for any other hypothesis.

Thus it is clear that however many data are known to support an hypothesis, its posterior probability remains at the mercy of the prior probabilities assigned to that hypothesis and to its competitors. It is clear that part of what is involved in induction comes down to assigning vanishingly small prior probabilities to Goodman-type, mutant hypotheses.

6. PRESUPPOSITIONS AND PRIOR PROBABILITIES

In using Bayes' theorem Burks urges that we should use any available background knowledge in assigning prior probabilities, for instance that a coin is weighted on one face (p. 529).

Here the distinction between prior and posterior probability rests on the distinction between rather general knowledge and more precise, directly relevant data, for example the frequency of heads in forty throws of the coin. But this use of background information is itself an application of induction. If this inference is itself explicated in terms of Bayes' theorem, it is evident that we must appeal to a more basic prior probability of the hypothesis. This, in turn may still be relative to rather more general information; but we have entered on a regress which can only end with what might be called "pure prior probabilities". Here we are concerned with judgements about hypotheses which are not relative to any specific factual information; but which may reflect the general structure of our conception of the world.

According to Burks, there are presuppositions which lie behind induction itself and which guide and direct us in assigning pure prior probabilities. These presuppositions could be formulated as rules of procedure in the method; but, alternatively they may be expressed as principles which assert general properties of the universe. He adopts this latter option. His three presuppositions are:

A. *The causal uniformity principle*
If a causal connection between nonindexical proper-

ties holds in one region of space-time, it holds throughout space-time. (p. 572).

B. *The causal existence principle*

Some space-time region is governed by quasi-local laws at least for the most part. (p. 577).

C. *The limited variety principle*

The laws of nature are based on a finite set of nonindexical monadic properties each requiring a finite region of space-time for its exemplification. (p. 633).

Burks employs the notion of "causal system", which is a set of possible worlds. A particular causal system, in effect, consists of a "complete" set of basic laws, so the assignment of a probability to a causal system determines a minimum probability for a law it includes. Given the three presuppositions it is possible to show that the number of possible causal systems is finite. So, using the principle of indifference, it is possible to derive finite values for each of these systems (pp. 624–626). Then via Bayes' theorem it will be possible to determine the probability of any given causal law relative to a body of empirical data. (Burks believes that the only adequate version of our concept of causation is deterministic. (pp. 574–580). But this is unnecessary. There is no reason why he should not allow that probabilistic laws, expressing propensities, may be basic. In this way he would avoid an embarrassing confrontation with modern physics.)

The third presupposition clearly plays a crucial role in determining that there are a finite number of possible causal systems; so do the spatio-temporal stipulations of the second and third. The first presupposition plays an essential role in excluding laws which assign different property-patterns to different space-time regions. Allowing such laws as possibilities would rule out standard induction. Since such laws may mutate across any space-time boundary, there would be an infinity of possible causal systems embodying them; hence falsifying the above argument for non-zero prior probabilities. The same point may be put differently. Given this possibility there would be an unlimited number of incompatible laws each entailing any given set of evidence. So the requirement of compatibility with data will not select between these alternatives. What does discriminate between them is the different prior probabilities of these hypotheses; and these, in turn, are determined in Burks' system by his presuppositions.

7. PRESUPPOSITIONS AND NORMS

It is clear that according to Burks the presuppositions play a crucial role in the development of the system of inductive logic. What is the status of these presuppositions? Do they in any sense serve in place of a justification? These are the two questions to which we must now turn.

The presuppositions are said to express a "complex innate and inter-subjective conceptual structure" (p. 632). This innate structure places limits on which structures of the universe we accept as possible. The concept of causal necessity, and presumably the conceptualisation of the universe determined by the presuppositions, is a priori. An a priori concept is one which "can only be obtained by reflecting on one's experience and seeing what is involved in them" (pp. 608–609). On this view the most basic workings of the mind are innate and determine the range of structures of the universe which we recognise as possible. So the way that we integrate our data and project them will be determined by these innate mental structures. Burks draws an analogy between automata (or computers) and humans; suggesting that the latter, like the former, have what may be called an innate structure-program (pp. 610–612). Thus, that we apply certain concepts to the universe is due to experience; but that we employ other more basic concepts is not.

The parallel with Kant is obvious; but there are differences, one of which at least we should note. Kant takes the a priori principle which served as a basis for inductive reasoning to be certainly true (see Burks, p. 645). But Burks, in agreeing with Hume that no non-circular justification of induction is possible, disagrees with Kant (p. 647). He holds that our disposition to think of the universe according to the three presuppositions is innate, and therefore prior to experience; but he acknowledges the possibility that the universe does in fact not have this structure. So it cannot be known a priori that the use of induction will continue to be successful – and this is what he believes would be necessary for a justification.

Burks does not hold that the presuppositions constitute a justification but he does hold that they explain a virtually unanimous agreement concerning the use of induction. Unfortunately, however, this alleged unanimity concerning induction is rather tenuous. Bad inductive reasoning is all too common, and much superstition and prejudice is based on barely any inductive evidence. Now bad induction is not in accord with the principles of induction. It might be argued that consistency precludes a policy of sometimes employing proper induction and sometimes generalizing from biased examples. But there is no literal contradiction here; it would be a simple matter to formulate a single, albeit complex, principle which prescribed any combination of different procedures on different occasions. If there is a procedure which we should always employ then this is presumably because it is the proper method. The "inductive method" is by no means "*the* way people form expectations"; it is the method employed by some people some of the time. Those who consciously use it are convinced that it is the best way of proceeding, try to live up to its standards and urge others to do so.

But even if the inductive method were much more widely employed than it is, it would still be clear that the human mind is not constrained to think in only this way. Indeed Burks himself makes this clear in supporting Hume's thesis by specifying ways of making predictions which differ from standard induction. That such methods are specifiable shows that it would be *possible* for the human intellect to follow them. Thus we cannot hold that induction needs no justification because there is no possibility of doing otherwise. There clearly is such a possibility. The best that can be said for induction is that it comes naturally to us. Indeed, it is by no means an implausible thesis that the central nervous system is so structured that we have an innate *tendency* to form expectations in a way which is *broadly* inductive. But that something comes naturally is no guarantee that it is the best way of proceeding. There are examples enough to hand of innate tendencies which divert us from the path of virtue.

There are alternatives to induction which are possible. If we do not think of employing them, that is because we are convinced that induction is better. If it is truly better we should be able to identify what makes it so. And if we should discover that we were mistaken in thinking it better we would have no reason to try to avoid sloppy induction which comes even more naturally than good inductive thinking.

8. WHAT IS A PRESUPPOSITION?

We still need a justification for induction. I will propose that one is possible by considering the possibility of adopting presuppositions different from Burks' and seeing what would be wrong with such a procedure. But first we must consider exactly what it is that we do when we make presuppositions about the universe.

Let me first explicate one sense of "acting on a presupposition" or "an activity presupposing something" and point out why Burks' presupposition theory of induction gets into trouble if so interpreted. The most natural way to take the phrase "presupposing a proposition" is as synonymous with "accepting that proposition as true as a starting point in some other activity". So in saying that someone acted on presupposition X in doing Y we would be saying that this person in fact took X to be true in so acting. This makes no claim as to whether X is true nor whether the person was correct in thinking that holding this belief was a necessary or sufficient condition for deciding to do Y. However, if we speak of a certain proposition, say X, being presupposed in a certain *kind* of activity, we will not be making any claim about particular persons believing X to be true. It also seems quite implausible to interpret such locutions as making general claims about the beliefs of all people so acting. Rather it seems that in using this expression we make a claim about what would need to be accepted as true for the action to be appropriate or rational. Thus "Doing Y presupposes X" would seem to mean "Any person doing Y acts rationally only if they believe X". (And, of course, one only acts rationally in believing X if one has rational grounds for doing so.)

If "Using induction presupposes the uniformity of nature" is interpreted in the sense under consideration we are surely in trouble. We might hope, on this basis, to account for the normative aspect of induction; but unfortunately we are led into circularity: the very same circularity which vitiates what according to Burks is the only possible approach to the justification of induction. The uniformity of nature is a very general structural property which may be true of the universe; but we know that it is logically possible that this is not so. A claim that the universe has this property could only be based on induction. But it is held that an acceptance of this proposition underpins the use of induction.

There is, however, a different analysis of "acting on a presupposition" according to which it makes much more sense to claim that the use of induction presupposes a general property of the universe. It seems to me that an alternative, and perhaps more natural way of expressing the notion in question is to speak of "acting on the assumption that X"; and I will use this form of words to distinguish the interpretation which I am proposing from any other. Suppose it can be established that if it is in fact the case that X then action Y is rational in a quite direct way. Then we may say of someone knowing this and doing Y that they act on the assumption that X. There are circumstances in which such a person may be far from certain that X is the case yet do Y and act rationally in so doing. Reichenbach has reminded us of the possibility of acting rationally in pursuit of a goal by doing an act which we know may not succeed. He cites a surgeon who operates knowing that he may fail to cure and not even being able to establish the probability of success; but knowing that medical science holds out no hope of any other way of saving the patient (1949, p. 474). In this case we might say that the surgeon acts on the assumption that the operation will produce such and such a result; yet knows that it may not.

For Burks the crucial thing about his three presuppositions is that if they are true then the standard system of induction follows from the probability calculus and nothing more than the principle of indifference. Thus *given* this kind of uniformity of nature the rationality of induction follows straightforwardly. But I would argue that this does not imply that people using induction in fact believe nature to be uniform. (If it were given that the patient's condition is such that the operation is going to succeed then doing the operation is rational in a quite direct way. But when the surgeon has no way of establishing whether this is the case but knows it may be and has no other way of treating the patient, it is rational to operate in any case.) Thus it remains possible that it is rational to act on the assumption of the uniformity of nature even though we know that this may not be the case. If we take the uniformity of nature to be a presupposition of induction *in this sense*, then it is not necessary to have reason to believe that it is so in order to establish that induction is a proper norm. Nor is it necessary to hold that everyone is constrained to act on this assumption. However it remains necessary to provide a justifying argument. We cannot hope to do this by showing that induction will succeed or will probably succeed. But there are other ways of establishing the rationality of a policy; perhaps along the lines of the surgeon's justification of the operation.

9. JUSTIFYING THE PRESUPPOSITIONS

In critically evaluating that policy of predicting which assumes that nature is uniform we will naturally explore alternative policies. We may suppose that nature does not have this property and consider what policy would then be appropriate.

Let us consider the causal uniformity principle, which is, perhaps, the most important of Burks' three presuppositions. Firstly, we must note that this principle is a good deal stronger than necessary for the job it does. The predictions we are concerned with typically deal with the more-or-less immediate future and with spatial regions not too remote from us. Consequently generalisations limited in this way would be adequate for our purposes and they could be based on a presupposition which only asserted that causal connections holding in one space-time region extend into adjacent ones. However it is natural to accept the stronger version, for we place no limit on the possible locations in which we would act on this assumption. And accepting the general principle is equivalent to accepting the conjunction of all possible limited principles.

Let us consider how we should proceed if the causal uniformity principle were false. In this case different causal laws would hold in different parts of space-time. Given only the falsity of the strong version of this principle, it would be possible that the causal laws which apply in our part of the universe only fail in regions utterly remote from us; and this need make no difference to any practical action. So let us consider the consequences of the weaker version being false. Unfortunately even this assumption does not determine any specific perdicting policy. If we were assured of its truth we could conclude that there must be some method of predicting which would give better results than induction. We would know that some causal laws which have applied within recorded human experience will be replaced by other laws at some time in some region concerning which we may be making predictions. But we would not know which laws will change nor where nor when, let alone what laws will replace them.

Is it possible to act in any way on the negation of the causal uniformity principle? Since this negative principle would assure us that things are going to be different in some respects somewhere sometime, we would be warned that any prediction at all might go wrong. So perhaps acting on this would consist in being cautious about our predictions. But what does this amount to? Being cautious about one prediction compared with others results in different behaviour on that occasion. But if we are uniformly cautious about all predictions what will we do differently? Will we refrain from all those actions that might go wrong? But we do not need to negate the uniformity principle to know that any action *might* go wrong and the negated principle gives no help in identifying which predictions may be wrong. Being cautious makes sense in some cases when we have a basis for discriminating in the confidence we assign to different expectations. But if I were to be cautious about everything, would I give up shaving on the grounds that my shaver might explode and blow my head off? Or should I remain clean-shaven on the grounds that beards may develop the property of spontaneous combustion?

It is, of course, possible to formulate assumptions such that if we were willing to act on them would give specific predictions which differ from those we make by induction. If we not only assume that the nature of some causal connections change from some space-time regions to others but also make assumptions about what the unobserved causal connections are and where and when they operate, we would make different specific predictions. The obvious trouble is that once we consider any such assumption we are confronted with an unlimited range of alternatives and have no way of choosing between them. Even if we retain the limited variety principle there will be an infinity of space-time boundaries over which a causal law may mutate (not that it is necessary to establish an infinity of options to make the point).

If, in order to decide on a method of predicting we consider how the world may be, we must first ensure that the possibilities we consider are compatible with what we know of the world. Given this constraint we may first recognise that there is one possible general characterisation of the universe, namely that it is uniform, such that if we act on the assumption that it is true we do induction. In this case the assumption makes no reference to specifics: to space-time locations or particular properties. The specific reference of predictions is determined by our experience, not by the method we employ. This characterisation of the world is certainly one of an infinite set, all compatible with what we know, but it is unique in an important respect. All other characterisations which are such that they lead to particular predictions when we act on them in predicting do make specific reference. And if we consider any one such possible universe we will see that there are a large number of alternatives each making a different specific claim and hence leading to different predictions. Thus if we consider acting on any specific assumption about the universe we see that there are always an unlimited number of options and there is nothing at all to indicate which we should use in predicting.

The method which predicts on the assumption that the world is uniform is justified not because we can know that the universe has this character but because predicting on any other assumption is not a viable procedure. Competing assumptions, at the same level of generality, about the structure of the universe simply fail to give predictions. Assumptions which do give competing predictions are in effect simply unbased assumptions that those very facts are the case. To employ a method based on such an assumption would be simply to make guesses about as-yet-unobserved facts. It would be completely different from induction which requires that predictions are based on the specific facts already known by experience. The point is that there is only one assumption which systematically gives such an objectively-based method.

10. IN CONCLUSION

Burks' presupposition theory of induction is, I have argued, unsatisfactory for a number of reasons. In the first-place it seems paradoxical to claim that induction is based on presuppositions about the universe when induction is the only way that we can know any such generalisation. However this paradox can be resolved. It is not necessary to interpret the theory as meaning that we must know that the universe is as we presuppose it to be in order for induction to be legitimate. We can take the proposition "Induction presupposes the uniformity of nature" to mean "In using induction we act on the assumption that nature is uniform". And, as we have seen, this does not imply that we know, or even are confident, that the universe has this structure.

The more basic problem that I have discussed is that the theory fails to account for the normative character of induction. This defect can be corrected by supplementing the theory with a justification of induction; but a justification of a kind which Burks does not consider and is of the same general kind as that proposed by Reichenbach. I have elaborated the nature of such a justification in more detail elsewhere (1977 and 1982). However, it can be usefully expressed in terms of the presupposition theory.

Whether or not we have an innate tendency to think inductively is neither here nor there as far as a justification goes. And an affirmative claim does not show that a justification is unnecessary. We can formulate alternative ways of predicting and we must consider which is the best way of doing so. The justification is as follows. If we act on the assumption that the universe will be, at least in the near future, generally similar to the way it has been within recorded experience, then we have a way of employing directly known facts in predicting. It is clearly possible that this assumption is true but we certainly do not know that it is. Whether it is rational to act on this assumption depends on what the alternatives are. A contrary assumption of the same level of generality fails to provide a basis for making any predictions. While there are specific contrary assumptions that would provide a basis for predicting, any such assumption would be on no different footing to an unlimited number of alternatives. Thus it would be an arbitrary, and therefore irrational, choice to act on any one of these. In short, there is no rational defect in using induction, but there is in using any of the possible alternatives. Hence it is rational to use induction.

The rationality of an action or policy is relative to what we can know; not to what is in fact the case. We recognise that the world may or may not continue to be as regular as it has been. In the latter case at least some inductive predictions will fail. But there is no way to allow for this possibility; for this general possibility breaks down into an unlimited set of alternatives one of which must be assumed if we are to make predictions. And we have no way of selecting between these alternatives. Thus it is rational to act on the assumption that the limited regularities we know extend further into space-time even though we cannot know that this is so.

APPENDIX

(A.) For any three propositions h, b and e the axioms of probability calculus give directly $\text{Prob}(h/b\&e) = \text{Prob}(h/b).\text{Prob}(e/h\&b)/\text{Prob}(e/b)$.

If we take h as some hypothesis, e as a body of data which is directly relevant to h and b as a statement of all additional, relevant, background knowledge, then this formula expresses Bayes' Theorem. $\text{Prob}(h/b\&e)$ is the probability of h given all relevant data and is called the "posterior probability of h". $\text{Prob}(h/b)$ is the probability of h given only the background information and is called "the prior probability of h". Note that this term is relative to the division of relevant information into evidence and background information; and this choice is arbitrary. If e should be selected so as to include all empirical data then b would become empirically empty and Prob(h/b), if meaningful at all, would either express a purely subjective judgement (as is held by the personalists) or the way h is related to our general conceptual structure (as is held by Burks).

Since Prob(e/b) can be expressed in terms of the prior probabilities of hypotheses (see B below), Bayes' theorem determines the probabilities of hypotheses relative to empirical data provided the prior probabilities of all hypotheses are available.

(B.) If h_1, h_2----, h_n are mutually exclusive hypotheses, all of which deductively explain e (that is, Prob(e/h_i& b)=1 for $1 \leq i \leq n$), and h' is the negation of the conjunction of h_1 to h_n, then

$$e = (e\&h_1)v(e\&h_2)v.......v(e\&h_n)v(e\&h')$$

and Prob(e/b) = Prob(e&h_1/b) + Prob(e&h_2/b)----------
$$-----Prob(e\&h_n/b) + Prob(e\&h'/b)$$

Since Prob(e&h_i/b) = Prob(h_i/b). (Prob e/h_i&b),

And Prob(e/h_i&e) = 1 for $1 \leq i \leq n$,

Prob(e/b) = Prob(h_1/b) + Prob(h_2/b)---------Prob h_n/b + Prob(e&h'/b)

So Prob(e/b) $\geq$ sum of Prior Probabilities of h_1 to h_n.

h_1 to h_n and h' are a mutually exclusive and jointly exhaustive set of statement, so their prior probabilities sum to unity.

So if all of h_1 to h_n have the same prior probability it will be a value u, say, such that $u \leq 1/n$.

The posterior probability of h_i, on the assumption of equal prior probabilities, will be given by:

$$u.1/n.u + d \text{ (where } d = Prob(e\&h'/b))$$

So Prob(h_i/e&b) = $1/(n+d/u) \leq \frac{1}{n}$.

F. John Clendinnen
Melbourne University

NOTES

* All page references to Burks' work will be to this book unless otherwise indicated.

REFERENCES

Burks, Arthur W.: 1943, 'Peirce's conception of logic as a normative science', *Philosophical Review,* **LII**, pp. 187–193.

Burks, Arthur W.: 1963, 'On the significance of Carnap's system of inductive logic for the philosophy of induction', pp. 739–759 in Shillp (1963).

Burks, Arthur W.: 1977, *Chance, Cause, Reason*, University of Chicago Press, Chicago.

Carnap, Rudolf: 1950, *Logical Foundations of Probability*, Routledge and Kegan Paul, London.

Carnap, Rudolf: 1952, *The Continuum of Inductive Methods*, University of Chicago Press, Chicago.

Carnap, Rudolf: 1963, 'Replies and systematic expositions' pp. 859–1013 in Schlipp (1963).

Clendinnen, F. John: 1977, 'Inference, practice and theory' *Synthese,* **34**, pp. 89–132.

Clendinnen, F. John: 1982, 'Rational expectation and simplicity' in McLaughlin (1982) pp. 1–25.

Colodny, Robert G.: 1966, *Mind and Cosmos*, University of Pittsburgh Press, Pittsburgh.

Goodman, Nelson: 1965, *Fact, Fiction and Forecast*, (Bobbs-Merril Co., Indianapolis (First edition 1955).

McLaughlin, R. (ed.): 1982, *What? Where? When? Why?*, D. Reidel, Dordrecht.

Reichenbach, H.: 1949, *The Theory of Probability*, University of California Press, Berkeley.

Salmon, Wesley, C.: 1966, 'The foundation of scientifc inference', pp. 135–275 in Colodny (1966).

Schillp, P. A.: 1963, *The Philosophy of Rudolf Carnap*, Open Court, La Salle.

Searle, John R.: 1970, *Speech Acts*, Cambridge University Press, Cambridge.

Strawson, P. F.: 1952, *Introduction to Logical Theory*, Methuen, London.

ARTHUR BURKS ON THE PRESUPPOSITIONS OF INDUCTION

1. INDUCTIVE STATEMENTS: AN OVERVIEW

A plausible picture of induction holds that (1) repeated conjoint occurrences of two event-types lead us to expect one in the presence of the other, (2) these anticipations are most often successful, and (3) their success is owing to discoverable facets of nature, above all, its 'uniformity'. Arthur Burks develops this line of thought in *Chance, Cause, Reason* (to which otherwise unspecified page numbers refer), but incorporates essential modifications.

First, he notes that the laws of nature to which induction leads us appear to go beyond mere uniformities. Our modally-tinged language suggests that they embody a physical necessity. His central problem is then to show how the methods of induction can establish such causal necessities given that we observe only successions of events. "There must be some difference in the data", he writes (p. 621), "to warrant saying one [generalization] is causally true, the other only contingently true". This poses a direct challenge to the Humian view that there are no necessary connections of matters of fact, or none, at any rate, which observation can disclose.

The traditional picture holds that the probability of event-type B (say, rain) in the presence of A (storm clouds) increases with the number of occasions on which B is found to follow hard upon A. Enumerative induction and the element of repetition figure centrally for Burks, too, but we have, in addition, the powerful tools of the calculus of probability. Bayes' theorem is the big weapon in this arsenal, and Burks duly emphasizes its relevance to the process of verifying a hypothesis by attesting its consequences. Yet, he insists (p. 89) that Bayesian conditioning does not tell the whole story. He thinks it incapable of accounting for the additional confirmatory force of varying the instances, and he also worries about the status of the initial prob-

Merrilee H. Salmon (ed.), *The Philosophy of Logical Mechanism*, 233–250,

abilities that must be plugged into Bayes' formula to obtain inductive probability statements of the form, $P(H/E) = p$, as outputs (i.e., claims to the effect that one statement renders another probable to a specified degree, which Burks calls "atomic inductive probability statements"). One way around the difficulty is to declare inductive probabilities measures of purely personal opinion, but Burks thinks the personalist theory too thin, for opinions based on evidence seem to him patently more reasonable than those which ignore that evidence (p. 327).

At the same time, he rejects the "logical" theory, though for more subtle reasons. Consider a Carnapian "model universe" described in terms of a finite number n of individual constants and a single predicate P. As Carnap noticed, the recommended assignment of equal probabilities to all 2^n possible states of this universe has the disquieting consequence that the initial probability of $1/2$ that an unsampled individual is a P is not altered by a homogeneous sample of P's, however large. Burks thinks it unreasonable that this predictive probability should not increase, and, in consequence, he thinks that probability calculus needs to be supplemented by the principle of enumerative induction. As he notes, this principle is tantamount to a directive that enjoins the assignment of higher probabilities to more homogeneous state descriptions (those in which the considered individuals are mostly alike in the properties they have or lack).

But Burks is also aware of the many, real or apparent, counterinstances of the principle of induction. His conception is one of a hierarchy of inductive rules and inferences at every level of which enumerative induction figures importantly. Thus, the methods of eliminative induction pre-require enumerative induction to determine what properties are likely to be relevant or causally interconnected and for the relative probabilities of uneliminated hypotheses (Burks, 1980, p. 184). Hence, enumerative induction occurs at a more fundamental level of the hierarchy than eliminative induction. But, in addition, we must recognize that enumerative inductions at lower levels are mediated or modulated by inductions at higher levels, and it is this fact that explains away the apparent counterinstances. The more often a metal wire is bent without breaking, the *less* likely it is to resist breaking if bent another time. Or, again, long runs of heads with a balanced coin do not increase the probability of heads on subsequent throws. The relatively sophisticated concepts of fatigue and randomness occur,

Burks says, "several levels up", and "at this level we have learned by enumerative induction that metal wire is subject to fatigue and that certain elaborate processes generate random sequences" (1980, p. 184).

Up to now, we may regard Burks as merely filling out the first two components of the traditional picture sketched at the outset. But he deviates from the third component, the empirical discernibility of the very features of the world that account for our inductive success. He argues that the Humian circle is inescapable: experience can tell us, at most, what methods have worked in the past, but it requires another inductive inference to conclude that those methods will continue to work. While the amalgam of probability calculus and enumerative induction Burks labels "standard inductive logic" is, he thinks, both reasonable and true, we cannot convince ourselves of this, without circularity, on empirical grounds. Rather, we must arrive at the concepts and principles of induction *a priori*, by reflecting on our own mental functioning and on what is involved in our ever more sophisticated constructions of reality. His approach is Kantian in spirit: ask not what inductive methods will work best in the world as we experience it; ask, rather, what properties that world of experience must have to account for the conspicuous successes of standard inductive logic, the refinement of commonsense reasoning employed throughout the sciences.

Since one can consistently assign state descriptions probabilities so as to favor inhomogeneous ones, standard inductive logic is not logically true. And, we have seen, there can be no empirical court of appeal, either for standard inductive logic or its putative presuppositions. Both have, therefore, a *synthetic a priori* status (Burks labels them "inductive statements"), and, as such, there can be no question of justifying or correcting them. As he bluntly puts it (p. 162): "an inductive logic constitutes a standard of evidence and of probable predictive success, and so a person who uses it can never find empirical grounds for modifying or abandoning it". What emerges, then, is a picture of closed (delusional?) systems passing each other like ships in the night. And it is in a rather Pickwickian sense that standard inductive logic turns out to be "reasonable".

As even this brief overview will suggest, Burks has woven a rich and complex tapestry. I have singled out only those threads on which I would like to focus critical attention, namely, the modal character of nature's laws, the central place of enumerative induction in probable reasoning, and, finally, the synthetic *a priori* status of both inductive logic and its presuppositions, and their inaccessibility to empirical reform.

2. THE DUAL ROLE OF FINITE AUTOMATA

Burks' presupposition theory makes significant use of cellular automata. These are discrete space-time systems. Picture an infinitely extended grid (in two dimensions, an infinite checkerboard) whose individual cells occupy different states as the system progresses through discrete time. The transition law is assumed to be spatio-temporally *contiguous* in the sense that the state of a cell at the next instant of time depends only on current states of neighboring cells (which may or may not be taken to include the given cell itself). A *finite-state* (or, briefly, *finite*) automaton is one in which the states a cell can enter are all drawn from a single finite set. If every cell is governed by the *same* transition law, the automaton is *homogeneous*. A *probabilistic transition law* maps the current state of a neighborhood (i.e., the states of neighboring cells) to a probability distribution over the possible next states of the given cell, while a *causal transition law* maps the current states of neighboring cells to a subset of the states the given cell can enter at the next step. When that subset is single-membered, the causal law is *deterministic*. For a homogeneous deterministic automaton, the entire future history is uniquely determined by the current states of its cells. It should be clear that almost any conceivable natural system can be realistically modelled by a finite cellular automaton. But Burks, we will see, makes far stronger claims.

Man is also conceived as a finite automaton. In his great pioneering work on self-replicating automata (von Neumann, 1966, edited and completed by Burks), von Neumann showed how to embed a Turing machine in a finite automaton of 29 states (the number has since been greatly reduced), and thereby to arrive at a machine capable of constructing any automaton if supplied with a blueprint, and, in particular, of self-replication when supplied with instructions for constructing itself. In principle, then, a finite homogeneous deterministic automaton, like von Neumann's, is capable of carrying out any effective routine. And if we think of intelligent behavior in terms of hierarchies of routines or adaptable plans, we arrive at the "man=machine thesis" (Burks, 1979).

In suggesting that deterministic automata can duplicate human intelligence, Burks argues (sec. 9.4) that the kinds of randomness such automata can produce (e.g., "pseudo randomness") are a sufficient source of chance to account for human behaviors and adaptations. And, of course, every bit of non-trivially intelligent or adaptive behavior deterministic computers or perceptrons are shown capable of exhibiting adds to the plausibility of the man=machine thesis.

It is not surprising in this connection that Burks makes frequent reference to the checker-playing program developed by Arthur Samuel (1959, 1967). Given the "combinatorial explosion" of moves and possible countermoves, even a large modern computer cannot hope to explore more than three or four moves ahead and must therefore prune the decision tree. Samuel's program evaluates board positions by using a weighted function of such parameters as king advantage (K), piece advantage (P), mobility advantage (M), and so forth. Thus, $5K + 3P + M$ would be a very simple evaluation function of this compound type. The really interesting feature of the program, however, is that it learns from experience by adjusting the weights of its evaluation function. One obvious way to do this is to compare the performance of different evaluation functions over successive games, but that method of learning is not terribly efficient. Samuel hit upon a continuous short-term alternative that compares the direct evaluation of the board position resulting from a given move with the evaluation of that move obtained by looking several moves ahead (and assuming that the opponent always selects what the evaluator judges to be the optimum countermove). The sign of the difference between these two "direct" and "extensive" evaluations is then used to re-adjust the weights.

Apart from exhibiting adaptation of a non-trivial kind, Samuel's program nicely illustrates Burks' hierarchical conception of induction. The example also serves to remind us that a computer's storage and computational facilities are finite, and any learning strategy or adaptive plan must work within these limitations, using a specified performance measure (like an evaluation function in checkers) as feedback. (The terminology "adaptive plan", "performance measure", etc., is borrowed from Holland, 1975, which offers a general mathematical formulation of the adaptation problem with many interesting theorems and biological illustrations.)

Given a fairly well-defined and relatively simple learning task and the idiosyncracies of a given learning machine or digital computer, one might even hope to show that a given adaptive plan (which that learning machine can carry out) is optimal. A theorem of that sort would solve the problem of induction for the given learner and problem situation. But in attempting to apply such an approach to human learners, we run up against the difficulty that the idiosyncracies and limitations of human automata are not given at the outset, nor is the exact nature of the learning task – save in rather contrived textbook cases. To find these things out, the presupposition theorist maintains, requires inductive learning of a rather sophisticated kind. Consequently, it would be viciously circular to hold that a given "adaptive plan", like standard inductive logic, is optimal, given the kind of automata we are and the kind of natural automata we are called upon to solve, for that inductive logic is presupposed in our conceptions of world and self. The only fruitful tack is to invert the problem and ask: given that standard inductive logic leads us to the establishment of causal necessities, what must be true of human automata and the natural automata whose transition laws they unravel? And it should be possible, ideally, to answer this question by a sort of transcendental deduction that has no recourse to alleged empirical facts.

Burks begins by assuming that induction presupposes a repeatable element and a space-time framework in which repetition occurs. Non-indexical properties are plausible candidates for the former and cellular automata for the latter. If we accept the thesis that humans are *finite* automata, then it follows (p. 634) that they can distinguish only a finite number of different qualities in any perceptually minimal space-time cell, and so the cellular automata man is equipped to solve are necessarily finite. And we may as well assume that all naturally occurring automata draw the states we can experience from a single vast but finite set of non-indexical properties. This "principle of limited variety" is the first of the Burksian presuppositions.

In addition, he argues that nature must be reasonably uniform and simple for the human mind to unravel its laws. More precisely, at least some accessible region of space-time must be governed by contiguous causal laws, and the causal systems that present themselves must be homogeneous. Or, more crudely put, we must assume that at least some of the locally manifested causal laws govern spatio-temporally remote regions of the universe.

To summarize, the three presuppositions are:
1.   the finiteness of the quality differences occurring in any

perceptually minimal region of space-time (the "limited variety
principle");
2. the existence of contiguous causal laws in some accessible region
    of space-time (the "causal existence principle");
3. the uniformity of causes, or the causal irrelevance of mere location
    in space-time (the "causal uniformity principle").

One further caveat is called for. I said that if we assume that man is
a *finite* automaton, the limited variety principle follows. But, surely,
that assumption is empriical? Not really, for, as we will see shortly, the
possibility of verifying causal laws presupposes the possibility of assigning equal (or, at any rate, non-zero) prior probabilities to the different
possible causal systems, and dividing these among the different causal
histories they admit. This, in turn, requires that the number of possible
causal systems be finite. And that constrains the number of possible
states to be finite, and also entails that the transition law be contiguous
and homogeneous, or very nearly so, to keep the number of possibilities finite. Burks is willing to weaken these assumptions somewhat,
but his point, I take it, is that there must be some upper bound on the
complexity of the automata we are called upon to solve if the empirical
confirmation process is ever to get rolling.

3. CERTIFYING CAUSAL NECESSITIES

We come now to the central problem of *Chance, Cause, Reason*: to
show how partial histories of the world can lead, via Bayes' theorem,
to the certification of causal laws. These, you will recall, are conceived
as irreducibly modal, "since they hold for sets of possible boundary
conditions most of which are nonactual" (p. 653). Burks develops a
formal logic of physical necessity (interpreted as truth in every physically possible world) and argues (p. 457) that the convincing way in
which this modal logic models ordinary and scientific discourse is
reason to think that "cause-effect relations are necessary".

To illustrate his conception, consider a two-state automaton in two
dimensions, and call the two states "empty" and "occupied". Neighbors
of a cell are taken to be the four adjacent squares, and the transition
rule is the following (contiguous and homogeneous) *parity rule* (due to
M. Fredkin): *each cell with an even number of occupied neighbors at
time t becomes or remains empty at t + 1; each cell with an odd number*

of occupied neighbors at t is occupied at t + 1. This automaton is 'self-replicating' in the sense that any finite configuration of occupied cells will replicate itself four times in two steps (or in one step, if no two adjacent squares are initially occupied). Moreover, if all and only cells in a given (infinite) row are occupied at some time, no configuration in which a cell has three or more occupied neighbors will subsequently arise. But configurations of the latter type are causally possible in the sense that they would arise given an appropriate initial configuration. And so the actual history of the automaton, which is uniquely determined by its initial configuration, is only one among the many causally possible histories its transition law admits, and there will be many true "*de facto* – generalizations" which are false in causally possible histories other than its actual history.

Given the Burksian presuppositions, there are only finitely many possible causal systems to consider. He advocates assigning them equal probabilities and then dividing their probabilities equally among the different histories they admit. The rub, of course, is that if discrete time is assumed infinite, each causal transition law will admit infinitely many histories, and not all of these can be assigned equal probability. To overcome this difficulty, Burks restricts the competing causal automata to a large but finite region R of space-time. So restricted, each causal law then admits just finitely many histories. Finally, because the causal transition laws are assumed homogeneous and time infinite in both directions, the result of shifting any causally possible history of an automaton in space and time must be itself causally possible (p. 629). Hence, any transition in violation of one of the competing causal laws which occurs outside R will occur inside R in an alternative possible history of that causal system, with the result that if R is chosen sufficiently large – and Burks recommends taking it large enough to compass any conceivable human interest – violations of a false causal law are almost certain to be detected in R. (But that does not mean, of course, that they will be detected in our lifetime or in our galaxy!)

Now, at least as I have presented it, the argument has a slight flaw. The different causal transition laws are not mutually exclusive. A deterministic law will be compatible with various causal laws which map some state of $N(c)$ to a proper superset of the unit set to which the given deterministic law maps it. But nothing I have to say really turns on this point, and so let us imagine that we have to deal exclusively with deterministic laws. For these are mutually exclusive.

What Burks offers, then, is an essentially eliminitive procedure: we observe (or, perhaps, bring about) all possible states of the neighborhood N(c) of some cell c, and then record which state of c ensues. At the end, we will be left with a single deterministic law compatible with all the observed transitions.

This procedure has, however, two weaknesses, one practical and one theoretical. The practical difficulty concerns the number of possible states. Even if we grant that the qualities that can be detected in a perceptually minimal cell are finite in number, the set of these properties remains an enormously large and rather amorphous collection. I don't see how one could ever hope to list all of its members. And, besides, as Burks is well aware, we only deal with a few properties at a time when searching for causal relationships. We deal with very tiny partial subsystems of the grand *"systeme du monde"*. But part of the problem, as we have seen Burks himself emphasize, is to decide which properties are likely to be causally relevant, and this determination presupposes a higher order enumerative induction. Hence, we are never really in a position to know that the set of alternative causal systems under consideration is exhaustive of the possibilities. Indeed, when natural laws are viewed as causal necessities, the situation is much worse. Various properties may be ruled irrelevant to a causal relationship by a standard statistical analysis, but no such analysis can ever assure us that the excluded properties have no influence on that relationship in some causally possible alternative to the actual universe.

And this is, in essence, the second difficulty. What if certain logically possible states of N(c) fail to occur, even after our own persistent efforts to bring them about? How can we determine whether those states of N(c) are causally impossible or merely absent in the actual history of the system? The difficulty is pervasive and fundamental and goes to the heart of the claim that empirical data can distinguish causal necessities from casual uniformities.

To see the difficulty more clearly, we turn to Burks' own formulation of the certification process. Consider the causal laws:

$$\text{(L)} \quad (c)(t)[\emptyset \, (N(c), t) \text{ npc } \psi(c, t')]$$

which says that the presence of property $\emptyset$ at neighborhood N(c) of cell c at time t non-paradoxically causes (npc) the appearance of property ψ in c at the next instant, $t' = t + 1$. Causal implication means causal necessity of the corresponding material conditional:

$$(U)\,(c)(t)[\emptyset(N(c),\ t) \rightarrow \psi(c,\ t')]$$

and *non-paradoxical causation* requires, in addition, the logical independence of antecedent and consequent, the causal possibility of the
antecedent, and causal non-necessity of the consequent. Using a
Bayesian (or positive relevance) criterion, (U) is confirmed by ob-
served transitions from $\emptyset$ at $N(c)$ and time t to ψ in c at t'. And since
(U) is a consequence of (L), these transitions will also confirm (L). At
any rate, this will be true if we think of (L) as belonging to a partition
of alternative causal laws. The question Humians will raise[1], however,
is whether confirmation of (L) adds anything to confirmation of the
corresponding contingent universal (U)? What they will especially fail
to see is how the observed transitions could ever discriminate (L) from
(U). I purposely framed that parity transition rule informally to under-
score the immateriality of the distinction. The transition law is the
same whether we express it as a conjunction of laws of type (L) or a
conjunction of laws of type (U). Barring cases where antecedent and
consequent are logically dependent (and it is not clear how they could
be), both versions of the law would admit and exclude precisely the
same transitions.

If sound, this criticism sharpens the dilemma, the other horn of
which is the apparently essential role of modal concepts in science. My
own inclination is to think that there are several ways, acceptable to
Humians, of giving a good sense to the kind of distinction Burks wishes
to draw in modal terms. The distinction between what *must* happen
and what merely *does* (but need not) happen is often just the distinction between a model which fits only what is actually observed and one
whose additional parameters (or "extra wheels") allow it to accomodate a good deal else besides. In an earth-centered astronomy,
for example, it is mere happenstance that the frequency of retrogression decreases in the sequence, Saturn, Jupiter, Mars, for the
periods of epicycle and deferrent are independently adjustable. In
other words, any order in the frequencies of retrogression could be ac-
comodated by suitable parameter adjustment. Not so, however, in a
sun-centered model, for the number of retrogressions is equal to the
number of times the earth overtakes the planet in one circuit of the
sun, and, consequently, a planet whose period is longer *must* retrogress
a greater number of times. We can, of course, evidentially discriminate
between two theories which differ in this way, but the difference is in

the theories themselves, not in the data. A Bayesian analysis shows that theories which fit less are more strongly confirmed when the data fall into the region of those which that theory can accomodate.[2]

4. STATUS OF THE PRESUPPOSITIONS

It does seem viciously circular to claim inductive support for the presuppositions of induction. At the same time, the three presuppositions are clearly not logically true. It is not logically inescapable to suppose that the world is a *homogeneous* finite automaton. What does seem inescapable, therefore, is the conclusion that the presuppositions of induction are statements of a third, synthetic *a priori*, kind.

Yet, in persuading us that the presuppositions are not logically true, Burks almost succeeds too well. He shows, not only that nature could violate the causal uniformity principle, but that we could discover this. In admitting (p. 635) that this principle is overly restrictive, he says that "the laws of nature could vary slowly over space and time, and these variations might be discovered by inductive inference". And this admission raises a suspicion that causal uniformity is an empirical claim afterall. In considering the possibility that Newton's inverse square law alternates with an inverse cube law over great distances, he writes (p. 638):

Note that if causal connections reversed periodically, and one did not know the length of the period, the correct rule of induction would be somewhat like inverse inductive logic. One should reason: the more often this "effect" has followed this "cause", the less likely is it to do so next time. In all these cases one could correctly infer counterfactual subjunctives from these hypothetical causal laws and theories even though they violate the causal uniformity principle.

Again, such reversals could be discovered. Burks does insist, nevertheless, that "if the variation were too great, man would never be able to start the empirical confirmation process" (p. 635). Even this dilute claim may be doubted, however, and one is left wondering just how much the presupposition of causal uniformity can be weakened and still serve to distinguish standard inductive logic from "inverse inductive logic".

In any event, one may query the status of the presuppositional claim itself, the claim that, broadly speaking, nature must be reasonably uniform and simple for man to have developed the sciences. Is this claim an empirical one? It *sounds* very like similar claims put forward by Fred Hoyle[3] to the effect that man's first tentative steps towards the explanation and prediction of natural phenomena would never have been taken if the earth were a cloud-bound planet like Venus. He speculates that our development of geometry and of a sense of time is owing to the fact that, by peering into space, we were able to develop the concepts of east, west, north and south, and so orient ourselves on the earth's surface, while the periodic motions of sun, moon, and planets enabled us to develop a sense of time. We can imagine facts that would countermand these claims, such as the existence of creatures of a cloud-bound planet with a fully developed geometry and sense of time. Burks' claim seems to be disconfirmable too, say, by the existence of finite automata no more complex than the human brain but capable of solving non-homogeneous automata, or, for that matter, by our own discovery of far more complicated natural laws involving action-at-a-distance and the like. (That Newton's gravitation law has been supplanted by Einstein's contiguous law is hardly an argument that we cannot unravel non-contiguous laws.)

Perhaps what Burks hankers after is a theorem which places an upper bound on the complexity of finite automata which a finite automaton of given complexity could solve. And this is not so very far-fetched. A finite automaton is essentially a formal system, and as Chaitin (1975) observes, the information content of the axioms and rules of a formal system can be measured and designated the *complexity* of that system. And in his 1974 paper, Chaitin remarks:

There are situations in which it is possible to reason as follows: if a set of theorems contains t bits of information, and a set of axioms contains less than t bits of information, then it is impossible to deduce these theorems from these axioms.

Consequently, I do not rule out theorems of the envisaged sort. (An interesting question here would be whether any finite automaton could solve automata as complex as itself.)

Yet, even if such theorems were at hand, the complexity of the human brain, qua finite automaton, would be a matter for empirical investigation to disclose. There can be no question, then, of determining an upper bound on the complexity of the finite automata human learners can solve in *a priori* fashion. But an empirical determination of our own complexity would, on Burks' own showing, involve the use of an inductive logic, and would thus be circular. "Since the causal uniformity, causal existence, and limited variety principles are presupposed by standard inductive logic", he writes (p. 638), "the use of this logic to verify them would be circular". And what of the man = machine thesis itself? What are the grounds for assuming that man is, in fact, a finite deterministic automaton? Perhaps the moral to be drawn here is that if induction has presuppositions, we can never known them.

5. INFIRMITIES OF STANDARD INDUCTIVE LOGIC

A demerit of the presupposition theory is that by elevating the various canons of induction to standards of criticism and rationality themselves immune to criticism, it bars the way to inquiry. Even if one takes higher order inductions into account, the principle of enumerative induction is far from self-evident. Attempts by Hempel and others to clarify the notion of instancehood on which this principle turns have only succeeded in plunging it into deeper obscurity. And I would emphasize that the difficulties about instancehood are in nowise diminished by excluding indexical predicates, as Burks proposes. For Hempel's satisfaction criterion (the only widely countenanced elucidation of 'positive instance' in the field) leads straightaway to the "anti-inductive" inferences proponents of enumerative induction are most at pains to exorcise. For green emeralds examined before time t are already Hempelian positive instances of "All emeralds not examined before time t are blue". And so if we understand the principle of induction in Hempel's way, finding emeralds of a given color strengthens the conclusion that unsampled emeralds are of a different color! And, notice, we have reached this counterinductive conclusion without introducing "grue" or any other indexical predicate.

Even if one can squirm out of the gruesome counterexample by appealing to higher order inductions, I doubt if such appeal can help with the following example. A slightly tipsy hatcheck person returns N hats to their owners in random order. Let H be the hypothesis that no man receives his own hat. As you can see most dramatically in the case N=3, the information that two of the men received each other's hat *lowers* the probability of H for all odd N, even though it is again a Hempelian positive instance of H. The difficulty here seems to be that when conforming cases of a law are specified in greater detail, as when a case of neither of two men receiving his own hat is further specified as a case of each receiving the other's hat, relevant conditional probabilities are changed in ways that can turn confirmation into disconfirmation. We cannot forthwith conclude, therefore, that finding FG's confirms (i.e., raises the probability of) "All F are G". What can be concluded is that finding of a known F that it is G confirms "All F are G". For by Bayes' theorem alone, it follows that hypotheses are confirmed by their consequences. The upshot of this discussion, which I have presented elsewhere at greater length (Rosenkrantz, 1981), is that a Bayesian analysis averts the paradoxes of confirmation to which Hempel's analysis is prey. Presumably, we are willing to part with paradox. What, then, of Burks' contention that a purely Bayesian analysis must be supplemented by the very principle which generates those paradoxes?

As I see it, the sound kernel of enumerative induction is this: lacking pertinent prior information, our estimate of a population proportion should be close to the observed sample proportion. Such estimates follow from Bayes' theorem if we input a diffuse or "uninformative" prior, viz., one that is easily swamped by the anticipated data. And the resulting objectivist Bayesian position is not all that far removed from Burks' approach (cf. pp. 326–327), as in his recommended procedure for certifying causal laws. But Burks, we also saw, lends his assent to the familiar Carnapian objection that the "uninformative" assignment of equal probabilities to the state descriptions of a population about which we have no information precludes learning from experience. The probability that an unsampled population element is a Q given a sample free of non-Q's will remain at its initial value, however large that sample. Burks labels the rule for assigning predictive probabilities on the basis of a uniform distribution of state descriptions "random inductive logic". Distributions which favor inhomogeneous state descriptions lead to "inverse inductive logic", which makes the probability of a Q decrease with the number of Q's already sampled. His position is that standard inductive logic gives reasonable results (though we cannot support this claim, either logically or empirically), while inverse and random inductive logic do not.

I am unconvinced. If we know nothing more about a population of n items than that each member of it is either a Q or a non-Q, the assignment of equal probabilities to its 2^N state descriptions seems apt. To realize standard inductive logic, the prior distribution must favor homogeneous state descriptions. Yet, Burks is willing (even anxious) to allow that higher order inductions may, on occasion, lead us to believe that our considered population is a heterogeneous one, and in that case, the indicated asymmetric distribution of probability over state descriptions will lead to inverse inductive logic. To my mind, the conclusion seems inescapable that where the background knowledge favors neither homogeneity nor heterogeneity of the population, our prior distribution should favor neither type. And a neutral distribution of this sort will, if it has other obvious symmetries, lead to random inductive logic. By insisting that predictive probabilities always increase in accordance with standard inductive logic, Burks pre-empts the mediating role of higher order inductions, as when we notice that an urn was drawn from a shelf full of mostly color-heterogeneous urns. His insistence on the need to supplement probability calculus with a principle of enumerative induction is therefore at odds with his own hierarchical conception, which allows enumerative inductions at a low level to be overridden by inductions at a higher level. In practice, of course, background knowledge will often license the predictive inferences of standard inductive logic. This is so if only because properties are usually projected over 'natural kinds', like chemical elements, metals, or biological species. And to qualify a class as a natural kind is already to suggest its relative homogeneity for a relevant class of properties.

6. CONCLUSION: TAKING 'INDUCTIVE LOGIC' SERIOUSLY

Two critical turns led Burks to his presupposition theory: (1) the belief that the principles of "standard inductive logic" are not logically true, and (2) the conviction that no empirical grounds can be given for preferring one inductive logic to another. In concluding this essay, I want to at least point out the existence of alternatives to these two tenets.

The evident consistency of random and inverse inductive logics led Burks to conclude that standard inductive logic is not logically true. But this argument does not apply to probability calculus *per se*, nor even to the principle that enjoins us to adopt a diffuse prior when pertinent data are lacking. The arguments of the preceding section suggest it is only the latter principles we should be concerned to justify. Indeed, from the perspective of Burks' hierarchical conception, we can see that the reasonableness (in appropriate contexts) of "inverse" or "random" inferences is not at all incompatible with holding (with Burks, p. 327) that extreme opinions based on little or no data are, *pace* de Finetti, quite unreasonable.

Moreover, the relevant principles of inductive probability may all be viewed as requirements of consistency in a relatively straightforward sense. One who believes both that E is more probable than not and that it is less probable than not has embraced a rather palpable inconsistency. And that remains true of anyone who assigns the members of a partition of propositions probabilities that do not add up to the probability assigned to their disjunction, the "certain" proposition of the algebra. Bayesian conditioning may be viewed in the same light.[4] Finally, the writings of E. T. Jaynes (e.g., Jaynes, 1968) have made a persuasive case for viewing uninformative priors in a similar light. And while I am firmly convinced that Jaynes' group theoretic reconstruction of the classical principle of indifference resolves all of the difficulties associated with the latter, this is not the place to defend that thesis in detail.[5]

Here, too, we are not all that far removed from views expressed in *Chance, Cause, Reason*. Burks regards the Dutch book theorem (p. 215) as establishing the equivalence of two kinds of consistency: "inductive consistency" (the consistency of one's partial beliefs) and "gambling consistency" (the invulnerability of one's betting odds to a Dutch book). The Dutch book theorem is a piece of pure mathematics; yet, it has a certain pragmatic force. But its force is diluted somewhat by the looseness of the conceptual ties that bind beliefs to actions and beliefs to betting odds. Given one's knowledge of prospective bettors, it may be entirely reasonable to offer incoherent odds, even if one's degrees of belief are coherent. For this reason, unlike Burks, I lay greater stress on the agent's purely cognitive goals, above all, the goal of getting closer and closer to the truth. And that brings me to the second critical turn mentioned above: the conviction that inductive methods cannot be adjudicated on empirical grounds.

Burks' reasons for holding this are Humian. He argues that different inductive logics will license different conclusions about the probable future success of the competing methods based on their past efficiency (pp. 130–131), and that their past efficiency is itself an inductive reconstruction of our records of the past (p. 132).

But there is an escape from the Humian circle if we look at the expected truthlikeness of a given method's outputs. Using a least squares measure of closeness to the truth, de Finetti (1972, ch. 1) is able to show that any incoherent assignment, $x = (x_1, \ldots, x_n)$, of degrees of prediction x_i to the elements e_i of a partition is *dominated* by a coherent assignment, $y = (y_1, \ldots, y_n)$, in the sense that y incurs a smaller penalty (is closer to the truth) than x whichever element e_i of the partition obtains. And one might hope to extend his result to any reasonable penalty function. (One property of a 'reasonable' penalty function is that it provides no incentive for the agent to assign degrees of prediction different from his actual degrees of belief, or, more precisely, the expected penalty should be minimized when the relative degrees of prediction are equal to the relative degrees of belief. I call penalty functions with this property *non-distorting*, and have shown that de Finetti's least squares function has this property.) Similarly, it can be asked how close to the truth you can expect to find yourself after sampling, using various different rules of conditioning. I have shown that Bayesian conditioning is optimal in this sense for all reasonable penalty functions, but the further elaboration of this line of thought lies beyond the scope of this paper. The point is that we can prove cognitive analogues of the Dutch book theorem which have, in a sense, even greater pragmatic force, given the closer conceptual ties that bind forecasts or degrees of prediction and degrees of belief, provided that non-distorting penalty functions are employed.

In contrast to the presupposition theory, I am proposing that the principles of inductive probability be viewed as constituting a proper logic of partial belief or inductive consistency. And I am suggesting, moreover, that these principles are optimally efficient means to the prosecution of a purely cognitive end.

Professor R. D. Rosenkrantz
Rt 1, Box 1070
Bradford, VT 05033
U.S.A.

NOTES

[1] Brian Skyrms raises essentially this point in his review of *Chance, Cause, Reason*
[2] Cf. Rosenkrantz (1977), chs. 5, 7.
[3] Hoyle (1962), pp. 10 ff.. HIs thesis, in a little more detail, is that science owes its beginnings to a number of lucky circumstances: (1) the earth is not cloud-bound, (2) space is locally Euclidean, (3) light is nearly corpuscular, and (4) the stars are nearly fixed. These fortunate approximations are paralleled in Burks' willingness to grant (p. 635) that the causal existence and uniformity principles could be slightly relaxed. It is

enough that nature be nearly homogeneous, that its laws be nearly contiguous, and so forth.

4. To illustrate, let your prior distribution for the experiment of rolling dice be uniform, where one die is red and the other white. By marginalization, your distribution for the white die (as well as for the red) is also uniform. Hence, the outcomes of the two dice are (for you) independent: P(white=j, red=k) = P(white=j) P(red=k) for all i, j. If, however, you see any alternative to Bayesian conditioning, P(white=j|red=k) ≠ P(white=j) for some pair (j, k). And, in that case, you believe both that the dice are independent and that they are not independent.

5. Jaynes' consistency requirement is that we assign a given proposition the same probability in two equivalent formulations of the same problem. A precise criterion of equivalence leads, in any application, to a group of admissible transformations, and a prior which correctly represents the prior information must be invariant under the admissible group. For a trivial example, suppose we are given a bare list of alternatives. Then the problem is unchanged by any re-labelling of the alternatives. Hence, the admissible group is the full symmetric group and the invariant prior is clearly uniform.

REFERENCES

Asquith, P. and H. Kyburg, (eds.): 1979, *Current Research in Philosophy of Science*, Philosophy of Science Association, East Lansing, Michigan.

Burks, A. W.: 1977, *Chanch, Cause, Reason*, University of Chicago Press, Chicago.

Burks, A. W.: 1979, 'Computer science and philosophy', in Asquith and Kyburg (eds.) (1979), pp. 399–417.

Burks, A. W.: 1980, 'Enumerative induction versus eliminative induction', in Cohen and Hesse (1980), pp. 172–189.

Chaitin, G. J.: 1974, 'Information-theoretic limitations of formal systems', *Journal of the Association for Computing Machinery* **21**, pp. 403–424.

Chaitin, G. J.: 1975, 'Randomness and mathematical proof', *Scientific American* **232** (May, 1975), pp. 47–52.

Cohen, L. J. and M. Hesse (eds): 1980, *Applications of Inductive Logic*, Oxford University Press, Oxford.

DeFinetti, B.: 1972, *Probability, Induction, Statistics*, John Wiley & Sons, New York.

Holland. J.: 1975, *Adaptation in Natural and Artificial Systems*, University of Michigan Press, Ann Arbor.

Hoyle, Fred: 1962, *Astronomy*, Rathbone Books, London.

Jaynes, E. T.: 1968, 'Prior probabilities', *IEEE Trans. Systems Sci. & Cybernetics*, SSC-4, pp. 227–241.

Rosenkrantz, R. D.: 1977, *Inference, Method and Decision*, D. Reidel, Dordrecht.

Rosenkrantz, R. D.: 1981, 'Does the philosophy of induction rest on a mistake?', *Journal of Philosophy* **KXXIX**, pp. 78–97.

Samuel, A. L.: 1959, 'Some studies in machine learning using the game of checkers', *IBM Journal of Research and Development* **3**, pp. 211–232.

Samuel, A. L.: 1967, 'Some studies in machine learning using the game of checkers II – Recent progress', *IBM Journal of Research and Development* **11**, pp. 601–617.

Von Neumann, J.: 1966, *Theory of Self-Reproducing Automata*, edited and completed by Arthur Burks, University of Illinois Press, Urbana.

PETER RAILTON

TAKING PHYSICAL PROBABILITY SERIOUSLY

INTRODUCTION

Arthur W. Burks has been prominent among those who have taken
seriously the inadequacies of narrowly empiricist construals of causal
relations and who have gone on to propose an alternative approach
based upon the idea of causal necessity. At the same time, Burks has
expressed dissatisfaction with another important component of various
empiricist views, the frequency interpretation of physical probability –
i.e., the kind of probability that figures in physical laws –, and has advocated a dispositional analysis of physical probability statements.[1] I
find myself in agreement with Burks' positions thus abstractly
described, but I would like to explore here a disagreement over the
kind of dispositional analysis to be given of physical probability statements. (I have a corresponding disagreement with Burks over the kind
of necessitarian analysis to be given of causation, but will not explore
that here.) In taking issue with Burks on the question of physical probability, I will also be taking issue with a range of views recently discussed by Brian Skyrms, David Lewis, and others who, like Burks,
have investigated analyzing physical probability in terms of epistemic
probability.[2] Although one might argue against epistemic interpretations of physical probability in a number of ways, I will concentrate in
what follows on questions related to possible *explanatory* roles for
physical probability. I will argue that physical probability should be
taken more seriously – more realistically, if you will – than epistemic
interpretations permit.

The form of the paper is as follows. In section 1, I will sketch Burks'
proposal for carrying out an interpretation of physical probability in
terms of inductive probability. In section 2, I will discuss a well-known
model of probabilistic explanation which, as it turns out, is well-suited
to Burks' approach to physical probability – not in spite of, but because of, his commitment to determinism. In section 3, I will suggest

251

Merrilee H. Salmon (ed.), *The Philosophy of Logical Mechanism*, 251–283,
© 1990 *Kluwer Academic Publishers*.

that broadly Humean subjectivist interpretations of physical probability may be formulated without a Burksian commitment to determinism. In section 4, I will ask whether a broadly Humean subjectivism can capture the *fundamental* explanatory roles of physical probability, express doubt on that score, and sketch a more realistic treatment of physical probability. Finally, in section 5, I will consider several arguments for and against such realism about physical probability. I should add that I do not consider any of the arguments made below on behalf of realism about physical probability to be more than preliminary, much less decisive. My aim is to raise some, but by no means all, of the questions I take to be important concerning this general issue, and to suggest reasons why certain answers to these questions might be preferred over others.

1. PHYSICAL PROBABILITY AS INDUCTIVE PROBABILITY: BURKS' APPROACH

For various reasons – some stemming from doubts about the observability or mere intelligibility of probabilistic connections "in the world", others from nothing more than frugality about fundamental ontological commitments – philosophers have attempted to analyze physical probability in terms of other notions. The two most prominent reductive analyses have been frequency interpretations and epistemic interpretations, which historically have held sway among philosophers in that order. Before discussing epistemic approaches, let us first say something about the reasons for the decline in enthusiasm for frequentism.

Very broadly, on a (non-finite) frequentist interpretation to say that a given physical system of type F has probability r to yield an outcome of type G is to say that in a sequence of N independent "trials" in which individual systems are prepared in condition F, the relative frequency of outcomes of type G will approach r as N approaches infinity. However, there may not exist infinite series of trials of the appropriate kind for some (or even most) kinds of chance systems, and thus to the extent that the interpretation is an *actualist* limiting frequency interpretation, the probabilities of such chance systems will be undefined. On the other hand, if the frequentist abandons actualism and speaks instead of *possible* or *virtual* infinite sequences, then the Humean purity of the original view – which treats probability extensionally much as regularism treats law and causation – may be compromised.

From an epistemological standpoint one might wonder whether virtual infinite sequences are any more accessible or well-defined – just *which* infinite sequences *would* occur?[3] – than some notion of physical probability as a modality "in the world". Indeed, if it is said that we have knowledge of certain features of virtual infinite sequences – e.g., their limiting relative frequencies – in virtue of what we know about the features of actual systems, we might wonder whether physical probabilities are being located in the chance set-ups generating sequences rather than in the sequences themselves.[4] Setting aside problems about the non-existence of certain actual infinite sequences, we may list several other persisting difficulties that have beset frequentism: (1) even when an infinite series of trials exists, it remains obscure how to justify applying what happens in an infinite limit to individual cases or to finite cases generally, and yet it is characteristic of scientific practice to apply physical probabilities to finite cases; (2) since a given trial of a physical system may belong to numerous actual or possible infinite sequences, not all of which have the same limiting relative frequency for the outcome in question, there are difficulties about identifying the appropriate sequence or "reference class" for individual probability attributions; (3) for reasons already mentioned, among others, there are problems about linking limiting relative frequencies to the actual, finite tests carried out in science or to epistemic probabilities in general (e.g., infinite repetitions of betting-like situations – a possible link to epistemic probabilities – are even less common than infinite repetitions of chance set-ups in nature); and so on. Frequentists are not without responses to these difficulties, but it has become increasingly hard to find philosophers cherishing high hopes for these responses.

Burks was ahead of his time in rejecting frequentism, and the kind of alternative he proposed has won increasing favor. He began by accepting Peirce's criticism of actualist frequentism: Peirce argued that probability attributions are *dispositional* in character, assigning "a certain habit or . . . a *would be*" to physical systems.[5] But Burks went on to reject what he calls Peirce's "dispositional-frequency" interpretation of physical probability for roughly the same reason he rejects other limiting frequency theories: Burks holds that any acceptable theory of physical probability must have application to possible *finite* sequences, but Peirce's account attributes a disposition only to produce possible *infinite* sequences.[6] In order to get direct application to the finite case, Burks develops an inductivist interpretation of physical probability; in order to encompass the notion of *possible* finite cases, his interpretation is dispositional.

Passing over some details, we can characterize Burks' inductivist interpretation in the following way. The appropriate analysis of a statement to the effect that a system *e*, given that it is of type *F*, has physical probability *r* to yield an outcome of type *G* is (1), below:

> (1) The system *e* has some, possibly complex, physical property φ – where φ is an appropriately complete description of certain features of system *e* – such that the inductive probability of *Ge* given knowledge of *Fe* and φ and given our general knowledge about the casual laws involved in systems of this kind, is *r*.[7]

It is said that φ is an "appropriately complete description of certain features of system *e*" because, depending upon context, there will be variation in the detail and depth of analysis of the system that a particular probability attribution presupposes. Thus, in an ordinary assertion that a doctored US penny has probability .7 of landing heads when shaken and tossed,

one is concerned mainly with gross, easily measurable properties (such as a hole or the presence of a large amount of a metal other than copper), and what one intends is that the description be complete with regard to such gross properties, not with regard to such difficult-to-measure properties as the turbulence of the air in which the coin falls, minor variations in the surface of the coin, etc.[8]

A probability attribution regarding system *e* may presuppose an appropriately complete description of certain physical properties of *e* even though *e* is an embedded subsystem of a deterministic system, such that if a yet more complete description of the embedding system were known, the laws of physics would imply one and only one outcome in each particular trial of *e*. Plainly, depending upon the thoroughness and kind of analysis given of a particular system, different probabilities may be assigned to the same outcome in a given trial. For example, knowing that the doctored penny has a hole dug on its obverse side may lead us to assign probability .7 to heads in light of what else we know about the coin, the way it is tossed, etc. But knowing the further information that this modification has altered the aerodynamics of the coin in a certain way may lead us to revise that probability estimate upwards or downwards. There is thus no one physical probability of the outcome heads on a given toss of a particular coin – there is instead an array of probabilities, varying according to the evidence upon which the relevant inductive probability is based. There are two ways in which this evidence might vary that are important here: we may have more or less complete knowledge of relevant facts, especially the initial and boundary conditions for a given trial; and we may have more or less complete knowledge of the laws governing the physical system that underlies the probabilistic disposition. Burks does not, in defining physical probability, require that knowledge of facts or laws be complete, only that it be "based upon all the information available" at the time.[9] And even so, a given probability attribution may be based on a description that is complete only up to a point, so that the description of the "underlying system ... is complete with respect to properties and laws of a certain kind and a certain level of specificity".[10]

For Burks, then, physical probability is inductive probability relative to a certain kind of description of a physical system. The existence of physical probability is compatible with determinism on this view, and in the end Burks rejects the idea that the world is tychistic or even indeterministic to a degree – what he calls "near-determinism". Hence he rejects the idea that there are "basic" probabilistic laws: "All basic laws," he writes, "are causal".[11]

Burks thus joins a long line of philosophers and scientists who have held that physical probability is not *ultimate*: if physical probability plays a role in our descriptions of the world or predictions of its behavior, that is only because our knowledge is partial or our descriptions incomplete; ultimately, at least in principle, physical probability can be done without even predictively. Now not all of those advocating an epistemic interpretation of physical probability need accept determinism, as we will see, but an acceptance of determinism removes at least one of the tensions involved in giving an epistemic interpretation of physical probability: it might seem odd to hold that the probability of physical systems to yield particular outcomes should be constituted by a fundamentally epistemic notion such as degree of confirmation or rational degree of belief; however, if all physical probabilities disappear when physics gives its fullest and deepest analyses, then it seems much less objectionable to link physical probability to epistemological notions in an essential way.

Before going on to discuss epistemic interpretations of physical probability that do not presuppose determinism, let us look briefly at a question that naturally arises for a Burksian account: If probability plays an ultimately eliminable role in scientific description and prediction, what role can it play in scientific explanation? Clearly, it will not have any ultimate explanatory role, but I hope to show that a familiar analysis of probabilistic explanation fits surprisingly (disturbingly?) well with a deterministic inductivist account of physical probability.

2. A VIEW OF PROBABILISTIC EXPLANATION UNDER DETERMINISM

In this section I will be arguing that Hempel's inductive-statistical (I-S) model of probabilistic explanation has the peculiar property that it fits nicely with a possible view about probabilistic explanation under determinism. Indeed, although I will not argue the point in detail here,[12] the I-S model is if anything *more* apt from a deterministic perspective than it is from a perspective admitting of genuine physical indeterminism. I am not claiming that Hempel himself was a determinist when he proposed this model – we have his own word to the contrary;[13] I mean to suggest only that a determinist like Burks has better reason to be happy with the I-S model than someone prepared to accept ultimately probabilistic physical laws.

The I-S model is perhaps well enough known to need only a brief restatement here. Essentially, Hempel claims that probabilistic explanation of particular facts takes the form of an inductive argument showing that the explanandum had high epistemic probability relative to the most specific probabilistic laws applicable in the current state of science. Hempel bases his model of probabilistic explanation on a Cramérian interpretation of probability, according to which:

> (2) [If] F [is] a given kind of random experiment and G a possible result of it; then the statement that $p(G, F) = r$ means that in a long series of repetitions of F, it is practically certain that the relative frequency of the result G will be approximately equal to r.[14]

Although Cramér and Hempel label this a frequency interpretation of probability, it is a rather distant relative of orthodox frequency interpretations, and can be seen not to be in fundamental conflict with Burks' approach. First, it applies directly to finite cases – (2) speaks only of "a long series of repetitions", not an infinite series – and even to the single case, since Cramer holds it to be a corollary of (2) that if $p(G, F)$ is very close to one, then we can be "practically certain" that result G will occur even if F is realized only once.[15] Second, according to Hempel, physical probabilities are to be viewed as dispositional: a "probability statement attributes ... a certain *disposition*" to chance systems and does not merely describe past or future *actual* relative frequencies.[16] Third, it defines physical probability in terms of inductive probability – the "practical certainty" of particular outcomes – and is not logically incompatible with determinism.[17] Thus it is not wholly contrary to the spirit of the Hempelian I-S model to use it under Burks' view.

Let us consider an example used by Burks, a biased coin-toss, and analyze the relevant explanation in Hempelian terms. Suppose that a coin c has been tossed in a particular way T and has come up heads (outcome H). Suppose further that c is a coin of type D, with a highly asymmetric distribution of mass, and that "with respect to certain properties and laws of a certain kind and a certain level of specificity",

$$p(H, D\&T) = .9.$$

What explanation may be offered of c's having come up heads on the toss in question? On the I-S model, the appropriate explanation would be:

(3) $p(H, D\&T) = .9$
$Dc \ \& \ Tc$
========== [makes practically certain]
Hc.

The double lines separating the explanandum from the explanans are meant to suggest that the inference is inductive rather than deductive; its inductive (and explanatory) "strength" is indicated by the expression in brackets – according to Hempel, the inductive probability of 'Hc' given the two premises should be .9. "Of course," Hempel writes, "an argument of this kind will count as explanatory only if the [inductive probability of the explanandum relative to the explanans] is fairly close to 1".[18] There appear to be two reasons for this *requirement of high probability*. First, on the Cramérian interpretation of probability used by Hempel, probabilities do not have application to the single case (e.g., to a particular coin toss) unless very near one or zero.

Second, Hempel holds the general view that explanations must be potentially predictive arguments, and that explanations succeed only when they establish the *nomic expectability* of their explananda. Clearly, I-S arguments establish the nomic expectability of their explananda only when the probability involved is close to one. Just *how* close the probability must be to one Hempel leaves open, but it is clear from his examples that .9 is close enough.[19]

To the requirement of high probability Hempel adds a second condition on I-S explanations, the "requirement of maximal specificity", devised to meet the following sort of problem. Suppose that coin *c* has, in addition to the asymmetry already mentioned, a small magnet inlaid in one side – or '*Mc*' for short –, and suppose further that the coin was subject to a magnetic field during the toss – '*Fc*' for short –, such that $p(H, D\&T\&M\&F) = .1$. (We may imagine that the south pole of the magnet in the coin points outward from the tails side, and that the area of the surface on which the coin landed is also a south pole.) We are now in a position to construct an I-S argument concerning the negation of the explanandum:

(4) $p(-H, D\&T\&M\&F) = .9$
     $Dc \& Tc \& Mc \& Fc$
     ══════════ [makes practically certain]
     $-Hc.$

If we construe (3) and (4) as nothing more than statements of evidential relationships among sentences, then there is no contradiction or competition between them. Yet, if we construe them as *explanations*, we have the anomalous result that the very same fact, *Hc*, both is and is not explicable. Hempel terms this the problem of "explanatory ambiguity":

any statistical explanation for the occurrence of an event must seem suspect if there is the possibility of a logically and empirically sound probabilistic account for its non-occurrence.[20]

We must, then, decide which of the two inductive arguments to accept for explanatory purposes. We cannot do so merely by following criteria of the kind used for assessing Hempelian deductive-nomological (D-N) explanation: *both* (3) and (4) have true premises, make essential use of a law, and have proper (in this case inductive) logical form. To resolve the explanatory ambiguity that may arise in I-S (but not in D-N) explanation, Hempel imposes the *requirement of maximal specificity*, which instructs us to refer any explanandum fact to the most specifically-defined class in which it has – for other than purely logical reasons – a characteristic and distinctive probability.[21] In the case at hand, (4) gets the nod, since properties M and F partition the class $\{x: Dx \ \& \ Tx\}$ in a statistically significant way. Hempel argues that application of the requirement of maximal specificity must be relativized to epistemic context: we are to pick the most specific reference class according to the totality of currently accepted beliefs. Thus, I-S explanation is fundamentally unlike D-N explanation, for whether an argument is a genuine I-S explanation depends upon our current epistemic situation and may change when beliefs change, while a D-N argument is a genuine explanation if its premises are true and its logic valid, whether or not this is currently believed to be the case and regardless of what we might come to believe.[22]

These last two requirements of the I-S model – the requirement of high probability and the requirement of maximal specificity – have generated a great deal of controversy. Why should we say that improbable facts, such as the occurrence of heads on three consecutive tosses of a fair coin, are inherently inexplicable? Doesn't the same explanation exist for probable and improbable facts alike, namely, that a chance process was at work and that the outcome observed had such-and-such a probability (high, middling, or low) of coming about by chance? Admittedly, in cases where the explanandum's probability is low or middling we cannot state a sufficient reason for the explanandum's occurrence, but we cannot state a *sufficient* condition in high probability cases either. And why should probabilistic explanation be epistemically relative when non-probabilistic explanation is not? I think the I-S model *as an account of probabilistic explanation under indeterminism* fetches up on these criticisms, and thus have attempted to develop an alternative covering-law account of probabilistic explanation that does not run afoul of them.[23] But *as an account of probabilistic explanation under determinism* the I-S model sails over these objections unperturbed.

To see why, suppose that one were a determinist, believing that a full physical analysis would, in principle, enable us to give D-N explanations of all facts now explained probabilistically. Thus, one would believe that there are hidden variables and associated non-probabilistic laws such that all apparently random processes could be shown, in principle, to obey an underlying determinism. In a familiar Laplacian

sense, probability would reflect our ignorance about, or unwillingness
to give a complete account of, this underlying determinism; probability, then, would disappear in the face of perfect before-the-fact information and calculation. Probabilistic explanations, on this view,
would never be ultimate, but only way-stations along the road to improved explanation, where the end of the road is outright D-N explanation. What would one do at such a way-station? One would try to
cite as many as possible of the factors believed to be relevant (at least
statistically) to the occurrence or non-occurrence of phenomena of the
same type as the explanandum. Hence, probabilistic explanation would
be inductive in form – i.e., would have the form of an evidential
relationship between explanans and explanandum – and moreover
would be governed by a requirement of maximal specificity relativized
to our epistemic situation. If the point of probabilistic explanation were
to argue inductively (on the basis of what we currently believe) to the
occurrence of the explanandum, and if we were to assume that every
explanandum could in principle be demonstrated outright in a D-N
fashion, then it would only be natural to require that acceptable probabilistic explanations confer high probability upon their explananda. If
we were unable to establish a high probability for an explanandum,
then this would show that we were ignorant of something quite important in the underlying deterministic process. Without an assumption of
underlying determinism, the requirement of high probability might
seem odd, since it would appear to dictate that all chance phenomena
that have low or middling probability are in principle inexplicable *even
probabilistically*. Under the assumption of determinism, the oddness
would be greatly reduced: with sufficient knowledge, we should always
be able to produce an I-S explanation establishing high probability,
since we could bring in a more and more extensive list of relevant factors and exclude as many "disturbing factors" as need be to get a high
probability for the explanandum. As our knowledge of laws and facts
grows, we should be in a position to have greater certainty about outcomes. In the ideal epistemic limit, we should be able to have deductive certainty about outcomes; as we approach that limit, we should
be able to achieve inductive "practical certainty". On a deterministic
view of this kind, then, probabilistic explanation would by its nature be
epistemically relative – there simply would be no empirical probabilities apart from particular epistemic contexts. Moreover, on such a
view it would be reasonable to require that an acceptable explanans be

maximally specific with respect to current beliefs and capable of establishing a high inductive probability for the explanandum. Of course, it may be possible to establish a high probability for the explanandum even without well-developed knowledge, as argument (3) shows, and not all increases of knowledge lead to increases in the explanandum's probability, as argument (4) shows. The most one can say is that, under deterministic assumptions, an *inability* to establish high probability would evidence an important gap in our understanding of the phenomenon to be explained, and, *in the epistemic limit*, we should be able (in principle) to establish a probability for the explanandum arbitrarily close to one (and thus no facts would be in principle inexplicable for reasons special to the I-S model).[24] In sum, the weaknesses of the I-S model as an account of probabilistic explanation under indeterminism become its strengths as an account of probabilistic explanation under determinism.

Again, I do not claim that Hempel held this deterministic view when he formulated the I-S model, but perhaps we may speculate that to an extent this view held him.[25] That would help explain why the anomalous features of the I-S model in comparison with his D-N model did not trouble him more than they did. For Burks, an avowed determinist, the I-S model ought to seem very plausible, its differences from D-N explanation quite appropriate. A determinist cannot allow physical probability to play a fundamental explanatory role apart from epistemic context, and the I-S model gives no account of probabilistic explanation independent of epistemic context. A believer in physical indeterminism and ultimate probabilistic laws, on the other hand, would for the same reason be likely to dissent from the I-S model, since he presumably would hold that at least some probabilistic explanations play a fundamental role that does not depend upon epistemic context.

In section 4 we will return to the question of giving physical probabilities a fundamental explanatory role, but first we should briefly discuss an epistemic interpretation of physical probability that does not involve a Burksian commitment to determinism.

3. PHYSICAL PROBABILITY AS SUBJECTIVE PROBABILITY: A HUMEAN APPROACH

Burks holds that causal relations are ineliminably modal and cannot be reduced to aspects of immediate experience. He holds a similar view of inductive probability, and as we have seen he reduces physical probability to inductive probability. Both concepts, causal necessity and inductive probability, have *a priori* elements according to Burks, and play a role in the presuppositions of induction.[26] It is open to a Humean to reject traditional frequentist interpretations of physical probability for much the same reasons as those considered above. Like Burks, he may attempt an epistemic interpretation, but, unlike Burks, he may reject the ideas of irreducible physical necessity and determinism.

In what follows I will use 'Humean' in a broad sense, meant to pick out a certain style of philosophical analysis that owes a great deal to Hume: the reduction of what might appear to be inherent features of the external world – such as causal and nomological relations – to complex patterns of particular facts lacking "intrinsic" connections, where these patterns include in a central way facts about human psychology. Although I must leave aside the historical question whether Hume himself was a Humean in this sense, I take it that it is clear enough why I wish to attach his name to this approach; after all, it was Hume who wrote:

... 'Tis certain that [the] repetition of similar objects in similar situations *produces* nothing new either in these objects, or in any external body. ... Tho' the several resembling instances, which give rise to the idea of power, have no influence on each other, and can never produce any new quality *in the object*, which can be the model of that idea, yet the *observation* of this resemblance produces a new impression in the mind, which is its real model. ... Upon the whole, necessity is something, that exists in the mind, not in objects. ... [The] contrary bias is easily accounted for. 'Tis a common observation, that the mind has a great propensity to spread itself on external objects ...[27].

Thus, for example, a Humean analysis of physical law would reduce the notion of nomologicality to: (1) the existence of certain patterns of particular facts – i.e., the truth of certain universal factual generalizations; and (2) the selection from among the true universal factual generalizations of those that are *laws* on the basis of human tendencies to project or protect some universal generalizations rather than others, or tendencies to employ certain standards of simplicity or strength rather than others (where, of course, these human tendencies themselves are ultimately matters of patterns of particular facts, e.g., the regular[28] co-occurrence of certain ideas with certain experiences of objects). Similarly, a Humean analysis of causation would reduce causal relations to: (1) the existence of certain patterns of particular facts – specifically, patterns of constant conjunction, spatio-temporal contiguity, and invariant succession; and (2) the selection from among these patterns of those that are *causal* on the basis of human tendencies to form expectations, or to make associations, or ..., concerning them (and again, these tendencies are themselves reducible to patterns).

For a Humean in this broad sense, nothing would be more natural than to attempt to view *physical probability*, too, as something "that exists in the mind, not in objects", and that the mind "spreads" on objects on the basis of experience. That is, physical probability would be reduced to: (1) the existence of certain patterns among particular facts – such as statistical and non-statistical regularities; and (2) the degrees of rational expectation or subjective probabilities that humans would form concerning outcomes on the basis of the evidence in (1).

The reduction could take a form akin to Burks': to say that the physical probability of an outcome of type G is r for system e of type F at time t is to say that a relatively complete description of system e, combined with the best available information at t concerning relevant laws (ultimately, of course, a matter of patterns of particular facts), would make rational a degree of expectation r for the outcome Ge.

Alternatively, a considerably higher standard of information might be required: for example, one might define the physical probability at time t of Ge given Fe as the degree of rational expectation one would have of Ge given *ideally complete* information concerning the history of particular facts up to t and ideal computing powers.[29] This second, idealized subjectivist account would not have the epistemic relativity of Burksian versions, since the physical probability of a fact at time t would be independent of what we happen to believe at t (or ever after), although what we *take to be* the physical probability of a fact at a time will of course vary with epistemic context. However, such an idealized account will yield a *unique* physical probability distribution at a given time only if all reasonable initial credence functions conditionalized on the same total history of facts up to a time would agree about what subjective probabilities to assign at that time. It is not obvious that this *uniqueness condition* will be met, and if it is not, then there will be a degree of ambiguity about what the physical probability distribution is at a given time. To eliminate this ambiguity it would be necessary to add a further element to the definition of physical probability, namely, some way of selecting a subclass of the class of reasonable initial credence functions that is appropriate for the purpose of determining physical probabilities. (We will discuss this problem below.) However the details are spelt out, both idealized and non-idealized subjectivist interpretations manage to avoid treating physical probability as a non-Humean modality "in the world".

It is important to observe that such subjectivism need not involve any claim that there are no basic probabilistic laws or that nature is *au fond* deterministic. On the contrary, it is perfectly compatible with reducing physical probability in a Humean way to claim that some or even all basic laws are probabilistic, since for a Humean whether a statement is a law is always a matter of what patterns of particular fact are found in nature, what expectations we associate with those patterns, etc. Thus probabilistic laws are just as capable of being fundamental as non-probabilistic laws. A number of theorists have discussed subjectivist accounts of physical probability along these broadly Humean lines, including Richard C. Jeffrey, Brian Skyrms, and David Lewis.[30]

4. A FUNDAMENTAL EXPLANATORY ROLE FOR PHYSICAL PROBABILITIES

What reasons might we have for accepting the idea that there are basic probabilistic laws? Various reasons might be given, but surely the most compelling reasons are furnished by contemporary physical theory. The outstanding successes of quantum mechanics, the prevalence of the "probabilistic" interpretation of quantum theory, and the existence of so-called "no hidden variable" proofs for the quantum-mechanical formalism (which establish that any "hidden-variable" theory compatible with quantum mechanics must suffer serious deficiencies, such as non-locality), have all conspired to give great credence to the notion that at least some fundamental physical laws are probabilistic, in the sense that even the most complete description of the physical state of a system may determine – in conjunction with all true physical laws – no more than a probability distribution over future states of the system. Burks agrees that the most formidable case against his assumption of determinism is made by the quantum theory, but he holds that the difficulties that continue to accompany the interpretation of this theory are sufficient to leave the possibility of determinism open. Indeed, it cannot be said that the success and development of the quantum theory resolve the issue of determinism, but it at least seems plausible at this time to consider in some earnest what questions would arise for an interpretation of probability and for an account of probabilistic explanation on the assumption that fundamental probabilistic laws exist.

We saw in section 3 that one could adopt a subjectivist interpretation of physical probability without commitment to an underlying determinism. For example, Skyrms has attempted to characterize a notion of the *resiliency* of certain epistemic probabilities, that is, their stability and invariance with respect to the addition of certain kinds of information. Thus, knowing general features of a given tossing apparatus, we might arrive at a subjective probability of .5 for getting heads by tossing the coin in that way – the coin's shape is symmetrical, its mass is uniformly distributed, the tossing apparatus is designed so that even a skillful person cannot use it to reliably produce a desired outcome, and so on. We might hold to this probability attribution despite learning that the coin has been tossed ten times by this apparatus and come up heads seven times. Similarly, we might hold to this probability attribution despite learning that the next toss will be initiated by John, and six of the seven times heads has come up it was John who initiated the toss, while in the other four tosses Jim did the initiating. The subjective probability of one-half for heads is thus *resilient* with respect to certain kinds of additional information. This is not to say that it would be resilient with respect to all further information: certain after-the-fact information could quickly alter the subjective probability of heads on a given toss to very near one or very near zero; and before-the-fact information concerning details of the physics and microphysics of a given toss – the exact initial angular momentum of the coin, the air resistance it encounters in its fall, the precise distance it falls, etc. – might also radically alter the subjective probability for heads from one-half. But with respect to a certain class of factors $F_1, \ldots, F_n$, a subjective probability may be quite resilient in the face of further information concerning $F_1, \ldots, F_n$, and Skyrms advances the hypothesis that physical probabilities (as characterized here) ought to be reduced to resilient subjective probabilities.[32] Skyrms' account clearly resembles Burks' treatment of physical probability as inductive probability based upon a description of a system that is complete only in certain respects and at a particular level of analysis and exactness, but for Skyrms there is no essential reliance of probability upon incompleteness of description.

Skyrms uses the range of factors concerning which information may be admitted without altering a given subjective probability as a measure of the *scope* of that probability's resilience, and in turn uses scope of resilience as a measure of the *lawlikeness* of a probability attribution.[33] He allows that there may be cases in which a subjective probability is (rationally) resilient even in the face of the most complete description of before-the-fact conditions possible according to our physical theory. Subjective probabilities concerning certain quantum-mechanical phenomena, Skyrms suggests, may have resiliency of universal scope in this sense, and thus are the strongest candidates for playing the role of lawful physical probabilities. But various subjective probabilities concerning the behavior of thermodynamical systems in classical statistical mechanics and the behavior of so-called "classical" gambling devices (coin-tossing, roulette, etc.) will have considerable scope of resiliency despite the assumed underlying determinism of these systems at the microphysical level. Skyrms is satisfied that the scope of resiliency of certain subjective probabilities associated with these classical systems is sufficient to warrant treating them as lawful physical probabilities, too.[34]

The attraction of this sort of solution to the problem of interpreting physical probability is that it enables us to treat quantum-mechanical and classical probabilities in much the same way: both are matters of the strongest predictive statements that can rationally be made at a certain level (and degree of completeness) of physical analysis. But attractiveness is all too often alloyed with vice, and in this case the vice is that a subjectivist interpretation seems incapable of taking adequate account of the *differences* between quantum-mechanical and classical probability. Are these differences merely quantitative (e.g., a matter of the extent of scope of resiliency), or are they qualitative? The question is not whether we can speak of probabilities in connection with classical deterministic systems, for we evidently can; the question is whether we can simply extend a notion suitable for analyzing probability under determinism to the indeterministic case without blurring important distinctions. The important distinctions I propose to attend to have to do with possible *explanatory* roles for physical probability.

To give the discussion focus, let us briefly consider an alternative account of physical probability, a *propensity interpretation* according to which physical probability is not to be reductively analyzed into any sort of epistemic probability, but instead is viewed as an objective

property of physical systems. Physical probability, on this interpretation, is a measure of the dispositional tendencies of indeterministic
physical systems to yield certain outcomes. One way these chance dispositions manifest themselves is in the relative frequencies obtained in
sequences of trials (for trials of the appropriate kind, the relation of
propensities to relative frequencies is established by the law of large
numbers). Moreover, knowledge of a chance disposition makes rational particular expectations concerning outcomes (if one is certain that
the propensity of a to be G at t is r, then one should assign subjective
probability r to Ga at t, regardless of whatever else one believes). But
a propensity is not reducible to its manifestations or to the expectations
it warrants, just as dispositions in general are not so reducible. Quantum mechanics seems to afford us plausible examples of propensities.
For instance, in virtue of its physical constitution a radioactive nucleus
has a certain dispositional tendency (if unperturbed) to spontaneously
emit an alpha-particle during a given time interval. This tendency may
or may not be realized during that time interval; whether it is or is not
is a matter of chance, but the chance is a lawful one, lawfully tied to
other physical characteristics of the nucleus – e.g., atomic number and
atomic weight. Needless to say, fully deterministic systems do not have
probabilistic propensities in this sense since they have no *chance* dispositions.[35]

Now if we should ask why some rational degrees of belief (other
than zero or one) are resilient even when conditionalized on the most
complete before-the-fact information about a physical system, while
others are not, the propensity theorist has an answer: when we are
dealing with genuine chance dispositions, no amount of before-the-fact
information could conclusively establish the outcome; when we are
dealing with statistical characteristics of deterministic systems, then additional before-the-fact information could do just that. That is, the
propensity theorist offers the existence of real chance in the world as
an objective foundation for this difference in resiliency. Skyrms claims
that epistemic interpretations, too, can offer objective foundations:

epistemic . . . views of [physical probability] need not deny that resiliency has an objective
foundation. . . . The gambler will find the physically symmetric dice, the vigorous shake,
. . . etc., sufficient objective foundation for the sort of resiliency he requires.[36]

But this is not adequate. What was wanted was an objective foundation for the difference between cases in which before-the-fact

information exists that would (if known) lead reasonable men to alter their probability assignments and cases in which such information cannot, even in principle, be found; what underlies *this* difference? Here it seems the subjectivist faces two alternatives, both problematic. He might simply say that all there is to it is the difference in resilience, nothing further. But then he is without any *explanation* or "objective foundation" for ultimate resiliency. Alternatively, he might claim that there is indeed an objective difference between deterministic and in-deterministic processes that explains the ultimate resiliency with respect to before-the-fact information of subjective probabilities concerning the latter. But then we might wonder why he is in the business of reducing physical probability to subjective probability. It is as if someone said that what *explains* the difference between conductors of electricity and non-conductors is the presence in the former but not the latter of easily dissociable electrons permeating the substance's microstructure, and yet went on to *reduce* the dispositional property of electrical conductivity to our *expectation* of conduction based upon existing (or ideal) physical knowledge. Surely this would be putting things back to front: when we have knowledge of the underlying physi-cal reasons why a substance is a conductor of electricity, this will serve to ground expectations about that substance's dispositional behavior were a potential difference applied across it, but if there is to be any reduction one would expect it to be of the disposition to the underlying physical reasons, not to an expectation based upon them. Moreover, we would like to be able to say that the existence of the disposition, when known, makes rational certain expectations about future be-havior, but this claim is blocked if we have reduced the disposition to a rational expectation. I can see no relevant difference distinguishing probabilistic from non-probabilistic dispositions in this regard; the order of *explanation* seems to be as follows: an underlying physical basis explains the disposition, and knowledge of this disposition or its basis explains the rationality of certain expectations. In the process of *discovery* we may first become aware of certain expectations about the behavior of a given phenomenon, then attribute a corresponding dis-positional property to it, and then later discover a physical basis for this disposition. But if reductive analysis is to follow an order, should it not be the order of explanation rather than the order of discovery?

It might be objected that reductive analysis should always proceed in the direction of experiential rather than explanatory basicness. But this is a recipe for phenomenalism, and I take it that relatively few subjectivists taking the broadly Humean line sketched above find *this* aspect of Hume appealing. Or, it might be objected that it simply is no explanation of resiliency to invoke indeterministic propensities "in the world" – after all, this purported explanation predicts no new observations, seems to be in principle unverifiable, and so on. But again, the harder this line is pushed the more difficult it is to resist phenomenalism. Many philosophers like to say that we may posit physical objects "in the world" as an explanation of the stability and coherence of our experience. But this posit seems to have just the faults lately attributed to positing propensities to explain the stability and invariance of certain epistemic probabilities: it predicts no new observations, it is in principle unverifiable, and so on. In the following section I will consider further the question of positing entities that do not make certain kinds of difference in experience. For now I simply want to note that certain initial attempts to disqualify propensities as explanatory are very costly if pursued, so that a *prima facie* claim can be made for entertaining propensities for explanatory purposes as long as we are prepared to entertain the external world for explanatory purposes, as I presume most of us are.

Propensity interpretations thus offer physical probabilities an explanatory role of a fundamental kind not available to them under subjectivist interpretations. In discussing Burksian determinism, we noted that physical probability was denied a fundamental role because of the assumption that no fundamental probabilistic laws exist. The subjectivist may hold that there are basic probabilistic laws, but in analyzing the probabilities figuring in these laws he argues in a broadly Humean way that they are not "in the world" independent of mind, and thus denies himself the possibility of using their objective existence to explain why the mind runs up against certain ultimate limits in attempting to discover hidden variables. I take it to be unproblematic to say, rather informally, that someone willing to accept the idea of basic probabilistic laws is prepared to take physical probability seriously in a way that the determinist is not. Somewhat more problematically and informally, we may say that the propensity theorist is prepared to take physical probability seriously in a way that the subjectivist is not. Intuitively, both the determinist and the subjectivist manifest forms of *idealism* about physical probability, while the propensity theorist espouses a form of *realism* on the same subject. I will not attempt to give a non-intuitive characterization of this difference here, and it will suit my purposes equally well to situate the determinist, subjectivist, and propensity theorist at successively higher points along a scale marked off in *degrees* of realism about physical probability. For my point here is that the more thorough-going realism of the propensity theorist permits him to make more explanatory use of physical probability than his less realist competitors. The price paid for greater explanation is precisely greater realism, and so the question is whether, as in the case of being realistic about physical objects, the price is right.

Before discussing that question further, let us say what probabilistic explanation of chance outcomes amounts to if one posits the existence of propensities. Very briefly, probabilistic explanation is a matter of showing that there existed a propensity for the explanandum fact and noting that, by chance, this propensity was realized. An ideally full probabilistic explanation would provide an accurate measure of the propensity for the explanandum fact as well as a derivation of this propensity from the most basic elements of physical theory; but less complete probabilistic explanations might still be perfectly adequate contextually.[37] There is a notable shift from the basic picture of I-S explanation: probabilistic explanation is no longer inductive in form and does not require epistemic relativization; probabilistic explanations can be true or false, like D-N explanations generally, depending entirely upon the truth of the premises and the validity of the explanation's logic; and probabilistic explanations need not show that the explanandum was highly probable, since relevance to the single-case exists in virtue of the single-case character of propensities, and since the point of probabilistic explanation is not to simulate deterministic explanation as nearly as possible, but to characterize the actual chance process responsible for the explanandum, whatever its probability of yielding the explanandum.[38] Because he recognizes the possibility of basic probabilistic laws and accepts the application of probabilities to the single case, the subjectivist could offer a formally similar account of probabilistic explanation, although independence of probabilistic explanation from particular epistemic contexts could only be achieved on *idealized* subjectivist interpretations. Furthermore, unless the uniqueness condition is met – i.e., unless subjective probabilities based upon ideal information up to a time do not vary across reasonable credence functions – there may exist a form of explanatory ambiguity for probabilistic explanation, as there could be competing probabilistic explanations. Further restrictions that might be imposed upon reasonable initial credence functions for the purpose of eliminating this ambiguity may well seem arbitrary from the standpoint of physical explanation, and yet without such further restrictions it may be more than we have any right to hope that the uniqueness condition will in general be met.[39] Finally, it would remain true for any subjectivist account that all probabilistic explanations, even those concerning such events as the chance mutation of a gene during the Miocene era or the spontaneous disintegration of a nucleus on a distant asteroid, involve a mind-dependent phenomenon – subjective probability – in an essential way. Once again, I find this degree of idealism about the constituents of physical explanation disturbing. It is a very basic intuition that wholly physical systems should be explained wholly physically – a detour through what is, on analysis, a rational degree of belief, seems quite irrelevant to such explanations.[40]

5. OBJECTIONS TO GIVING PHYSICAL PROBABILITY A FUNDAMENTAL EXPLANATORY ROLE

In the previous section I suggested that positing chance physical dispositions to explain probabilistic behavior and the stability and invariance of certain epistemic probabilities is in some respects comparable to positing enduring physical objects to explain the stability and coherence of our experience. In both cases, we posit things "in the world" independent of mind in an effort to get an underpinning or "objective foundation" for epistemic phenomena even though we cannot have immediate epistemic access to the posited objects.[41] In both cases, too, the explanations made available by the posits permit us to reduce the amount of coincidence and disunity in the phenomena, as, for example, when a succession of similar experiences may be accounted for in terms of a single, enduring, objective thing or disposition. Moreover, both posits enable us to avoid an intuitively objectionable idealism.

There are a number of objections to this parallelism, however. For example, it might be said that daily life and scientific life could get along quite well without positing propensities, while it is hard to imagine how either daily life or science could continue without positing an external world. This is a difficult question, which should be discussed in connection with a more general survey of modalities "in the world" – e.g., causal necessity. I cannot pursue that general survey here, but let me say that it is at least doubtful that everyday life and scientific life could go on as they do without positing some "powers in the world" of a causal and (where appropriate) probabilistic kind. Burks and others have stressed the role of presuppositions about causal necessity in the process of induction and in the sphere of practical action, and the vocabulary and mental set characteristic of causal discourse so pervade our thought and activity that it is difficult for me to think that we could dispense with modalities "in the world" as readily as the objection under consideration suggests. But again, this is too large a matter to discuss here.

A second objection is that we have a much clearer notion of the concept of a physical object existing "without the mind" than we do of the concept of a propensity "in the world". I can only agree, and submit that propensities ought to be thought of as theoretical entities akin to other posits of physics. We may lack a clear notion of force, of particles as described by the quantum theory, and so on, but we recognize that relatively unfamiliar posits may be called for in extending our understanding of the world. Clearly, the justification for making a given posit will depend upon the virtues of the theory into which it fits, and while it is unquestionable that quantum mechanics has great virtue, it is a vexed question how we should interpret that theory. *Does* quantum mechanics furnish good reason for believing that there are single-case chance dispositions in the world? The answer will depend upon the tenability of a realist probabilistic interpretation of that theory. The issue is far from resolved, but it would seem unreasonable to claim that propensities ought to be denied potential explanatory status on grounds of unfamiliarity.

The complaint that propensities are not well-understood might have another origin, however. A Humean might hold that the idea of any sort of modality "in the world" is merely incoherent. Put this way, the complaint may succeed only in begging the question. Of course, propensities cannot be reductively analyzed in a Humean fashion, and if such reductive analysis is what is required for the Humean to concede the intelligibility of propensities, then no such concession will ever be forthcoming. But the issue is: Should we require that all acceptable concepts be reductively analyzable in a Humean fashion or not? I have attempted to suggest that we should not adopt this requirement because it would prevent physical probabilities from playing an important explanatory role. Any countermove must at least offer a further *defense* of Humean requirements, and not merely presuppose their legitimacy. Let us look again to our parallel. The phenomenalist might complain that positing an external world for explanatory purposes is fruitless since the notion of an object existing "outside the mind" in some irreducible sense is incoherent. Presumably, realists about the external world would respond that if the phenomenalist is simply insisting that the only acceptable notions are those capable of being phenomenologically reduced he is begging the question against realism. If the issue is to be joined, the phenomenalist must offer good reasons for accepting his criteria of intelligibility.

Now the Humean might claim that he can give good reasons for his criteria, reasons of an epistemological kind. After all, he could argue, we have no direct awareness of propensities and we can imagine that our experience would be the same if they were not "behind it". Thus it is hard to see how we could ever have *evidence* for them. But again, if the Humean presses his point he will find he has backed into phenomenalism: the same epistemological strictures count in favor of phenomenalist criteria of intelligibility and against positing an external world.

Still, aren't there *some* epistemological advantages of Humean subjectivism? Yes, but they come at a cost. Consider first the form of Humean subjectivism that reduces physical probabilities at a time to the subjective probabilities we would assign on the best available evidence at that time. This approach has the clear epistemological advantage of making physical probabilities readily accessible. *Too* readily, perhaps, for on this view we may know the physical probability of an outcome even on the basis of *very* limited information – as long as this limited information is the best available. As Popper has often asked: How could physical probabilities be such magical things that they enable us to coin ignorance into knowledge? Moreover, if physical probabilities depend upon our epistemic context, then they will vary as beliefs change, and surely it is somewhat counterintuitive that (for example) the physical probability of Uranium 238 to alpha-decay has actually changed whenever decay laws have been made more precise. (Note that requiring contextual resilience will not eliminate this problem, since an unwillingness to alter a probability attribution in the face of new evidence could be due to ignorance of the bearing of that evidence.) Further, if physical probabilities change with available information, so will probabilistic explanation, and while it may be acceptable to view probabilistic explanation as epistemically relative if one assumes the world to be deterministic (as we saw in section 2), if one thinks there *are* basic probabilistic laws it would be much more unpalatable to make *correctness* (and not merely *acceptability* or *degree of confirmation*) of probabilistic explanation a function of what we happen to believe at a time. Are we in the enviable position of being able to coin ignorance into explanation, too? If we think that explanation has an important role to play in epistemology, as it does in causal theories (and perception presumably involves some probabilistic links) we may find ourselves in a circle if we claim that what counts as an explanation depends upon our epistemic condition. Hence I do not find the epistemological advantages of this sort of Humean subjectivism compelling: it fits too ill with intuition and may make trouble for a systematic epistemology.

We may avoid this alleged dependence of physical probabilities and probabilistic explanation upon epistemic context if we adopt an idealized form of Humean subjectivism and interpret the distribution of physical probability at a time as the distribution of subjective probability that would be arrived at that time given ideal information about before-the-face conditions, i.e., given the whole history of particular facts up to the time in question. Unless qualified, this approach would eliminate any physical probabilities associated with classical deterministic systems, but that may be a point in its favor. Now this approach lacks some of the epistemological advantages of less idealized approaches, since we may have no access, even in principle, to the whole truth about the universe's history up to a time. Idealized subjective probabilities are certainly no more accessible than propensities. However, while this approach lacks certain epistemological advantages over more realistic propensity views, it might be thought to enjoy the epistemological advantages of greater conceptual clarity and economy, since it gives an account of context-independent physical probability that avoids positing propensities "in the world" as conceptually basic modalities over and above the world's history of particular facts.

But conceptual clarity and economy are not bargains if they are achieved at the expense of being able to make conceptual distinctions we think we ought to be able to make. A brief example may suggest that idealized Humean subjectivism is an expensive bargain in just this way. It seems quite possible to imagine two worlds completely alike in history of particular fact – including of course all psychological facts – but which differ with respect to physical chances. For example, world w_1 and world w_2 may agree on all points concerning the history of particular facts, and thus it may be true in both that all particles of kind K decay into one particle of kind L with spin s and one particle of kind M with spin $-s$. But in world w_1 this universal regularity is the result of a deterministic law, while in world w_2 it is a matter of chance whether a K–particle decays in this way or not, there being a probability of roughly 1^{-100} that a K–decay will produce an L–particle with spin $-s$ and an M–particle with spin s. (If it aids the imagination, one may think of these worlds as containing only finitely many K–particles, and perhaps as differing with respect to chance – although not with respect to actual history of particular fact – in various other ways, too.[42]) Notice that whatever subjective probabilities would be assigned on the basis of total factual history alone in world w_1 would also be assigned on the basis of total factual history alone in world w_2, so that the two could not differ in idealized subjective probability distribution at a time. (We cannot include facts about which physical probabilities have obtained in the history of particular facts without spoiling the Humean reduction; any difference in chances must be traced to a difference in manifest history of particular facts not about chances as such.) Thus on an idealized Humean subjectivist interpretation, there simply could not be any such difference between w_1 and w_2 as we have imagined here – one could not have a physical probability that the other lacks. Yet intuitively, this possibility seems real enough – no conceptual impossibility would appear to be involved. It should further be noted that there could not exist, according to an idealized Humean subjectivist account, two worlds w_3 and w_4 alike in history of particular fact but differing in the following way: in both, the relative frequency of (say) G given F is .7, but in w_3 the physical probability of G given F is .69 while in w_4 the physical probability of G given F is .71. Again, this seems to be a genuine possibility; in fact, the Humean subjectivist must admit that in a world where the physical probability of G given F is .69 there is some chance (even probability of [standard] measure zero is not impossibility) that the relative frequency of G's among the F's is .7, and similarly for a world in which the physical probability of G given F

is .71. That is, he must admit that there is some chance that two worlds with different physical probabilities might look alike with respect to relative frequencies of particular facts.[43]

Now the idealized subjectivist might attempt to make room for such intuitions about different chances being compatible with the same histories by noting that the uniqueness condition (section 3) may not be met, but this maneuver will not ease his difficulties. For disagreement among credence functions that have come from reasonable initial credence functions by conditionalization on the total factual history up to a time could only leave physical probabilities at that time undefined, ambiguous, or partially constituted by a choice from among reasonable initial credence functions. The idealized subjectivist would be well advised to hope that the uniqueness condition will in fact be met, and to attempt instead to convince us that any intuitions we might have about the possibility of worlds w_1 and w_2, or w_3 and w_4, differing as described above are bogus. Hence, even if we charitably ignore important problems that arise for the subjectivist – such as making it plausible that all reasonable credence functions would agree on the basis of the same total factual histories, or giving a subjectivist characterization of extremely small physical probabilities and extremely small differences in physical probabilities – we arrive at the unhappy result that differences we thought could exist among worlds have been excluded for reasons of conceptual clarity and economy.

But don't we often have to give up *prima facie* distinctions for the sake of conceptual clarity and economy? Indeed we do; the question is whether we should do so in *this* case. Rather than attempt to answer this question conclusively, I will finish by appealing to two rather deep intuitions that I hope a Humean in our broad sense will share.

First, let me return to the parallel with phenomenalism. What, after all, are the origins of philosophical disaffection with phenomenalism? Part of the reason is the well-known failure of phenomenalists' attempts to work out in any detail a reduction of the world of things to the world of appearances. But surely part of the reason is the enduring conviction that there really is a world "out there" independent of experience that helps to explain experience, and it seems obvious from the start that this intuitive idea cannot satisfactorily be captured in phenomenalist terms. There are corresponding reasons for being disaffected with subjectivist interpretations of physical probability. In the first place, we may be dissatisfied with existing attempts to carry out such an interpretation. But there may also be an enduring conviction – at least among those who have accepted the possibility of irreducibly probabilistic laws – that there may well be chance "out there" in the world independent of mind, and that the existence of such physical chance would help to explain why certain outcomes are distributed statistically and why we are unable to find hidden variables governing certain processes. It seems obvious from the start that this intuitive idea cannot be captured by a subjectivist analysis, however cleverly it "objectifies" subjective probability,[44] and that it can only be accommodated if we take physical probability more seriously – more realistically – than subjectivism allows. I trust that Humean subjectivists share the intuition that phenomenalism renders the world too "ideal", and share a belief that contemporary physics involves us in irreducibly probabilistic laws. Is it satisfactory to say that physical probability is more "ideal" than the mere *stuff* physics has discovered?

Second, I take it that Humeans even of the broad sort do not want to endorse the rationalistic view that knowledge of physical reality could be arrived at *a priori*. And yet if we interpret physical probability as a form of subjective probability, we will have thereby "established" that physical probability obeys the classical probability calculus. Perhaps it will be countered: How could physical chance be otherwise? Some have proposed, as one component of an interpretation of the quantum theory, a non-classical theory of the physical probabilities of quantum mechanics.[45] Do we want to exclude this possibility *a priori*? Given the interpretative successes achieved in physics using non-classical geometries and other non-standard mathematical theories, we would do well to be cautious. There would be a Hegelian irony in it if interpreters of physical probability eager to preserve their Humean virtue were to derive conclusions about the nature of physical chance *a priori* from the requirements of a reasonable credence function, thereby obeying Hegel's dictum that the rational will become real, and the real, rational.

Peter Railton
The University of Michigan

NOTES

*I would like to thank David Lewis and Richard C. Jeffrey for valuable discussions of the issues considered in this paper and many related matters. I am also indebted to Daniel Hausman for criticisms of an earlier version of some of the material included here. I am grateful to the Society of Fellows at the University of Michigan for research support.

1. See Burks (1955), and further developments of his views in Burks (1977).

2. For some references, see note 30. I use the expression 'epistemic probability' to encompass both inductive probability of a Carnapian or Burksian kind, on the one hand, and subjective probability as developed in the tradition of Ramsey, De Finetti, and Savage, on the other.

3. For a forceful presentation of this line of criticism of the frequentist's appeal to virtual infinite sequences, see Jeffrey (1971). See also Sklar (1970).

4. Again, for further discussion, see Jeffrey (1971).

5. Peirce (1958), vol. 8, p. 225.

6. Cf. Burks (1977), p. 537.

7. Cf. Burks (1955), pp. 188ff. Burks advocates a "pragmatic" theory of inductive probability, according to which atomic inductive probability statements are not logically true (or false), but have both a logical and a pragmatic element: they obey not only the classical probability calculus, but also what Burks calls "standard inductive logic"; however, in addition, "An atomic inductive probability statement does and should express a disposition to act in certain ways in conditions of uncertainty" (Burks [1977], p. 319). They differ from orthodox personalist probabilities in that they must obey standard inductive logic, "norms held in common by all rational investigators" (Burks [1977], p. 531); they differ from orthodox inductive probabilities in that they are not logically true (or false) and they are essentially tied to dispositions to act in situations akin to betting.

8. Burks (1955), pp. 188ff.

9. Burks (1977), p. 513.

10. *Ibid.*

11. *Ibid.*, p. 657.

12. For further discussion, see Railton (1978).

13. Cf. Hempel (1965), sec. 3.

14. *Ibid.*, p. 387. See Cramér (1946), pp. 148–149.

15. Cramér (1946), p. 150.

16. Hempel (1965), p. 378.

17. Indeed, the appeal to relative frequencies in schema (2) is not essential. One could equally well treat the corollary of (2) – that if $p(G,F)$ is very close to one, we can be "practically certain" that result G will occur even if F is realized only once – as the basic definition, and then get application to relative frequencies in finite sequences as follows. Take as the chance set-up F a long series of independent trials of a given chance experiment, each trial of which has probability q of producing a particular outcome O; take as G the obtaining of a relative frequency of outcomes of type O very nearly equal to q; then by the law of large numbers, $p(G,F)$ is very close to one, so that we can be practically certain of observing a relative frequency nearly equal to q the very next time the long series of trials F is realized.

18. Hempel (1965), p. 390.
19. More precisely, the probability of the explanandum proper must be high – an I-S argument with *low* probability may count as an explanation if its conclusion is the *negation* of the explanandum.

   By way of contrast, Jeffrey (1969) claims that I-S arguments are successful explanations only when the probability of the explanandum is "so high as to allow us to reason, in *any* decision problem, as if its probability were 1" (p. 105). Although he naturally does not think this establishes a sharp threshold, he indicates that a probability of $1 - (.5)^{100}$ would be sufficiently high (p. 108).
20. Hempel (1965), p. 345.
21. For further discussion of the requirement of maximal specificity, see Hempel (1965) and (1968).
22. Hempel (1965), p. 403.
23. See Railton (1978).
24. Of course, some facts may be in principle inexplicable for perfectly general reasons. For example, if there are any such, the merely factual initial conditions of the universe would be brute facts relative to even a complete true theory of the universe.
25. This speculation receives some support from the manner in which Hempel applies the parallelism of prediction and explanation to probabilistic explanation. He invokes the following "*condition of adequacy for any rationally acceptable explanation of a particular event*":

Any rationally acceptable answer to the question 'Why did event X occur?' must offer information which shows that X was to be expected – if not definitely, as in the case of D-N explanation, then *at least* with reasonable probability. [Hempel (1965), p. 368; emphasis added.]

   Here the suggestion seems to be that probabilistic explanation ought to be concerned with showing as nearly as possible that the explanandum *had to* occur. A more natural reading of the parallelism between prediction and explanation for the probabilistic case might be that a probabilistic explanation ideally would establish the potential predictability of the explanandum with exactly the probability that the explanandum actually had under the circumstances, be it high or low (see Railton [1978], p. 224).
26. Burks (1977), sec. 10.4.
27. Hume (1734), pp. 164–165, 167.
28. See, for example, the analyses of nomologicality given by Ayer (1956) and Goodman (1955).
29. For a discussion of idealized subjectivism, see Lewis (1979) – where he considers the thesis that the "complete theory of chance" is the same for every world – and Lewis (1980) – where he considers idealized subjectivism the less plausible of two strategies for defending Humean supervenience, a strategy requiring the revival of logical probability but one nonetheless not refuted. Although Lewis finds idealized subjectivism attractive, he produces several serious but not conclusive objections to the view, and in both papers suspends judgment.

   I have glossed over some complications in stating the required restrictions on admissible information on Lewis' view. Clearly, one could not permit perfect *after-the-*

fact information, since it ordinarily would be possible to use non-probabilistic laws to reconstruct chance outcomes on the basis of their aftermaths, and thus to collapse physical probabilities (on this analysis) to zero or one. Skyrms (1977) and Lewis (1980) rightly worry about building temporal asymmetry into the very analysis of chance; it does seem possible, for example, for worlds in which backwards causation is present to contain in the information available before a time *t* sufficient evidence to establish the outcome of certain chance processes at *t*. (Does this count against an analysis of determinism in terms of predictability?) I will ignore this problem in what follows. I will also ignore the distinction between Lewis' *supervenience* version of Humeanism and an explicitly *reductivist* version, in part because I believe many of the problems discussed below arise for both. Unfortunately, I received a copy of Lewis' illuminating "Subjectivist's Guide to Objective Chance" (Lewis [1980]) too late to discuss it in any fullness here.

[30] See Jeffrey (1965), Skyrms (1977) and (1978), and Lewis (1979) and (1980). For a related approach, also broadly Humean, see Mellor (1971). There is some awkwardness about calling Lewis a Humean, even in a broad sense, in light of his realism about possible worlds. However, in the restricted sense in which the expression is used here, it does seem applicable to Lewis' views about physical probability and about laws in general (see Lewis [1973], sec. 3.3). Lewis applies the expression to himself in Lewis (1980).

[31] See Skyrms (1977) and (1978).

[32] For more details, see Skyrms (1977) and (1978). Skyrms at times states his view in two stages: first, he reduces physical propensities to resilient objective probabilities, then he gives a subjectivist interpretation of objective probability. Thus he writes:

Propensities are *resilient* objective probabilities. . . .
 Notice that nothing we have said about "objective" probabilities implies that they are "out there". . . . . Indeed, for the purposes of what has been said so far, "objective" probabilities might just be conditional epistemic probabilities. [Skyrms (1978), pp. 557–558.]

[33] Skyrms (1977), p. 708.

[34] *Ibid.*, p. 709. Again, we assume that the resiliency is not irrational.

[35] For further discussion of a propensity interpretation of the sort under discussion here, see Giere (1973).

[36] Skyrms (1978), p. 559.

[37] For further discussion, see Railton (1981).

[38] A fuller analysis of probabilistic explanation using propensities may be found in Railton (1978).

[39] Because Burks' account of inductive probability requires that inductive probability obey principles of inductive logic as well as the probability calculus (see note 7), he claims there would be less room for variation in the inductive probabilities assigned by acceptable credence functions on the basis of a given set of evidence on his view than there is on prevalent subjectivist views. He therefore claims that his view of probability is less "personal and subjective" than the views of Ramsey and Savage and thus better able to account for the objectivity and definiteness of probabilities associated with physical phenomena such as coin tossing. Moreover, he claims:

Prima facie, it seems that we should examine the behavior of the coin, i.e., how it falls, rather than the behavior of a person, i.e., how he bets, to determine the empirical probability that the coin will fall heads. [Burks (1977), p. 540.]

With this sentiment I am in strong agreement, however it is arguable whether Burks' own view accords with it. After all, Burks does reduce physical probability to an epistemic probability partly constituted by "norms common to all men", and further "[t]he probability values determined by these rules may only be qualitative", so substantial variation is possible (Burks [1977], pp. 529, 326). At best, Burks is able to account for intersubjective agreement (within certain qualitative limits) on physical probability attributions, and he suggests that his interpretation might be termed *intersubjective* as opposed to *subjective*. But as the parallel with phenomenalism shows, *intersubjective* may also be contrasted with *objective* – crudely, existence independent of subjectivity in any form–, and it thus seems that Burks, too, fails to capture the idea of objective, definite physical probability. (See also note 40, below.)

40. It should be added that on a logical interpretation of epistemic probability, an analysis of physical probability in terms of idealized epistemic probability could permit probabilistic explanation to escape *mind*-dependence. In its place, however, would come dependence on *norms* – the canons of rational credence. The true Humean presumably would want to go one step further and reduce the norms in question to facts about human psychology; but a non-Humean might think that these norms are ultimately independent of mind. However, it seems no less troubling to say that physical explanations of physical phenomena involve norms essentially than to say that they involve mind essentially. Certainly, the issue of *idealism* is not mitigated by a shift from mind-and-particular-fact dependence to norm-and-particular-fact dependence. I am indebted here to David Lewis.

41. To be sure, there is a sense in which we may have direct experience of physical objects, causal relations, and so on. However, I assume that we are able to make sense of the traditional empiricist's concern that our relation to things "in the world" is mediated epistemically, at least upon reconstruction.

42. For the sake of those impatient with outlandish examples, it is worth noting that it is not mere example-mongering to imagine that there might be propensities so small that they never chance to be realized in a reasonably long-lived, well-populated universe. According to a contemporary form of (partly) unified field theory, protons have a probability of undergoing spontaneous decay that is sufficiently small to give them a mean life of approximately 10^{30} years. By comparison, the world is believed to be 10^{10} years old. And, of course, even events of standard probability measure zero *can* happen; indeed, if there are any actual infinite sequences of chance outcomes, measure zero events *will* happen, since any *particular* infinite sequence of chance outcomes will have probability zero in the standard measure.

43. Although I will not argue the point here, it seems to me easier to imagine that two worlds with perfectly matching histories of particular fact might differ in *chances* than in (non-probabilistic) *causal relations* or *deterministic laws*. If this is right, then the probabilistic case may afford the strongest intuitive test for Humeanism in our broad sense.

44. The term 'objectified' is Jeffrey's; see Jeffrey (1965), pp. 190ff.

[45.] For a discussion of what has been characterized as the non-Boolean character of quantum-mechanical probabilities, see Bub (1975).

REFERENCES

Asquith, P. and I. Hacking (eds.): 1978, *PSA 1978* 2, Philosophy of Science Association, East Lansing.

Ayer, A. J.: 1956, 'What is a law of nature?', *Revue international de philosophie* 10, pp. 144–165.

Bub, Jeffrey: 1975, 'Popper's propensity interpretation of probability and quantum mechanics', in Maxwell and Anderson (eds.) (1975), pp. 416–429.

Burks, Arthur W.: 1955, 'Dispositional statements', *Philosophy of Science* 22, pp. 175–193.

Burks, Arthur W.: 1977, *Chance, Cause, Reason*, University of Chicago Press, Chicago.

Butts, R. and J. Hintikka (eds.): 1971, *Basic Problems in Methodology and Linguistics: Proceedings of the Fifth International Congress of Logic, Methodology and Philosophy of Science*, Part III, Dordrecht, D. Reidel.

Cramér, H.: 1946, *Mathematical Methods of Statistics*, Princeton University Press, Princeton.

Giere, R. N.: 1973, 'Objective single-case probabilities and the foundations of statistics', in Suppes *et al.* (eds.), (1973) pp. 467–483 .

Goodman, Nelson: 1955, *Fact, Fiction, and Forecast*, Harvard University Press, Cambridge.

Hempel, Carl G.: 1965, 'Aspects of scientific explanation', in *Aspects of Scientific Explanation and Other Essays*, Free Press, New York.

Hempel, Carl G.: 1968, 'Maximal specificity and lawlikeness in probabilistic explanation', *Philosophy of Science* 35, pp. 116–133.

Hume, David: 1973, *A Treatise of Human Nature*, ed. by L. A. Selby-Bigge, Oxford University Press, Oxford. Reprinted from the 1734 edition.

Jeffrey, Richard C.: 1965, *The Logic of Decision*, McGraw-Hill, New York.

Jeffrey, Richard C.: 1970, 'Statistical explanation vs. statistical inference', in Rescher (1970), pp. 104–113.

Jeffrey, Richard C.: 1971, 'Mises redux', in Butts and Hintikka (eds.), (1971), pp. 213–224.

Harper, W. L., R. Stalnaker, and W. Pearce (eds.): 1980, *Ifs*, D. Reidel, Dordrecht. Reprinted from R. C. Jeffrey, ed. *Studies in Inductive Logic and Probability* II, University of California Press, Berkeley.

Lewis, David: 1973, *Counterfactuals*, Blackwell, Oxford.

Lewis, David: 1979, 'Causal explanation', The Howison Lectures in Philosophy at the University of California at Berkeley.

Lewis, David: 1980, 'A subjectivist's guide to objective chance', in Harper *et al.* (ed.) (1980), pp. 57–85.

G. Maxwell and R. M. Anderson, Jr. (eds.): 1975, *Induction, Probability, and Confirmation: Minnesota Studies in the Philosophy of Science* 6, University of Minnesota Press, Minneapolis.

Mellor, D. H.: 1971, *The Matter of Chance*, Cambridge University Press, Cambridge.

Peirce, C. S.: 1958, *Collected Papers of Charles Sanders Peirce* **8**, ed. by Arthur W. Burks, Harvard University Press, Cambridge.

Railton, Peter: 1978, 'A deductive-nomological model of probabilistic explanation', *Philosophy of Science* **45**, pp. 206–226.

Railton, Peter: 1981, 'Probability, explanation, and information', *Synthese* **48**, pp. 233–256.

Rescher, N. (ed.): 1970, *Essays in Honor of C. G. Hempel*, Dordrecht, D. Reidel.

Sklar, Lawrence: 1970, 'Is probability a dispositional property?', *Journal of Philosophy* · **67**, pp. 355–366.

Skyrms, Brian: 1977, 'Resiliency, propensities, and causal necessity', *Journal of Philosophy* **74**, pp. 704–713.

Skyrms, Brian: 1978, 'Statistical laws and personal propensities', in Asquith and Hacking (eds.), (1978), pp. 551–562.

Suppes, P. *et al.* (eds.): 1973, *Logic, Methodology and Philosophy of Science* **IV**, North-Holland, Amsterdam.

BRIAN SKYRMS

PRESUPPOSITIONS OF INDUCTION

We may be far from an adequate account of good scientific method, but however such an account might turn out, it is hard to believe that it would not have a significant contrast. There will presumably be logically consistent ways of assimilating evidence which count as bad science, or unscientific or outright mad. To this extent, at the very least, scientific method will have its presuppositions.[1] An examination of the presuppositions of scientific induction is nothing other than an analysis of scientific method itself.

One can believe that good scientific reasoning should be representable in a probabilistic framework without believing that representation must be unique. A probabilistic approach to the analysis of good scientific reasoning need not assume that a scientist ought to be able to come up with a precise number as the probability of each hypothesis, etc.; all that is needed is that scientific reasoning is, in a well-known sense, coherent. On such a probabilistic reconstruction, the presuppositions of induction may take the form of constraints on the probability measures which can legitimately represent scientific induction. This is essentially the approach that Arthur Burks takes in *Chance, Cause, and Reason*, and the one that will be assumed throughout this paper.

In the first part of this paper, I will discuss presuppositions which relate to learning from experience. The upshot of the discussion will be, *contra* Mill, that learning from experience does not require any presupposition of the uniformity of nature. Rather it is best viewed in the light of the representation of epistemic (or inductive) probabilities as weighted averages of probability measures. Discussion of such representations will naturally involve questions about the nature and interrelation of different varieties of probability: epistemic probabilities, relative frequency probabilities, and propensities. A special case, where the whole picture falls into place beautifully, is the situation in which DeFinetti's celebrated representation theorem applies. I will dis-

285

Merrilee H. Salmon (ed.), *The Philosophy of Logical Mechanism*, 285–319,
© 1990 *Kluwer Academic Publishers.*

cuss the epistemological status of the conditions of that theorem; and will survey more general mathematical techniques which allow analogous representations in cases where the conditions of DeFinetti's theorem do not apply. It will be seen that, when viewed in the proper light, ergodic theory and, from a slightly different angle, the theory of sufficient statistics; allows philosophically salient features of DeFinetti's analysis to be reproduced in a quite general setting. The discussion in part one is, I think, congenial to Burks' views, although somewhat different in emphasis.

In the second part of this paper, I will discuss causal presuppositions of induction. Two types of question are interwined. The first is whether any structural principles governing causation (e.g. asymmetry; locality) are presupposed by scientific induction. The second has to do with the modal force implicit in causal principles. In particular, we are led to ask in what sense empirical evidence can be brought to bear on the modal force implicit in causal principles. In part II, I will argue for a quasi-Humean alternative to Burks' position.

I. LEARNING FROM EXPERIENCE

I.1 Coin flips and the uniformity of nature.

Consideration of induction in a non-probabilistic setting might lead one to conclude that scientific inductive logic must presuppose that nature is *uniform* with the future resembling the past, the unexamined resembling the examined, rather than *random* like the outcomes of independent flips of a coin, or willfully *perverse* like a clever opponent in a zero-sum game.

A closer examination of objective uniformity reveals unexpected complications. Postulation of uniformity means nothing unless we specify the respects in which uniformity is postulated. Uniformity is relative to a system of classification. This basic insight was already there, submerged in the chorus line, with Hume and Mill.[2] Nelson Goodman brought it to center stage with such glamour[3] that it dazzled the eye.

There are complications even if we abstract away Goodman's problem. Suppose that we consider a sequence of outcomes of the same experiment, assuming that the notion of "same experiment" and the outcome space have been chosen with an eye towards projectibility. The question arises as to what we are to count as a uniform sequence; or better, as a sequence with some degree of uniformity. One is naturally led to the consideration of *random* sequences as the really non-uniform ones.

There is an extensive mathematical literature on the characterization of random sequences; it is by no means a trivial question either mathematically or philosophically. Without pursuing the matter in detail, I want to note two features common to the alternative analyses. First, as one would expect, a probabilistically independent sequence of trials will, with high probability,[4] produce a random sequence of outcomes. The second, and more surprising consequence is that random sequences *must* have a limiting relative frequency.[5] Put the other way around, a sequence whose relative frequency does not converge must exhibit highly non-random behaviour, and must fail natural tests of randomness. This is a rather spicy revelation in view of Reichenbach's taking the *existence of limiting relative frequencies* as the principle of the uniformity of nature! The most chaotic and disordered alternative we can find to uniformity *entails* uniformity-in-the-sense-of-Reichenbach.

Without going into the technicalities of these proofs, let me point out a simple fact that makes the conclusion more intuitive. Suppose we have a sequence of coin flips. We test for randomness by looking for patterns in the sequence. One of the simplest of such patterns consists of a run of consecutive heads and tails. As we look at longer and longer initial segments of the sequence, keeping track of the relative frequency in each initial segment, such runs correspond to fluctuations in the relative frequency of heads. Roughly speaking, the size of the fluctuation is for a run of given length, is inversely proportional to the length of the initial segment in question. (e.g. to take an extreme example, suppose that in an initial segment of length n there are no heads and on the n+1 th toss a run of m heads commences. Then the relative frequency of heads on the initial segment of length n + m, and indeed the change in relative frequency resulting from the run, is m/n+m.) If the frequency and distribution of runs remains roughly the same as we move along the sequence, the fluctuations in the relative frequency will approach zero simply because of the increase in the number of trials. For a sequence to fail to approach a limiting relative frequency, the frequencies of runs would have to behave in a fashion that would be counted as non-random on anybody's definition of randomness. The naive intuition of a dichotomy between randomness and order needs to be qualified. Randomness is indeed a kind of disorder, but it carries with it of necessity a kind of statistical order in the large. If randomness is taken as a standard of non-uniformity, then the postulation of this type of non-uniformity is the postulation of the existence of limiting relative frequencies.

I haven't said anything yet about the third *prima facie* possibility of nature as a wilfully perverse opponent. Certainly the world could be arranged so that its creatures with fixed finite intelligence could be frustrated at every turn. Put in this way there is simply no answer to Hume's problem. It is, however, worth noting that if the capabilities of Nature's opponents knew no bounds, then the best nature could do is to adopt a random strategy.

Now, what I have been building up to is that the whole question takes on a quite different complexion when considered in the context of rational degrees of belief (epistemic probabilities).

Consider one of the oldest illustrations in the business:[6] A biased coin is to be flipped a number of times. There are two hypotheses about the physical probabilities (chances, propensities) which characterize the process. BH: The coin is biased two to one in favor of heads and the tosses are independent. BT: The coin is biased two to one in favor of tails and the tosses are independent.[7] Suppose we believe that these are the only two possibilities with non-neglegible probability, and that they are equally likely; we assign them each epistemic probability 1/2. Then it is reasonable that we take as our epistemic probabilities of outcomes, an average of their physical probabilities on each hypothesis, with the weights of the average (in this case equal) being the epistemic probabilities that the corresponding hypothesis is correct. So, the epistemic probability of heads on toss one is $(1/2)(1/3) + (1/2)(2/3) = 1/2$, likewise for the epistemic probability of heads on toss two. But the epistemic probability of heads on toss two *conditional on* heads on toss one is not 1/2, but a bit more. Intuitively, this is because the information that the coin come up heads on toss one supports the hypothesis of bias towards heads a bit more strongly than the hypothesis of bias towards tails. Our elementary example has led us to a philosophical conclusion of fundamental importance. *We can have learning from experience without uniformity of nature!*

In our epistemic probability distribution, (heads on toss 1) has positive statistical relevance to (heads on toss 2) even though we are sure that the tosses are really (*vis a vis* the true physical probabilities) independent! Mathematically, this positive statistical relevance is an artifact of the averaging (mixture) of possible chance distributions to come up with our rational degrees of belief. The conditions for learning from experience are here not created by *knowledge*, but rather by *ignorance*. Here we are learning from experience not by presupposing that nature is uniform, but rather in the teeth of the conviction that it is *not* uniform, by virtue of our uncertainty about *how* it is not uniform. It is evident already from this most elementary example that the traditional problem of induction undergoes a profound transformation when reconsidered from a thoroughly probabilistic point of view.

With regard to the coin flip example, it is worth pointing out here that the epistemic probabilities can be represented as a mixture (weighted average) in more than one way. Consider the *possible relative frequency* distributions corresponding to the relative frequency of heads. Let us construct them so that outcome sequences giving the same relative frequency of heads are equally probable.[8] For simplicity, consider the case in which there are only two tosses. Then the relative frequency distributions corresponding to the realizable relative frequencies in such a sequence are:

Relative Frequency of Heads is:	1	1/2	0
pr(Heads on 1 & Heads on 2):	1	0	0
pr(Heads on 1 and Tails on 2):	0	1/2	0
pr(Tails on 1 and Heads on 2):	0	1/2	0
pr(Tails on 1 and Tails on 2):	0	0	1

(with each of the three columns representing a possible relative frequency distribution.) If we average these possible probability distributions using their epistemic probabilities as weights[9] we get back our epistemic distribution. Note that here the possible *relative frequency* distributions into which we factor the epistemic probabilities are radically different than the possible *chance* distribution in terms of which it was represented previously. In the chance distributions the tosses were independent. But consider the relative frequency distribution corresponding to the relative frequency of heads being 1/2. In this distribution, the probability of heads on toss 2 is 1/2, but the probability of heads on toss two conditional on heads on toss 1 is zero! We have here *negative* statistical relevance with a vengeance. Such distributions (hypergeometric) have been taken as a model[10] for counterinductive reasoning. That is, if I was *sure* that the relative frequency of heads was 1/2 in a sequence of two tosses, observation of a head on toss 1 would lead me to drop the probability of a head on toss two from 1/2 to zero. (Obviously the same sort of effect can be illustrated with respect to any finite number of tosses.).

We have, in our example, an illustration of three different conceptions of probability: epistemic probability; physical propensity or chance; and relative frequency. Our epistemic probability distribution could be viewed either as a mixture of possible propensity distributions, or as a mixture of possible relative frequency distributions, even though the propensity distributions made the trials independent while the relative frequency distributions made an outcome on a trial negatively relevant to an outcome on another trial. The interaction of these three conceptions of probability lies at the heart of inductive reasoning.

I.2 DeFinetti's theorem.

In the preceeding section, we took a weighted average to get an epistemic distribution. If we start with the epistemic distribution, we say it has that representation as a mixture of probability distributions. In general, such representations need not be unique. Indeed, we saw that the epistemic distribution in question also had a representation as a mixture of quite different relative frequency distributions.

Each of the distributions in the preceeding paragraph has the property of being invariant under finite permutations of trials (pr(Heads on 1 and Tails on 2)=pr(Tails on 1 and Heads on 2)). Probability measures with this property are said to make the trials *exchangeable*. Any independent sequence of trials is exchangeable and any mixture of exchangeable sequences (i.e. of measures that make the sequence exchangeable) are exchangeable. DeFinetti introduced the concept of exchangeability to prove a representation theorem. The statement of the theorem is neatest if we consider infinite sequences of trials (and assume sigma-additivity of the probability measures involved).[11]

Then we have:

Any exchangeable sequence of trials is representable as a mixture of independent sequences of trials. Furthermore, it has a *unique* representation as a mixture – of independent sequences.[12]

DeFinetti's theorem establishes a broad domain for the sort of model of learning from experience that we looked at in minature in the last section. A fact of particular interest that is pointed out in *Cause, Chance and Reason*, is that within the framework under consideration here, counterinductive inference cannot arise. If our epistemic degrees of belief are exchangeable, then we can treat them as an average of independent distributions. We can learn from experience, just as before. If we observe a sequence of outcomes, and conditionalize on that observation, then the subsequence of remaining outcomes is still exchangeable, but the weights of the independent sequences into which it factors will have shifted as a result of our observation. As the sequence grows, then with probability approaching one, the weight of the average will become concentrated on one of the independent sequences.[13] Positive statistical relevance between like outcomes of different trials will arise from the averaging, except in the extreme case of a degenerate average where we start out with our degrees of belief already concentrated on an independent sequence. In this case of infinite sequences of trials, DeFinetti appears to have replaced the presupposition of the *objective* uniformity of nature with the modest *subjective* condition of exchangeability. Thus DeFinetti's sanguine remarks about resolving the problem of inductive reasoning along Humean lines. How far DeFinetti's theorem goes toward resolving the problem of induction remains to be considered. But it cannot be denied that DeFinetti's ideas put the problem in a dramatically new light.

The second important thing that DeFinetti's theorem does for us is to establish the relationship between degree of belief, chance, and relative frequency in the case under consideration. To show this let me say a little more about how the DeFinetti representation works. *Any* probability for a finite sequence can be represented as a mixture of those distributions gotten by conditionalizing out on statements of relative frequency. Exchangeability of the original distribution assures that in the factors so obtained all outcome sequences with positive probability are equally probable. Decomposition of finite exchangeable sequences into relative frequency probabilities in this way yields hypergeometric distributions (the "counter-inductive", sampling from an urn without replacement type distributions that were the relative frequency distributions of the last section.) Every finite exchangeable sequence has a unique representation as a mixture of hypergeometric sequences. As the length of the finite sequence approaches infinity the hypergeometric distribution approaches independence. If we are willing to speak directly of infinite sequences, then we can regard the DeFinetti representation as a mixture of those distributions that can be gotten from the epistemic distribution by conditionalizing[14] on statements of *limiting relative frequency*.

If we believe that our infinite exchangeable sequence is governed by unknown propensities which make the trials independent, then by the uniqueness clause of DeFinetti's theorem, we cannot represent our degree of belief in two distinct ways: as a mixture of relative frequency probabilities and as a mixture of physical propensities. It comes to the same thing, whether we think of our sequence as governed by a mixture of propensities, or governed by the same mixture of relative frequency probabilities, or simply governed by an exchangeable probability measure.

DeFinetti regards this as a way of dispensing with metaphysically dubious propensities. (One might ask whether this is really any different from the familiar argument that propensities can be identified with limiting relative frequencies by virtue of the strong law of large numbers. Without yielding to the temptation for an enormous digression here, we can emphasize for the purposes of representation of the epistemic distribution as a mixture, a set of situations in which the single case propensity and limiting relative frequency of an outcome do not coincide which has epistemic probability zero can be disregarded. Thus, *if* our focal point is the epistemic distribution, then in the case under consideration, the postulation of propensities pulls no weight).)

DeFinetti's analysis for the situation described is so beautiful that we would almost like to forget that it is a special case. But although it is an important case, at the heart of probability theory, it is, after all, a special case. What if the number of trials is finite? What if the plausible "propensity" measures do not make the trials independent? What if the epistemic probabilities are not exchangeable? How far can the bold insights of DeFinetti's analysis be generalized?

Arthur Burks takes finite exchangeable sequences which are hypergeometric to be models for counterinduction: demonstrations of its consistency, and dramatic evidence that scientific inductive logic has presuppositions. Finite sequences are perhaps really the usual objects of our attention. And although all finite exchangeable sequences can be represented as mixtures of hypergeometric sequences, some can also be represented as mixtures of independent sequences: (as in the example of I.1) and thus permit inductive learning from experience. These issues regarding finite exchangeable sequences will be discussed in the next section.

What of physical sequences where the chances are not independent (e.g. where they are Markov)? Burks argues that we can still take epistemic probability to the epistemic expectation of chance (i.e. the average of the chance distributions weighted by their epistemic probability) and learn from experience *via* Bayes theorem. This is the standard Bayesian approach to statistical inference. Furthermore, he argues that we can take chances (propensities) to be epistemic probabilities conditional on a description of the experimental situation. Like DeFinetti, this allows us to think of chances as artifacts of a representation, but the representation need not have so frequentist a flavor.

I have put forward a view of propensities much like this myself, with the addition that the description of the possible experimental situation be chosen with an eye towards "*resiliency*". A rough description of what I mean by "resiliency" is that the probabilities conditional on a specification of an experimental situation should remain invariant on further refinement of the specification of that situation.

It might be objected against Burks and myself[15] that all this is mere philosophical posturing without some mathematics to back it up. "... even Don Quixote did not consider venturing forth upon the world astride a rocking horse."[16] Most of the rest of part one will consist of a quick check of the stables. Mathematics has been kind to us. Don Quixote can have a thoroughbred.

What if the epistemic probabilities do not make the sequence exchangeable? Should this be allowed to happen? It is possible to get oneself in a state of mind where exchangeability appears to be a natural *a priori* constraint on rational degrees of belief. The trick is to think of the names of the trials as utterly devoid of meaning. Suppose the numbers that index the trials do not mirror temporal order, but were merely assigned at random. Then, failing definite knowledge of any of the outcomes of the trials, an exchangeable epistemic distribution appears the most natural. This is the position that Carnap took regarding exchangeability (symmetry). But as soon as the numbers which index the trials carry some information as to physical order, as intended in the theory of stochastic processes, exchangeability is no longer a plausible *a priori* requirement. Already in 1937, DeFinetti suggested generalizations of exchangeability to "partial exchangeability" to deal with such cases (DeFinetti, 1937). This may be thought of as a move towards the more free swinging treatment of chance referred to above. How far it goes in that direction remains to be considered.

I.3. Finite exchangeable sequences.

In section I.1, the example was of a finite exchangeable sequence that could be represented as a mixture in two ways: as a mixture of hypergeometric sequences or as a mixture of independent sequences. Every finite exchangeable sequence has a unique representation as a mixture of hypergeometric sequences, with the elements of the mixture gotten from the original probability by conditioning out on statements of relative frequency, and the weights being the respective original probabilities of those relative frequency statements. (e.g. fixing a sequence of coin tosses at length n, we can take the appropriate statements to be; no heads, 1 head, ... n heads). Not every finite exchangeable sequence has a representation as a mixture of independent sequences. For instance, consider the hypergeometric distribution in the two toss case corresponding to one head and one tail.[17] (Pr(H on 1 and T on 2)=Pr(T on 1 and H on 2)=1/2)

Those finite exchangeable sequences which admit a representation as mixture of independent sequences are of some special interest because they are ones that permit learning from experience in the way illustrated in the example of I.1. Considering that example a little further, I want to point out the features of the situation that make this so. Let me concentrate on showing why Pr(Head on 2 given Head on 1)>Pr(Head on 2). By the definition of conditional probability:

$$\text{Pr(H2 given H1)} = \text{Pr(H2 \&H1)/Pr(H1)}$$
$$= \text{Pr(BH given H1)Pr(H2 given BH \& H1)} +$$
$$\text{Pr(BT given H1)Pr(H2 given BT \& H1)}$$

But by *independence* of trials in the distribution conditional on BH and on that conditional on BT, Pr(H2 given BH &H1)=Pr(H2 given BH), likewise for BT, so we have:

$$= \text{Pr(BH given H1)Pr(H2 given BH)} +$$
$$\text{Pr(BT given H1)Pr(H2 given BT)}$$

or, emphasizing the role of the probability measures that are elements of the mixture:

$$= \text{Pr(BH given H1)}\text{Pr}_{BH}(\text{H2}) +$$
$$\text{Pr(BT given H1)}\text{Pr}_{BT}(\text{H2})$$

Since the unconditional probability of heads on 2 is:

$$\text{Pr(BH)}\text{Pr}_{BH}(\text{H2}) + \text{Pr(BT)}\text{Pr}_{BT}(\text{H2})$$

So the change in the probability of heads on toss two upon observing heads on toss one come about by *remixing* the objective distributions for toss two with different coefficients, or as Richard Jeffrey says, by "probability kinematics". (Jeffrey, 1965) The remixing makes Pr(H2 given H1) greater than Pr(H2) because the shift in coefficients gives more weight to the objective distribution (Pr_{BH}) which makes H2 more probable. That the remixing gives more weight to the objective distribution that makes H1 more probable follows from Bayes theorem; that this is the same one that makes H2 more probable follows from exchangeability. The reader (to whom this is not already all old hat) should have no trouble seeing how this reasoning generalizes.

So we find models of straightforward inductive behaviour in the finite case when we find representability as a mixture of independent sequences. Under what conditions do we have this, or a good approximation to it? There is an illuminating discussion of this question in a recent article by Persi Diaconis (1977). Diaconis shows that a finite exchangeable sequence of length r, which can be extended to an exchangeable sequence of length k, approximates a mixture of independent sequences, with the error in the approximation going to zero as k approaches infinity.[18] To say that the sequence of length r can be extended to a sequence of length k just means that there is a probability measure on a sequence of length k which yields the original measure of the original subsequence of length r as marginal, by summing over the extra outcomes. We may think that it will be highly likely that an experiment will only be repeated a finite number of times, but not be absolutely sure. Even if we are sure that an experiment will only be repeated a finite number of times, we may be unsure as to what finite number of times it will be repeated. Even if we are sure that it will only be repeated n times, we may believe that it could have been repeated a greater number of times. Consequently, there is *wide scope* for the application of the DeFinetti analysis of learning from experience, in the case of finite sequences.

What does this tell us about finite hypergeometric sequences as a model for counterinduction? It should be reemphasized at the onset, that hypergeometric sequences are not, in themselves, perverse. If I know than an urn contains 50 red balls and 50 white balls and nothing else, and my experiment consists of random sampling from that urn without replacement, then the hypergeometric distribution is the one that I should adopt. If, after drawing 50 red balls, I change the probability that the 51st draw will be red to zero, I will have learned from experience in a quite unexceptionable way, notwithstanding the negative statistical relevance of like outcomes on different trials. It is adopting such a distribution in less appropriate circumstances, say in the coin flipping case, that is perverse.

Concentrating one's probability on a hypergeometric distribution in a case like the coin case which is more plausibly a mixture of independent distributions, can be taken as a model of counterinductive *behaviour*.[19] Finding a model for a counterinductive *method* may be another matter. Suppose someone believes there will be only 100 tosses of this coin and that exactly half will be heads, and adopts the corresponding hypergeometric distribution as his epistemic probability. What does he predict about the outcome of the 101st toss while the coin is spinning in the air? What probabilities does he assign after observing the 51st heads? There are falsifiable contingent assumptions which lie behind his assignment. They provide no clue as to how the assignment should be extended to a greater number of trials. One might require of a counterinductive *method* that it deal with any possible sequence of observations as input. Diaconis' result then shows just how tightly the counterinductivist is hemmed in by exchangeability.

(This is not to claim that in no sense is a model of counterinductive method possible. The counterinductivist might, at the moment of truth, simply leap to a different hypergeometric distribution consistent with the observations to date, e.g. the one corresponding to an urn with 51 red balls and 50 black balls.[20] This sort of counterinductive strategy would not respect extendability – life would be full of crises requiring leaps to entirely new probability distributions which are not extensions of the old[21] – but the counterinductivist is, after all, rather odd duck anyway.)

Scientific induction has it presuppositions. We do not really need the picture of "the counterinductivist" to dramatise the point. The difficulties encountered in painting the picture are none the less instructive. They emphasize the considerable role of mixing in the creation of statistical relevance in the finite case as well. The theoretical importance of mixtures of independent sequences leads, in the finite case, to a special appreciation of alternative ways of representing a sequence as a mixture. The canonical DeFinetti representation as a mixture of relative frequency distributions is, in the finite case, a representation as a mixture of non-independent sequences. In many cases, the same mixture can be naturally be represented as a mixture of independent chance sequences. Thus the study of learning from experience in the finite case tends to emphasize the distinction between relative frequency and chance.

I.4 Stochastic processes and partial exchangeability.

In the theory of stochastic processes, the numbers which index the events are taken to convey information as to some physical order (often temporal order.) Here, in the discrete case we have repetitions of an experiment indexed by the positive integers, or in a doubly infinite sequence indexed by the whole numbers ($\ldots -1, 0, +1 \ldots$), and in the continuous case indexed by a continuous "time" parameter. (The theory can be generalized so that the experiment is parametrized in several dimensions, in which case we have a *random* field. Random fields form the analytical framework for the study of the statistical mechanics of lattice gases.) Here it is no longer possible to flirt with the idea that exchangeability might be given some Kantean *a priori* justification.[22] Here, order can, and often does, make a difference. Exchangeability in the epistemic distribution must be considered a special case, and by the same token, so must independence in the objective distributions. A great many physical stochastic processes which are not independent, satisfy Markov's slight weakening of independence, where the chances of a outcome on a trial depend only on the outcome of the preceding trial. (The Markov property is obviously connected with causality, and an appropriate Markov property in a random field where the parameters represent space and time would represent a probabilistic principle of locality of causation.) It is therefore of great interest whether the DeFinetti analysis can be extended to Markov chains. Again, when we are dealing with arbitrary sequences of random variables, rather than just repetitions of an experiment, there is no reason to assume in general that the random variables in question should be exchangeable in the epistemic distribution, or independent in the possible objective distribution.

DeFinetti appreciated the importance of these questions from the beginning. In 1937 he writes: "To get from the case of exchangeability to other cases which are more general but still tractable, we must take up the case where we still encounter "analogies" among the events under consideration, but without their attaining the limiting case of exchangeability" (Jeffrey, 1980, p. 197).

To this end, DeFinetti introduces the notion of *partial exchangeability*:

"All the conclusions and formulas which hold for the case of exchangeability are easily extended to the present case of *partially exchangeable* events which we could define by the same symmetry condition, specifying that the events divide into a certain number of types 1,2, ...g, and that it is only events of the same type which will be treated as "interchangeable" for all probabilistic purposes. Here again, for this definition to be satisfied quite generally, it need only be satisfied in a seemingly much more special case: it suffices that there be a unique probability ω n_1, n_2,...n_g that n given events all happen, where n_1, n_2...,n_g belong respectively to the first, second,... gth types and $n_1 + n_2 + ... + n_g = n$ – and where this unique probability is independent of the choice of particular events from each type (Jeffrey, 1980, p. 198).

DeFinetti introduces this notion of partial exchangeability with a coin tossing example, where several odd shaped coins are tossed. Here it is not so plausible to expect exchangeability in the sequence of all tosses, but it is plausible to expect exchangeability within the sequence of tosses of a given coin. The example can be varied to illustrate various degrees of subjective analogy between the different types of toss. In one limiting case, the coins may be thought to be much the same, with the upshot of this judgment being global exchangeability. As the other end of the spectrum, the coins may be thought so different that the results of tossing one give no information about the bias of the others.

It may not be apparent how this notion of partial exchangeability is meant to be applied to the theory of stochastic processes. The leading idea, in the Markov case, is clear: "In the particular case of events occurring in chronological order, the division into classes may depend on the result of the previous trial; in such a case we have the Markov form of partial exchangeability (DeFinetti, 1970, II, p. 212)." That the probability of an outcome on toss 56 of a Markov chain depends on the outcome of toss 55 is not a reflection of any special status of toss 55 over and above its being the predecessor of toss 56. Thus, if we take the subsequence of tosses whose predecessor has a given outcome, renumber them 1,2, . .–. in order of occurrence (this suppressing some information as to position in the original sequence), we should expect the resulting sequence to be exchangeable.

Even though the leading idea is clear, this case is more complicated than the case of the odd coins, and DeFinetti's suggestion constitutes a broadening of the notion of partial exchangeability. To illustrate, consider an example of Diaconis and Freedman (1980a). Instead of flipping a coin we flick a thumbtack across the floor, and we play it as it lays. The probability of its landing point up (U) or not may plausibly depend on the result of the previous trial. It would be a mistake to think that we could get an exchangeable subsequence of trials just by conditioning on a statement which posits a fixed outcome (say, point up) for each of their predecessors. Conditioning on the statement that tosses 1, 5, and 6 have predecessors that land point up, makes it certain that toss 5 lands point up, but cannot be expected to make U on 1 or U on 6 certain. Here the very act of dividing up the trials into classes may presuppose information which destroys the exchangeability of the trials within those classes.

On the other hand, there is a generalization from DeFinetti's alternative characterization of exchangeability. We can almost give a frequentist characterization of a sequence, so that that characterization can plausibly be held to uniquely determine the probability of that sequence. That is, we can specify the frequency of transitions from state to state (up to up, up to down, etc.) in the sequence. I said "almost" because to this, we must add the non-frequentist information of the initial state of the system[23]. Given that we believe that the probabilities are governed by a Markov law, sequences with the same initial state and the same transition count should be equiprobable.

Freedman, (1962), brought exchangeability and various forms of partial exchangeability together under the concept of a certain form of sufficient statistic. Let a statistic here be a function from sequences of length n into some set of values (which may be a set of vectors.) We say that the statistic, T, is a *summarizing statistic* for the probability assignment, P on sequences of length n; or alternatively, that the probability assignment is *partially exchangeable with respect to the statistic* iff $T(x)=T(y)$ implies $P(x)=P(y)$ where x and y are sequences of length n. As Diaconis and Freedman (1980a) point out this covers all the intended cases of partial exchangeability. In the case of exchangeability, the summarizing statistic consists just of the frequencies (or relative frequencies) of outcomes. In the case of tossing several funny coins and analogous cases, the value of the summarizing statistic consists of a vector whose components are the frequencies (or relative frequencies) for each coin. In the Markov case, we have a vector whose components are the transition counts together with the outcome of the first trial.

Although these two applications are much in the spirit of DeFinetti's original analysis, the notion of partial exchangeability with respect to a statistic is a very substantial generalization of the notion of partial exchangeability. In particular, it should be noted that there is nothing in the definition of a summarizing statistic that requires it to be or involve a frequency count. Freedman's treatment moves us very far towards the higher levels of generality that will be discussed in the subsequent sections of part one.

If the statistic, T, is a *summarizing statistic* for two probability assignments P_1 and P_2, then by definition, for any value of the statistic, the sequences in its inverse image are given the same value by P_1 and likewise for P_2. Those sequences will therefore be given the same value by any weighted average of P_1 and P_2. That is, partial exchangeability with respect to a fixed statistic T, is a property of probability distributions preserved by mixing. If we consider the family of probability assignments (on a given set of sequences of length n) which are partially exchangeable with respect to T, it is closed under mixing.[24] Those probability measures which concentrate probability one on the set of sequences which are the inverse image of one value of the statistic, are the extreme points of the set of measures, i.e., the ones that cannot be represented as a non-degenerate[25] mixture of distinct measures in the set. If the set contains a regular assignment (one which gives each possible outcome non-zero probability), then all the extremal probability assignments can be recovered from it by conditioning on the possible values of the statistic. We now have the following generalization of DeFinetti's theorem for finite sequences:

Every probability assignment P which is partially exchangeable with respect to T is a unique mixture of the extreme measures P_i. The mixing weights are $w_i = P (x:T(x)=t_i)$. (Diaconis and Freedman, 1980).

(For a discussion of the extreme measures in the finite case for some of the special cases of partial exchangeability that have been mentioned see Diaconis and Freedman) (1977, p. 238).

Freedman (1962) proves a general representation theorem for partial exchangeability with respect to a statistic. Since this concept is defined with respect to finite sequences, it is necessary to find a way to relate the various summarizing statistics for subsequences of an infinite sequence. For our probability space with infinite sequence as its elements, let us take as our new sense of summarizing statistic, an infinite sequence of summarizing statistics in the old sense. That is, a statistic, U, in the new sense can be taken as a function from the integers to statistics in the old sense for finite sequences, and a summarizing statistic in the new sense is one such that for any n, U_n, is such that if two finite sequences have the same value of U_n the set of infinite sequences that have them as subsequences in a specified position must be equiprobable. Freedman introduced the notion of an S-structure to assure that the U_ns mesh together nicely. U has an *S-structure* if for any for any finite sequences A,A' of length n and B,B' of length m, if $U_n(A)$ = $U_n(A')$ and $U_m(B) = U_m(B')$ then $U_{n+m}(A @ B) = U_{n+m}(A' @ B')$, where A @ B is the result of concatenating the sequence A with the sequence B. The summarizing statistics mentioned specifically in this section (relative frequency, frequency within types, initial outcome together with transition counts) all have an S-structure. Freedman then proves that a probability P is partially exchangeable with respect to (is summarized by) a statistic U with an S-structure if and only if it has a representation as a mixture of ergodic (metrically transitive) probabilities which are partially exchangeable with respect to U. (I will postpone the discussion of ergodic measures until the next section.) He uses this result to attack the question of the characterization of mixtures of Markov chains. The answer when the stochastic process is stationary is in Freedman (1962). Diaconis and Freedman (1980b) show that when a stochastic process is recurrent[26] it has a representation as a mixture of Markov chains if and only if it is partially exchangeable with respect to the statistic of initial state and transition count. The representation is unique.

In the simpler case of partial exchangeability, (several different biased coins) a representation as a mixture of products of Bernoullian measures is given in DeFinetti (1938) Link (1980) and Diaconis and Freedman (1980 a). Link give proofs of representation theorems for such "k-fold partial exchangeability" using Choquet's theorem as the main analytical tool. Diaconis and Freedman indicate how the proof of one of these theorems can be gotten by passing to the limit in the finite case.

On the other hand, when random processes are generalized to random fields, there are generalizations of DeFinetti's theorem. This is an area of much current interest, largely because of the connections with statistical mechanics and thermodynamics. Two recent studies are Preston (1976) and Georgii (1979).

It is clear that methods in the spirit of DeFinetti can be used to give an analogous account of a much more general class of cases than those treated in DeFinetti's original theorem. The question as to how general that class in fact is has an answer which depends in part on ongoing mathematical research and in part in how generously we interpret "analogous" and "in the spirit of DeFinetti". The consequences of a generous interpretation will be discussed in the next two sections.

I.5 Invariance, resiliency and ergodicity.

Many of us would like to think of physical probabilities quite generally in the way that DeFinetti thinks of his "relative-frequency probabilities"; as an artifact of our representation of epistemic probabilities as a mixture. What is needed to back up this position is a general mathematical theory. Ideally, such a theory should do two things for us. First, it should prove a unique representation theorem. That is, given degrees of belief about the outcome of an experiment (which might be required to be invariant provided that the basic experimental arrangement is fixed) we would like to have a unique representation of our degrees of belief as a mixture of possible physical probabilities. Secondly, we would like an account of learning from experience. This could be provided if we could prove some kind of law of large numbers for repetitions of this experiment. If we had such an account, we could show how to learn the correct physical probabilities from experiment, and how it is possible to use these to predict the results of a sequence of experiments given *the correct* physical probabilities. Finally we, or at least I, would like to claim that the objective physical probabilities should have an objective resiliency, or invariance, which is connected with the subjective resiliency or invariance of the degrees of belief with which we started.

Let me formulate these desiderata for the mathematics of a subjectivist theory of chance precisely at a high level of generality. Let our degrees of belief be idealized to a probability measure, P, on a standard Borel space (Ω, F, P)[27] where Ω is a set, F is a sigma-field of measurable subsets of Ω, and P is a probability measure on F. We will think of a point in Ω as specifying not only the outcome of an experiment, but also the experimental arrangement and background conditions of the experiment. Although we may not know exactly what the physical probabilities are, we may know that some of the background conditions (e.g., time of the experiment) are irrelevant to them. That is, the physical probabilities should be invariant under variation of some background conditions, and so should be their mixture. We may typically expect that there may be some such symmetry, or invariance which characterizes our epistemic probabilities. We may express this formally by introducing a transformation (e.g. a shift in time of the experiment) which maps the probability space, Ω, into itself. (We should require that the transformation respect measurability, i.e. that the inverse image of a measurable set be measurable.) We will then say that our probability *P is resilient with respect to the transformation T* if for every measurable set A, a set which the transformation carries into it $(T^{-1}A)$ has the same probability that it does $P(T^{-1}A)=P(A)$. It is evident that resiliency with respect to a fixed transformation, T, is a property of probability measures that is preserved under mixing.

We know that some background factors are irrelevant to the probabilities, and this information manifests itself in resiliency with respect to some transformation, T. But we may not know the true values of the *relevant* variables, and so we are uncertain about the correct physical probabilities. We want to think of the possible physical probabilities as probabilities which are fixed by a knowledge of *all* the relevant factors. The "all" is important, for if we were uncertain as to some of them, we would again have our degrees of belief being a mixture of possible physical probabilities. We are thus led to the following characterization of the possible physical probability measures of which our epistemic probability measure is supposed to be a mixture:

> (Invariance) (I) A possible physical probability measure should be resilient with respect to T.
>
> (Indecomposability) (II) We should not be able to get "truer" physical probability from a physical probability by conditioning on a specification of further factors.

To get a "truer" physical probability by conditioning, we would have to condition on a set, such that conditioning would lead us to a different measure which was also resilient with respect to T. We can formulate (II) as the denial of the existence of such sets. By a *possible total specification of the additional factors* I will mean a set A such that $A = T^{-1}A$ mod O (i.e. that the probability of the symmetric difference between A and the set that T takes into A should be zero.) The idea here is that a total specification of factors should be invariant under the transformation that represents repetitions of the experiment. We might then replace (II) with

> (II')There is no possible total specification of additional factors, except those which the measure in question gives probability one or zero.

Conditioning on a set of measure one would leave the probability unchanged; conditioning on a set of measure zero is undefined. If a probability satisfies both (I) and (II') I will say it is *objectively resilient* with respect to T. (The "objective" comes from the motivating idea that these objectively resilient probabilities should be taken as our "objective" physical probabilities.) One might think of substituting for (II') in this definition, a condition which is at least equally well motivated: (II") The physical probability cannot be represented as a non-trivial mixture of probabilities which are resilient with respect to T. It so happens that the substitution gives an equivalent definition. The representation theorem that we would like is: *Every probability that is resilient with respect to T has a unique representation as a mixture of probabilities that are objectively resilient with respect to T.*

We also want our physical probabilities to have a connection with frequency. For a probability measure that is objectively resilient with respect to T, the transformation T is to be thought of as a repetition of the experiment with all the relevant factors fixed. For each point, ω, in the probability space, Ω, we can consider the sequence ω, Tω, T(Tω) ... T^nω, ..., and we can consider the relative frequency of any measurable set, A, in F, in any finite segment of that sequence, and the limiting relative frequency in the sequence provided that it exists. Call this sequence the *orbit* of ω. The nicest generalization of the law if large numbers that we could imagine here is that *for any set A in F, the limiting relative frequency of A in the orbit of ω exists and equals the probability of A almost everywhere* (i.e. for every ω in Ω excepting a set of measure zero.)

Of course, the theory that I have been hypothesizing already exists, although it is not usually described from this perspective.[28] It is none other than ergodic theory.[29] Where I said that *P is resilient with respect to the transformation T*, the more usual terminology is that *T is measure preserving with respect to P*. Where I said that *P is objectively resilient with respect to T*, it is more common to say that T is *ergodic* (or metrically transitive or indecomposable) with respect to P^{30}. The fantasized generalization of the law of large numbers is nothing other than Birkhoff's celebrated pointwise ergodic theorem (or rather a special case thereof.) It and the representation theorem are central facts of ergodic theory.

Actually, Birkhoff established the existence of the limiting relative frequency almost everywhere in general, when T is measure preserving with respect to P but need not be ergodic with respect to P. (More generally, under these conditions, if f is an integrable function, then its limiting empirical average:

$$\lim_{n \to \infty} 1/n \sum_{k=0}^{n-1} f(T^k \omega)$$

exists almost everywhere. Considered as a point function on Ω, the limiting empirical average is integrable, invariant, and has its expectation equal to the expectation of f.) If the transformation is, in addition, ergodic with respect to P, then the limiting relative frequency is almost everywhere constant and equal to the probability (and, more generally, the limiting empirical average is almost everywhere constant, and equal to the expectation of f.)

In light of the representation theorem,[31] this should be understood as follows. If the probability P is *resilient* with respect to T (T measure preserving), then the limiting relative frequencies will almost surely converge to a probability measure P' which is one of the *objectively resilient* (T ergodic with respect to P'). One is reminded of the De-Finetti analysis whereby the scope of the law of large numbers was extended from independent sequences to exchangeable ones. There the limiting relative frequencies almost surely converge to the probabilities of one of the independent sequences of which the exchangeable sequence is a mixture. Indeed, that case is a special case of the relationship, between resilient and objectively resilient measures.

Historically, ergodic theory arose out of statistical mechanics, where the transformation, T, was thought of as a time transformation specified by the dynamical law of the system. Then, taking M as the Lebesgue measure on phase space, the question arises as to whether one can prove that T is ergodic for a given dynamical system. This is a difficult mathematical problem for the physical systems for which the theory was designed and even for idealized approximations of them. For a survey from this point of view see Sinai (1976). However, the mathematics is independent of this interpretation, and is equally congenial to the one we have put upon it.

Before I indicate how the ergodic theory applies to stochastic processes as a special case, allow me to use the ergodic theorem to substanciate my earlier claim that (I) & (II') is equivalent to (I) & (II"); i.e. objectively resilient (ergodic) probability measures are the extreme points of the convex set of resilient (preserved by T) measures. That is, a measure is objectively resilient if it is resilient and cannot be represented as a non-trivial mixture:

$$P=(a/a+b)P_1 + (b/a+b)P_2 \; a,b > 0; P_1 \neq P_2$$

of distinct resilient measures. One direction is trivial. If P is not objectively resilient, then by condition (II') there is a possible total specification, A, such that $P(A) \neq 1, P(A) \neq 0$. Then P can be written as a nontrivial mixture of resilient probability measures, i.e. the measures gotten by conditioning on A and $-A$:

$$P(\bullet)=P(A)P(\bullet|A) + P(-A)P(\bullet|-A)$$

For the other direction, notice that if P is a non-trivial average of P_1 and P_2, and P is objectively resilient, the P_1 and P_2 must be also since under these conditions P gives a set probability 0 only if P_1 and P_2 give it probability 0, likewise for probability 1. Now suppose for reduction that P is objectively resilient and a nontrivial average of the resilient measures P_1 and P_2. Then, by the previous observation, P_1 and P_2 are also objectively resilient. By hypothesis, P_1 and P_2 are distinct, so there is a measurable set A to which they assign different values. By the ergodic theorem, the limiting relative frequency almost everywhere exists and is equal to $P_1(A)$, $P_2(A)$, and $P(A)$. This contradicts the assumption that $P_1(A) \neq P_2(A)$.[32]

Consider a discrete stochastic process consisting of a doubly infinite sequence of random variables, $..f_{-2}, f_{-1}, f_0, f_1, f_2...$, on a common probability space, S, and let Ω be the infinite product space whose elements ω consist of doubly infinite sequences of members of $S: \omega = (..\omega_{-1}, \omega_0, \omega_1, ...)$. Let T be the *shift* transformation, which takes $(...\omega_{-1}, \omega_0, \omega_1 ...)$. to $(... \omega_0, \omega_1, \omega_2, ...)$. Any statement about the random variables can be rephrased in terms of one of them and the shift transformation. To say that the shift is measure preserving is the same thing as saying that the stochastic process is stationary.

Suppose that the stochastic process is Bernoullian; that is, that the random variables are independent. Then the shift is *ergodic* with respect to the product measure on the product space, Ω. The Birkhoff ergodic theorem applied to this case gives the strong law of large numbers for Bernoulli sequences.[33] The Bernoullian measure is not the only sort of measure with respect to which the shift transformation is ergodic. Certain Markov measures corresponding to certain Markov processes do as well.[34] So the Bernoullian measures are not the only extreme measures corresponding to the shift transformation. However, if instead of the shift, we consider the group of transformations which take a sequence ω into another ω' just in case one can be gotten from the other by a finite permutation of its elements, the measures with respect to which that group of transformations is measure preserving are the exchangeable ones, and the extreme probability measures are the independent ones.

The DeFinetti analysis of exchangeable sequences as mixtures of independent ones is a special case of the theorems of ergodic theory. Freedman (1962) uses the ergodic representation theorem as a tool to show that any stochastic process that is partially exchangeable with respect to the statistic consisting of the transition counts and the outcome of the first trial, is a mixture of stationary Markov chains. In the course of establishing this, he proves a more general theorem with respect to partial exchangeability with respect to a statistic, to which I referred in I.4. However, at a higher level of generality, the ergodic theory itself *is* the generalization of DeFinetti's analysis. Any stationary stochastic process has a probability preserved by the shift transformation, in the manner indicated. That probability therefore has a unique representation as a mixture of ergodic (metrically transitive, "objectively resilient") probability measures. Birkhoff's ergodic theorem shows that for almost every sequence of observations, the relative frequencies will converge to one of the extremal ergodic probabilities. Here we have an account of learning from experience which preserves the main features of the DeFinetti account and which is applicable to *any* stationary stochastic process.

As indicated earlier, even this level of generality is a special case. We need not confine ourselves to stochastic processes, and to the shift transformation. Rather, we can consider an arbitrary probability space, and an arbitrary measure-preserving transformation on it. This is quite a generalization. One can think of it this way. In the theory of stochastic processes we already have a notion of the repetition of an experiment built into the probability space. (i.e. the elements of the product space, Ω, consist of repetitions of the experiment in the little space, S.) We merely recapture this notion of the repetition of an experiment by considering the shift transformation. In the general case, we define the notion of repetition of an experiment (choose a reference class) by choosing a transformation T. The choice is not entirely arbitrary. T must be measure preserving, or it would hardly provide a plausible idealization of the repetition of an experiment. But given this, we have the representation theorem and the ergodic theorem.

The result seems almost too good to be true. But we haven't really gotten something for nothing. It is the choice of the transformation T which determines what we count as a repetition of the experiment. If we chose the identity map as a trivially measure preserving transformation, then we would only count as a repetition of the experiment one with exactly the same conditions and same results. If we are interested in finding useful "possible objective probability measures" as factors of our epistemic distribution, we would not want to pick such a trivial measure preserving T, but rather one which in some sense captures the maximum symmetry that we can find in our subjective distribution. Then we can think of T as a way of varying those factors which are irrelevant to fixing the physical probabilities. On the other hand, we think of the objectively resilient (ergodic) probability measures, as the probabilities conditional on the relevant experimental variables being fixed in a certain definite way. (By virtue of the ergodic theorem we can think of "objective" probability as conditional expected value of the limiting relative frequency under repetition of the experiment with respect to the sigma field of invariant sets.) At this general level,[35] the ergodic theory can be thought of as an analysis of the relation between *epistemological symmetry* (invariance, resiliency) and *physical symmetry*. For those of us who believe that invariance lies at the heart of the modal force of laws, causes, and propensities, it is therefore a key to the interpretation of physical probabilities. It shows how generally the representation of a state of ignorance as a mixture of possible physical probabilities can be obtained, and how such a representation sets up the conditions for learning from experience.

I.6 Probabilistic presupposition of induction.

One can have one's prior degrees of belief be in accordance with the probability calculus, conditionalize on one's data and still refuse to learn from experience.[36] In this sense, induction has probabilistic presuppositions. Our conception of the nature of those presuppositions has been profoundly transformed by the line of analysis initiated by DeFinetti. I count the theory of measure-preserving transformations as a part of this line of analysis. Invariance under a group of automorphisms is a natural generalization of partial exchangeability, even though it was not first introduced with these concerns in mind.

The common theme of this line of analysis, at increasingly higher levels of generality, has been that the precondition of learning for experience is the adoption of an epistemic distribution which is a mixture of extremal distributions. The effect of conditioning on a finite number of observations will, in general, be to *remix* the extremal distributions with different weights. As the number of observations goes to infinity, the weight concentrates on one of the extremal measures. We say that we learn what the chances (propensities, physical probabilities) are. In the special case of mixtures of independent sequences, this means that we have positive relevance of outcome types between trials; e.g. observation of a head makes it more likely that the next trial will come up heads. We do not have this simplest type of learning from experience in general, and do not want it. As we approach concentration on physical probabilities answering to the following Markov transition matrix:

$$\begin{array}{cc} & S_1 \quad S_2 \\ S_1 & .1 \quad .9 \\ S_2 & .9 \quad .1 \end{array}$$

we want just the opposite. Learning from experience in general is best thought of as convergence to an extremal measure. The ergodic theorem clarifies both the nature of the extremal measures and their connection with limiting relative frequency. Once we fasten on a relevant notion of repetition of an experiment by choice of a measure preserving transformation, T, we find that the extremal measures are ones which satisfy strong invariance (or resiliency) properties, and we find that they satisfy the form of the law of large numbers with respect to repetitions of the experiment that finds expression in Birkhoff's ergodic theorem.

Induction has its presuppositions, but they turn out in surprisingly large measure not to be requirements that we feign knowledge of the world that we never had, but rather that we be forthcoming and honest in representing our ignorance.

II. CAUSAL PRESUPPOSITIONS

II.1 Locality.

A predisposition to conceive of causation as acting locally, is manifest in the learning behavior of humans and animals. This predisposition can be overcome not at all by some animals, and only with great difficulty by humans. Indeed, locality of causation is a principle that has been enshrined by philosophers of diverse schools.

Principles of locality need to be formulated against the background of some space-time framework. We can think of space-time as represented by either a discrete or continuous model; and of the relevant sense of causation as either deterministic or probabilistic. For the discrete model, the principle of locality is that causation acts on contiguous events. For discrete space-time and deterministic causation this is the condition that the state at some point in the space-time matrix be a function of the states of contiguous points. For probabilistic discrete models, the analogous condition is that we have a Markov random field; that the probability of a state at a point given the states at contiguous points, is independent of the specification of states at other points. With respect to continuous space-time we can formulate analogous conditions in terms of a closed hypersurface which the point in question lies inside. In the deterministic case the state of the point would be a function of the states of the points on the hypersurface; in the probabilistic case, the probability of a state at the point given the states on the hypersurface, would be independent of the states outside the hypersurface. (The probabilities here are intended to be physical probabilities. The principle should be expected to fail for epistemic probabilities, since knowledge of the states of the rest of the universe would constitute evidence as to the correct theory regarding the connection of the states on the hypersurface and the state of the point in question.)

The principles of locality formulated in the last paragraph ignore the direction of causation. They put past, future and simultaneous events on an equal footing. To go further requires assumptions about the causal structure of space-time. In the Minkowski space-time of special relativity, one might want to restrict the conditioning events in the foregoing definitions to those in the intersection of the closed hypersurface (or set of contiguous events) with the rear light cone of the conditioned event. Should one wish to construe special relativity broadly enough to admit tachyons (faster than light particles), then one could retrench to the position which only excludes events in the forward light cone from the class of conditioning events. On the other hand, as Gödel's famous example shows, there are solutions to the field equations of general relativity which preclude any such spatio-temporal restriction of the conditioning events, and in which the best we can do for locality is something along the lines of the preceeding paragraph.[37] If one is willing to consider variations in the topology, things get worse.[38]

Let us draw back from the bizarre possibilities suggested by the general theory of relativity, and suppose that for our purposes space-time is usefully approximated by a Newtonian or Minkowskian model (and that tachyons are forbidden). Then locality of causation as formulated appears to be an eminently reasonable condition. In fact, one might wonder how we could ever discover causes if they did not obey some such locality condition. Nevertheless, it so happens that the theory of quantum mechanics is non-local in the sense at issue, and every hidden variable theory which produces the probabilistic predictions of quantum mechanics is non-local as well. (I have discussed this matter in some detail elsewhere and will forgo an analysis here, (Skyrms (1979), Bell (1964, 1971), d'Espagnat (1978)). One can of course hope that quantum mechanics is seriously wrong, but failing that we may have to get used to the idea that the universe is non-local.

The most palatable kind of non-locality would be one whose effects diminished with distance (adopting here a Newtonian ST framework for simplicity). Arthur Burks calls this quasi-locality. Quasi-locality is an hypothesis *alternative* to quantum mechanics (as shown by Furry). Other forms of quasi-locality could be considered. Baracca, Bohm, Hiley Stuart put forward the hypothesis that the probability of spontaneous separation of quantum-mechanical systems depend on temperature as well as distance. Attractive as these hypotheses are from an intuitive point of view, they have not found experimental support.

What sort of presuppositional status can we maintain for a hypothesis which, on our best evidence, appears false. Burks suggests (1977, Ch. 10) that the sort of structural causal principles that we have been discussing should be presupposed *in the sense of having high prior probability*. Presuppositions in this sense are *revisable*; we start out assigning them high probability, but in such a way that they may be superseded given sufficiently weighty evidence. I think that this account is just right, not only in that it permits what ought to be allowed to happen, but also in that it permits what did happen.

It should be pointed out that the sort of presupposition that we discussed in part one also has this "soft" character. We will never learn from experience if we concentrate all our prior probability on one of the extremal distributions, so the presupposition has the character of requiring our prior to be a non-trivial mixture. But this does not prevent the posterior probabilities from converging to one of the extremal points with the accumulation of evidence. Quite the opposite is true.

A theory of corrigible presuppositions is so different from – and so much more attractive than – trrraditional theories of incorrigible presuppositions, that one wonders whether it does not deserve a different name: one which does not call up thoughts of demonstrations of the synthetic *a priori* status of Euclidian geometry and Newtonian mechanics. Again, it is the consideration of induction in a thoroughly probabilistic framework, that leads to Burks' theory of "soft" presuppositions for structural causal principles.

II.2 Modal force.

Laws of nature and statements of causality are widely acknowledged to have some kind of modal force. How is that force to find expression in a precise version of such statements – for instance in the statements of locality discussed in the last section – and how is that force to receive confirmation on the basis of empirical evidence? One approach is to build an explicit modal claim into the semantics of such statements. Then epistemic probability is taken as applying to sentences, some of which contain modal operators. A contrasting approach takes the modal force of laws and related statements as pragmatic: as arising from their epistemological status. On this view, the objects of epistemic evaluation are *de facto*, unmodalized statements; some of which may acquire a kind of modal force as a result of that evaluation.

Arthur Burks takes the first approach. If the *de facto* content of a law is p, than confirmation of the law is confirmation of the statement: *Necessarily p*, where the modal operator expresses an irreducible *a priori* concept. Burks argues that it is an immediate consequence of Bayes' theorem that such modalized propositions can be confirmed by non-modal evidence. Here is the basic idea in the simplest case. The probability of p conditional on *Necessarily p* is one; likewise for any empirical statement, e, which is a logical consequence of p. By Bayes' theorem, conditionalization on e will give a posterior probability to *Necessarily p* of l/pr(e) times its prior probability. Thus, if the modal, *Necessarily p* has non-zero prior probability, and the statement p has empirical consequences that do not have prior probability one, then the modal judgment can be confirmed by empirical evidence, where "confirmation" means that the posterior probability is higher than the prior.

There is no quarrelling with the theorem just stated, but there are questions to be raised as to its interpretation. It appears to prove too much, for it applies to any statement that is logically stronger than p; e.g. to *p and it is so by the will of the gods*. We appear to have a way of falling into the trap that Hempel kept warning us about – confirming any metaphysics whatsoever.

The problem lies not with the probability calculus, but with our interpretation of "confirmation". This can be seen as follows: The *de facto* proposition, p, is logically equivalent to the disjunction of (I): It is necessary that p. with (II): p, but it is not necessary that p. I want to ask how we confirm (I) as against (II). By Bayes' theorem, a bit of evidence, e, will render the ratio of posterior probabilities of (I) over (II) higher that the corresponding ratio of prior probabilities only if the conditional probability of e on (I) is greater than the conditional probability of e on (II); that is, only if the modal componant is positively relevant to the evidence. In the case where p entails e, (I) and (II) both entail e, so the ratio of posterior probabilities of (I) and (II) will be the same as the ratio of their priors. So, in the case under consideration, observation of e has the effect of transferring some probability from not-p to p, and distributing it between (I) and (II) according to their prior probabilities. So (I) is confirmed in the sense of having its probability increased, but not in the sense of being confirmed relative to (II). Indeed, if (I) accounts for a small proportion of the prior probability of p, then (I) can never achieve a high posterior probability by this means; it can never become "well-confirmed".

We could confirm (I) as against (II), and thus have prospects for having (I) being well-confirmed, only if we could discover evidence for which the modal componant of (I) is relevant; that is, evidence, e, such that the prior probability of e conditional on (I) is higher than the prior probability of e conditional on (II). The difficulty of imagining a non-modal e for which the modal componant of the law would make a difference engenders the suspicion that the modal component is empty metaphysics.

We do not have these problems on the other approach which takes the modal force of laws as arising from their epistemological status. On that approach, epistemic probabilities attach to *de facto* statements, and modal force is explicated in terms of the probabilistic structure. The difficulty for this approach is to give a convincing account of the nature of the necessity of laws.

The framework for probability is already modal, in a sense. Probabilities are defined over a space of possibilities. Probabilities that are invariant under conditioning on further specification of experimental conditions are modal in a stronger sense. In 1.5 I discussed how this leading idea allowed us an epistemic construction of propensities, or physical probabilities. Following these ideas through would lead to an account of probabilistic law, and taking universal laws as limiting cases, of laws in general. For this to be a convincing account of the modal force of laws, it must be shown that it illuminates areas of the pragmatics of laws where modal force appears important. I argue elsewhere that this is so (Skyrms, 1980).

The position just sketched has a great many points of contact with Burks' theory. The idea that inductive probability already has a kind of modal force is emphasized by Burks; as is the idea that physical probabilities should be conditional inductive probabilities. One main difference is the way in which the explicit modal operator gives way to the notion of resiliency. I believe that our understanding of physical necessity, like our understanding of uniformity, is radically transformed when we adopt a thoroughly probabilistic point of view.

Brian Skyrms
University of California
Irvine, California

NOTES

*Research on this paper was partially supported by National Science Foundation grant SES-8007884.

1. Whether we classify them as a special kind of knowledge is another question.
2. Thus Mill in *A System of Logic*: "The universe, so far as known to us, is so constituted, that whatever is true in any one case, is true in all cases of a certain description; the only difficulty is, to find out what description.'
3. As the "grue-bleen" paradox.
4. Probability one for an infinite sequence.
5. The result for infinite sequences and a powerful notion of randomness is in Martin-Löf, (1966). The analogous theorem for the finite approximation is in Fine, (1973).
6. One which is thoroughly discussed in (Burks, 1977).
7. So pr_{BH}(heads on toss 1)=2/3; pr_{BT}(heads on toss 1)=1/3; pr_{BH}(heads on toss 1 and heads on toss 2)=$(2/3)^2$, etc.
8. Exchangeability.
9. The epistemic probability of two heads = 5/8 = the epistemic probability of two tails. That of one head and one tail in any order = 4/9.
10. I.e., if our *epistemic* distribution took this form, our reasoning would, in a sense, be counterinductive.
11. "Mixture" now includes continuously weighted averages: $\int p \ du$.
12. I.e., the mixing measure, the weight of the weighted average, is unique. N.B. that this does not mean that it cannot have other representations as mixtures of nonindependent exchangeable sequences. For instance, it has a representation as a degenerate mixture whose weight is concentrated on itself.
13. This is basically a consequence of the law of large numbers for independent trials together with Bayes's Theorem.

[14.] A slightly, extended sense of conditionalizing.

[15.] And others who have advocated similar positions: J. M. Keynes, I. J. Good, Richard Jeffrey, David Lewis.

[16.] Bruno DeFinetti, (1970), p. 146. (See the discussion of the preceeding two pages.) But also compare vol. II, (1975), p. 214.

[17.] Or any of the finite extremal hypergeometric distributions except the ones corresponding to all heads and no heads.

[18.] The error being less than c/k where c is a constant that doesn't depend on the original sequence. See Diaconis (1977), pp. 277–278. For more details see Diaconis and Freedman, (1980c).

[19.] The "gambler's fallacy" in its usual form is a bit different. The gambler reasons that he has a fixed finite number of gambles in this lifetime and a fixed finite number of wins. He correctly sees that in the hypergeometric distribution corresponding to these numbers, a loss raises the probability of subsequent wins. Since he is uncertain about these numbers, his epistemic probability will be a mixture, and he incorrectly assumes that negative statistical relevance in each of the elements of the mixture implies negative statistical relevance in the mixture.

[20.] More generally, to predict the result of the next toss after observing which h have been heads, the counterinductivist might adopt the hypergeometic distribution corresponding to an urn which N+1 balls, of which h are heads if h>N-h; of which h+l are heads otherwise.

[21.] Some philosophers would not think this odd, but normal.

[22.] I should emphasize that DeFinetti himself never entertained such an idea. In 1937 he writes: "In other words it (Exchangeability) explains the psychological reasons which often force us to attach to the probability of certain events a value near the frequency observed in analogous events. But the case of exchangeability can only be considered as a limiting case: the case in which this "analogy" is, in a certain sense *absolute* for all the events under consideration." DeFinetti, "On the Condition of Partial Exchangeability" (Jeffrey), 1980, p. 197. The attractions of a quasi-Kantian position which tempted Carnap, did not tempt him; indeed DeFinetti has nothing but scorn for Kant: "No sooner had Hume begun to prise apart the traditional edifice, then along came poor Kant in a desperate attempt to paper over the cracks, and contain the inductive argument – like its deductive counterpart – firmly within the logic of certainty." DeFinetti, (1970, II, p. 201).

[23.] Here we have finite one-sided sequences in mind.

[24.] It is a convex set in the appropriate vector space of measures.

[25.] By a degenerate mixture, I mean a weighted average which puts all its weight on one element.

[26.] I.e., when the probability that the initial state recurs infinitely many times is one.

[27.] A standard Borel space is a Borel space that is isomorphic to the Borel space defined by a Borel subset of a complete metric space.

[28.] For a treatment whose motivation is close, though not identical to that given here, see Billingsley, (1965).

[29.] In addition to Billingsley, I have relied heavily on Mackey (1974) and Oxtoby, (1952).,

[30.] *More properly, with respect to (,F, P).

[31.] In the case where the probability space is a compact metric space, we have this analysis already in 1937 in the theory of Kryloff and Bogoliouboff. For an account of this and generalizations thereof see Oxtoby (1952). For statement of further generalizations see Farrell (1962) and Dynkin (1978).

[32.] Proof adapted from Billingsley (1965).

[33.] Let f ( ) be 1 if the random variable f_1 takes a value in s, a measurable subset of S, and O otherwise. Then

$$\frac{1}{m} \sum_{R=O}^{m-1} f\,(T^R\omega)$$

is the relative frequency of s in the sequence of trials $f_1, f_2, \ldots f_n$. According to the ergodic theorem this limit exists and equals the probability that the outcome of trial 1 is in s.

[34.] For a characterization of the ergodic Markov shifts see Billingsley (1965), pp. 31–33.

[35.] To carry this point of view to its logical conclusion, it is clear that we need also the generalization of ergodic theory to groups or semigroups of measure preserving transformations. Here again the role of the invariant sets in generating a sufficient subsigma field is the crux of the matter. See Farrel (1962) and Dynkin (1978).

[36.] In general terms, this can be done by concentrating one's prior epistemic probability on an extreme point.

[37.] There are causally degenerate space-times in which every event is causally connectible to every other. See Hawking and Ellis. (1973).

[38.] One must also consider the possibility that the fundamental theory may not be a space-time theory.

REFERENCES

Asquith, P. and I. Hacking (eds.): 1978, *PSA 78 II*, Philosophy of Science Association, East Lansing.

Baracca, A., D. J. Bohm, B. J. Hiley, A. E. Stuart: 1975, 'On some new notions concerning locality and non-locality in the quantum theory', *Il Nuovo Cimento* **28B**, pp. 453--464.

Bell, J. S.: 1964, 'On the Einstein, Podolsky, Rosen Paradox', *Physics* I, pp. 195–200.

Bell, J. S.: 1971, 'Introduction to the hidden variable question', in *Foundations of Quantum Mechanics*, ed. B. d'Espagnat, New York.

Billingsley, P.: 1965, *Ergodic Theory and Information*, Wiley, New York.

Burks, A.: 1979, *Chance, Cause, Reason*, University of Chicago Press, Chicago.

DeFinetti, B.: 1937, 'La prevision: ses lois logiques, ses sources subjectives', *Annals de l'Institut Henri Poincare* 7. Translated as 'Foresight: Its logical laws, its subjective sources' in Kyburg and Smokler, (1964).

DeFinetti, B.: 1938, 'Sur la condition d'equivalence partielle', *Actualities Scientific et Industrielles* no. 739, Hermann & Cie, Paris. Translated as 'On the condition of partial exchangeability', in Jeffrey (1980), pp. 193–204.

DeFinetti, B.: 1970, *Theory of Probability*, Wiley, New York.

DeFinetti, B.: 1974, *Probability, Induction and Statistics*, Wiley, New York.

Diaconis, P.: 1977, 'Finite forms of "Definetti's theorem on exchangeability"' *Synthese* **36** pp. 271–281.

Diaconis, P. and D. Freedman: 1980a, 'DeFinetti's generalizations of exchangeability', in Jeffrey (1980, pp. 233–249).

Diaconis, P. and D. Freedman: 1980b, 'DeFinetti's theorem for Markov chains', *Annals of Probability* **8** no. 1, pp. 115–130.

Diaconis, P. and D. Freedman: 1980c, 'Finite exchangeable sequences', *Annals of Probability* **8** no. 4, pp. 745–764.

Dynkin, E. B.: 1978, 'Sufficient statistics and extreme points', *Annals of Probability* **6** no. 5, pp. 705–730.

d'Espagnat, B.: 1978, *The Conceptual Foundations of Quantum Mechanics* 2nd ed., Academic Press, New York.

d'Espagnat, B. (ed.): 1971, *Foundations of Quantum Mechanics*, Academic Press, New York.

Farrell,. R. H.: 1962: 'Representation of invariant measures', *Illinois Journal of Mathematics* **6**, pp. 447–467.

Fine, T.: 1973, *Theories of Probability*, Academic Press, New York.

Freedman, D.: 1962, 'Invariants under mixing which generalize DeFinetti's Theorem', *Annals of Mathematical Statistics* **33**, pp. 916–923.

Furry, W. H.: 1936, 'Note on the quantum mechanical theory of measurement', *Physical Review* **49**, pp. 393–399; 476.

Georgii, H. O.: 1979, *Canonical Gibbs Measures*: Lecture Notes in Mathematics 780, Springer Verlag, Berlin.

Halmos, P.: 1956, *Lectures on Ergodic Theory*, Math. Soc. Japan, Tokyo.

Harper, W., R. Stalnaker, and W. Pearce (eds.): 1980, *Ifs*, D. Reidel, Dordrecht.

Hawking, S. W. and G. F. R. Ellis: 1973, *The Large Scale Structure of Space-Time*, Cambridge University Press, Cambridge.

Hewett, E. and L. J. Savage: 1955, 'Symmetric measures on Cartesian products', *Transactions of the American Mathematical Society* **80**, pp. 470–501.

Jeffrey, R. C.: 1965, *The Logic of Decision*, McGraw Hill, New York.

Jeffrey, R. C.: 1980, *Studies in Inductive Logic and Probability II*, University of California Press, Berkeley.

Kyburg, H. and B. Smokler: 1964, *Studies in Subjective Probability*, Wiley, New York.

Lewis, D.: 1980, 'A subjectivist's guide to objective chance' in Jeffrey (1980), pp. 263–294.

Link, G.: 1980, 'Representation theorems of the DeFinetti type for (partially) symmetric probability measures, in Jeffrey (1980), pp. 207–231.

Mackey, G. W.: 1974, 'Ergodic theory, and its significance for statistical mechanics and probability theory', *Advances in Mathematics* **12**, pp. 178–268.

Martin-Löf, P.: 1966, 'The definition of a random sequence', *Information and Control* **9**, pp. 602–619.

Oxtoby, J. C.: 1952, 'Ergodic sets', *Bull. American Mathematical Soc.* **58**, pp. 116–136.

Preston, C.: 1979, *Random Fields*: Lecture Notes in Mathematics 760, Springer Verlag, Berlin, Heidelberg, New York.

Ruelle, D.: 1969, *Statistical Mechanics*, Benjamin, New York, esp. Ch. 6.

Sinai, Ya.G.: 1976, *Introduction to Ergodic Theory* Princeton University Press and University of Tokyo Press, Princeton and Tokyo.

Skyrms, B.: 1977, 'Resiliency, propensity and causal necessity', *Journal of Philosophy* **74**, pp. 704–713.

Skyrms, B.: 1978, 'Statistical laws and personal propensities', in Asquith and Hacking (1978), pp. 551–562.

Skyrms, B.: 1979, 'Randomness and physical necessity', *Pittsburgh Lecture in the Philosophy of Science*.

Skyrms, B.: 1980, *Causal Necessity*, Yale University Press, New Haven.

Skyrms, B.: 1980, 'The prior propensity account of subjunctive conditionals', in Harper, et al. (1980) pp. 259–268.

Weyl, H.: 1952, *Symmetry*, Princeton University Press, Princeton.

Weiner, N.: 1949, *Time Series*, M.I.T. Press, Cambridge.

ROBERT AUDI

SCIENTIFIC OBJECTIVITY AND THE
EVALUATION OF HYPOTHESES

The natural sciences are often cited as paradigms of objective disciplines, and scientific method is widely believed to provide an objective way of acquiring knowledge of the world. The social sciences have been far less widely taken to be objective, but most philosophers of science who think they should be considered sciences regard them as objective in principle, typically on the ground that their subject matter is amenable to study using scientific method. Some philosophers and some scientists, however, have raised doubts about whether the scientific evaluation of hypotheses can be objective, even in principle. Perhaps the most common reason for this doubt is the idea that "value judgments" are unavoidable in scientific assessment. There have been other grounds for the doubt, of course, particularly in relation to the social sciences, and some of these grounds will also concern me in this paper.

Many philosophers and many scientists have defended what I shall call the objectivity thesis – the thesis that the sciences provide an objective way of acquiring knowledge (or at least justified beliefs) about the world. But most discussions of scientific objectivity have done less than they might have to clarify this notion. For one thing, a great many of them have been heavily preoccupied with refuting arguments or doubts about scientific objectivity. Another factor is that the tendency to take scientific method as one's paradigm of an objective method is so strong that many philosophers of science, and perhaps most scientists, writing about scientific objectivity have left unexplored some questions about objectivity in general and, in some cases, tended to conceive objectivity in general as having properties peculiar to scientific objectivity. It may be useful, then, to approach the topic of scientific objectivity by way of a discussion of objectivity in general. On the

321

Merrilee H. Salmon (ed.), *The Philosophy of Logical Mechanism*, 321–346,
© 1990 *Kluwer Academic Publishers.*

basis of an account of objectivity in general, we can see what it is for science to be objective and assess the question whether value judgments must occur in scientific assessments in a way that undermines their objectivity.

Part I will outline a conception of objectivity and, on that basis, suggest what is distinctive of scientific objectivity. Part II will explore the issue of whether scientists as such must make value judgments, with a view to seeing whether the case for that position undermines the objectivity thesis. Part III will take up some influential views that raise special problems for the putative objectivity of the social sciences. The last section will draw together some of the key points that emerge and identify some remaining problems important in assessing the thesis that science is objective.

I. OBJECTIVITY

In sketching a conception of objectivity my strategy will be to concentrate on what it is to say that a method is objective. We can then understand an objective discipline, such as a branch of science, as, roughly, one whose methods of answering its questions are objective. Thus, if scientific method is objective and used by all the sciences to answer the questions appropriate to them, we can say that "science" is objective. Regarding persons, we can say that an objective person is, roughly, one who, for at least most of his important judgments, accepts them on the basis of their being confirmed through an objective method. An objective claim, in one sense, would be a claim assessable by an objective method, and, in a stronger sense, one both assessable by an objective method and arrived at through an application of such a method. (I shall not define 'method'; our purposes will be served by using it broadly, to refer to any systematic way of answering a kind of question.)

In discussing objectivity, particularly in relation to science, my main focus will be on issues concerning the context of scientific confirmation rather than the context of scientific discovery. Questions of objectivity certainly arise concerning how scientific truths are or should be discovered, and some of what I say will be applicable to one or another aspect of scientific discovery; but clearly the most important domain in which scientists need to be capable of objectivity is in applying criteria of confirmation and of acceptance to scientific hypotheses. The objectivity thesis that chiefly concerns me – and has occupied most philosophers of science writing on scientific objectivity – is that scientific method embodies, or at least can be informed by objective criteria of confirmation and acceptance.

The notion of objectivity can be best characterized, I believe, if it is treated as one of those positive notions which, like that of free action, is to be understood by contrasting it with what it rules out. Let us consider, then, some subjective methods. Imagine a purported method of communicating with departed souls by going into a special kind of trance and waiting for certain kinds of signals, where the relevant kinds are not characterized specifically and each user of the method is allowed to interpret them in what seems a reasonable way. Suppose further that some people are not capable of going into the appropriate sort of trance, perhaps because they lack the required mystical temperament. We shall not be surprised to find that users of the method often disagree on whether certain signals, e.g. those perceived as a ringing in the ears, are relevant to the conclusion that the signaling soul is unhappy. Certainly users might disagree on how much these signals support that conclusion, even if they do agree that the signals have some positive relevance. They might also simply each get different signals, though each sets out to communicate with the same soul and they agree that they are getting signals from it at the same time.

This is a method it is natural to call subjective, in part because it does nothing to prevent prejudices or eccentricities of the user from biasing the results of its use. One reason it fails to prevent this is the apparent absence of any publicly accessible object to which it is applied: only the initiated can have access to the data received in the trance. But even if the uninitiated could use the method – as they can use a method of arriving at moral judgments by introspectively consulting their conscience – the privacy of the data called for by the method would make it subjective.

If we consider what conditions a method of inquiry must exhibit to avoid the subjective features of the one just described, at least five broad conditions emerge. An objective method for investigating a given subject matter is, minimally, one such that (a) it can be used by any competent investigator, (b) the data needed for its use are public; (c) independent competent investigators applying the method to answering the same questions concerning this subject matter tend to agree on what answers are relevant and how these answers are to be evaluated; (d) independent competent investigators who apply the method to evaluating the same hypothesis will tend to agree on the extent to which a given body of data confirms the hypothesis; and (e) independent competent investigators applying the method to obtaining data regarding the same hypothesis in the same (or relevantly similar) circumstances tend to discover the same data. Together, (d) and (e) imply that independent competent investigators applying the method to obtaining data regarding the same hypothesis in the same circumstances tend to exhibit a measure of agreement. But this tendency should not be exaggerated. For instance, (e) does not preclude significant differences in the data various investigators discover.

The central ideas are these: that of a competent investigator; that of public data; that of clear criteria of relevance; that of a high (but not necessarily perfect) intersubjective correlation in judgments of confirmation; and that of intersubjective agreement on what the data are, provided they are sought under appropriate conditions. Given the vagueness of the notion of objectivity itself, it should be no surprise that these concepts are vague. It is important to note, however, that the comparative notion, *being more objective than*, can be explicated along the suggested lines and, without arbitrariness, made fairly precise. To see this, and to get a better idea of what constitutes objectivity conceived non-comparatively, let us explore (a)-(e) further.

First, the relevant criteria of competence may presuppose specific knowledge, even a high level of knowledge, but not such things as extrasensory powers, or knowledge which, like supposed mystical knowledge, is possible for only the initiated. Related to this, an objective method must not be biased in favor of any particular view, but *doctrinally neutral*. This requirement is very hard to make precise. *Some* special knowledge, e.g. procedural knowledge, may be required for the use of an objective method, but the method must at least be neutral with respect to the sorts of hypotheses for whose assessment it is to be used. There is also a certain relativity built into the notion of objectivity: if it became normal for people to have a sixth sense, then the above characterization of an objective method would presuppose this additional capacity and being a competent investigator could presuppose having this sense. At present, however, even if someone had a sixth sense it is doubtful whether we should consider a method requiring it is unqualifiedly objective.

The publicity requirement regarding data might seem to follow from the requirement that the method be usable by any competent investigator. But a method of, e.g., predicting rain could call for each person to introspect his *own* sensations in the joints and judge by the felt quality of the sensations. In an arthritic society, this method might even give results better than chance, but it is surely not objective. The problem is not that qualities of sensations could not be good indicators of the weather, but that each user of the method seeks data not available to any other user. Thus, it need be no more than good fortune that different competent investigators using this method in the same circumstances succeeded in agreeing in their forecasting. Ordinarily, subjective influences might bias them so as to prevent consensus.

Concerning the requirement that an objective method cannot leave competent investigators in doubt about how to evaluate relevant answers, notice that this does not imply that *in practice* they always can obtain and evaluate such answers. The history of science shows that it can take a long time to determine how to test a hypothesis. But it would seem that if an objective method applies to a type of question, then in the long run competent investigators can evaluate any admissible answer to that sort of question. This is perhaps not a necessary truth, but it is at least a reasonable supposition.

Regarding the fourth condition, as already suggested it is more important that competent investigators agree on *comparative* matters of confirmation than on its absolute degree. If we agree that a hypothesis is better confirmed by the available evidence than its rivals, it is relatively less significant that some of us think that, say, its probability on the evidence is .90 while others put the figure at .80. We would, e.g., bet on the former if we had to choose just one to bet on, though some of us would doubtless be willing to bet more than others. Our cognitive tendencies – such as inferential tendencies – would also differ in a similar way; e.g., we would be more likely to believe a consequence of the former than of the latter, other things equal. (The *ceteris paribus* claim is important; some consequences of each – such as the disjunction of the two – are equally probable, and a consequence of the less probable one may be far more probable than a consequence of the other, e.g. where the former consequence is the disjunction of the entailing hypothesis with one having a probability close to 1.)

Fifth, regarding the requirement that independent competent investigators using the method to obtain data concerning the same hypothesis under the same (or relevantly similar) conditions discover the same data, this is needed to prevent subjective differences from yielding different data in a way that undermines objectivity. Suppose that a method of identifying important elements in a painting called for listing what one sees in the first sixty seconds of viewing. While this method might possibly satisfy the first four criteria, it allows idiosyncrasies to prejudice the data base. Similarly, imagine a method of hypothesis testing that required simply pulling from a hat a card listing a probability, where the hat contains 11 cards with probabilities 0, .1, .2, . . . 1 on them, respectively, and the probability on the drawn card is assigned to the hypothesis. Here two successive draws would tend to yield inconsistent results, even if no relevant change occurs between the two applications of the method. Granted, there may be no subjective influence operating; but the method also provides no rational way to resolve subjective disagreements. The method is haphazard, and the operation of chance here functions similarly to subjectivity in producing differences which it is unreasonable to take seriously.

So far, I have said nothing about an essential connection between the objectivity of a method and its tending to lead its users to truth. We do tend to conceive objective methods as meeting this condition, and I think that the paradigm cases do. One might try to explain this by saying that the relevant intersubjective agreement would be most unlikely apart from some causal connection between the facts discovered through applying the method and the beliefs arrived at through its use, e.g. between facts about the accelerations of balls on an inclined plane and beliefs about their accelerations, as well as general beliefs arising from the former by some kind of induction. There seems to be some truth underlying the suggested explanation, but this truth is not easy to articulate precisely. Intersubjective agreement would be very high indeed for a method of predicting world population trends by the number of cars crossing the George Washington Bridge on the Fourth of July. But though this method is objective it seems very unlikely to yield much truth about world population trends. The example suggests that while intersubjective agreement may not be possible except where there are appropriate connections between facts ascertained, or experiences had, in applying the method, and, on the other hand, the beliefs formed through its use, neither intersubjective agreement nor objectivity requires causal connections between these beliefs and what they are *about*, such that a significant proportion of the beliefs is true.

There are philosophers who would argue that the imagined method is not objective. Richard Rudner would doubtless have pressed this objection. For on his view (1966, p. 76), reliability is crucial to the objectivity of a method, where "Method *A* is more reliable than method *B* if, and only if, its continued employment is less liable to error (i.e., is less likely to result in our coming to believe, or continuing to believe, false sentences)."[1] I would agree that an objective method must be reliable in the sense that its use by different competent investigators in the same situation (or a relevantly similar one) will tend to yield the same (or relevantly similar) results. But the imagined method of predicting population is highly reliable in this sense, and certainly there is nothing subjective about it. It is not, however, truth-conducive, nor does it seem at all likely to help anyone avoid error. Rudner seems to be thinking of objective methods as satisfying both these criteria. I agree that at least typically the objective methods which interest us do satisfy them. But if I have been right, a method need not satisfy either criterion simply by virtue of its objectivity.

It is also important to notice that a subjective method can be truth-conducive. Using introspection could, under certain conditions, be considered a method of ascertaining what one wants. It might generally result in correct judgments; but it is clearly subjective, since (for one thing) only the subject of the wants can use it. To be sure, a psychologist may rely on the truth-conduciveness of this method in applying his own method of investigation, say one using overt, first-person reports of wants. But this does not imply that the former method is objective or that the latter is subjective. I am not taking the possible truth-conduciveness of introspection to show by itself that the truth-conduciveness of a method is not necessary for its objectivity. It does not show this; but by dissociating truth-conduciveness from objectivity in the way it does, it reduces the plausibility of taking the former to be necessary for the latter. It also shows that objectivity is not *in general* necessary for truth-conduciveness, though it may of course be necessary apart from cases in which a non-objective method succeeds by virtue of the user's having a kind of privileged access to the truths he seeks through using the method.

What about scientific method, however (which is what Rudner had in mind in the context)? It is quite plausible to say that it must be truth-conducive, at least in a conditional sense: roughly, if it is properly applied to the testing of hypotheses, then in the long run it will lead to greater and greater confirmation of any true one to which it is applied, and to greater and greater disconfirmation of any false ones to which it is applied. But it is not clear that truth-conduciveness in this sense is a conceptually necessary condition for a method's being scientific. No doubt some people are inclined to use 'scientific' in an evaluative sense such that scientific method *must* by its very nature be truth-conducive; they would thus not call a method scientific unless in the long run it is truth-conducive. But surely it is not obvious that there is an *a priori* guarantee that in the long run true hypotheses will be progressively confirmed, or false ones progressively disconfirmed, through repeated uses of scientific tests. In principle, e.g., the tests could remain inconclusive or oscillate between confirmation and disconfirmation. That this is logically possible is part of what gives the traditional problem of induction its bite: however well confirmed a true hypothesis is, we can never rule out *a priori* a change in the pattern of events, in the light of which it would be equally disconfirmed.

Let us suppose, however, that in fact scientific method is truth-conducive in the sense specified – not, of course, in the sense that it provides us with a "logic of discovery" which guarantees our selecting some true hypotheses for testing in the first place. Does this truth-conduciveness derive from the objectivity of the method? I think not. It is truth-conducive because it is scientific – e.g., in requiring empirical tests of hypotheses – not simply because it is objective. I think, then, that what distinguishes the objectivity of scientific method from that of other methods, including truth-conducive ones such as those of logic, is its integration with the distinctively scientific procedures for testing and evaluating hypotheses. These procedures generally maximize the likelihood that false hypotheses will be discovered to be erroneous through some of their false test implications, and that true hypotheses will be confirmed through the discovery of some of their true test implications. These test and evaluation procedures have been discussed very ably and at great length in the literature, and I shall not try to describe them further. My main concern is rather to formulate and assess some significant challenges to the view that scientific objectivity – the integration of objectivity as I have described it with the testing and evaluative procedures of science – is possible.

II. THE EVALUATION OF SCIENTIFIC HYPOTHESES

A major challenge to the objectivity thesis arises from various arguments to the effect that even in the proper application of scientific method scientists must make value judgments, either in a straightforwardly moral sense or in some other sense viewed as undermining the objectivity of the judgments. It is of course a controversial question whether there is an objective method for assessing moral claims. There is widespread doubt that such a method is available in practice. Here I want to leave that question open. Even if we suppose there is such a method, any candidate we choose – whether Kantian or utilitarian or whatever – will be sufficiently unclear to sufficiently many people so that, for them, at least, it will be quite important to know whether scientific evaluation requires making moral judgments. If it does, the need for an objective method in ethics is even greater than otherwise. If it does not, then at least one can, without begging moral questions, appeal to scientifically established statements in arguing for or against various moral views.

Richard Rudner has provided what is to date one of the most plausible statements of the case for the view that the scientist *qua* scientist makes value judgments, understood as implying that moral or other non-epistemic normative judgments are required for the scientific assessment of hypotheses. His central argument runs as follows:

1. The scientist *qua* scientist accepts or rejects hypotheses.
2. This requires deciding whether the evidence is sufficiently strong to warrant accepting the hypothesis.
3. The scientist's decision whether the evidence is strong enough to warrant accepting the hypothesis is "a function of the *importance*, in the typically ethical sense, of making a mistake in accepting or rejecting the hypothesis." Hence
4. The scientist *qua* scientist makes value judgments (1953, p. 2).[2]

Rudner illustrates this with the example of the hypothesis that a toxic ingredient in a drug is not present in lethal quantity. Here "we would require a relatively high degree of confirmation or confidence before accepting the hypothesis – for the consequences of making a mistake here are exceedingly grave by our moral standards," and "*How sure we need to be before we accept [such] a hypothesis will depend on how serious a mistake would be*," (1953, p. 2). Rudner is quite aware of the objection that a scientist's business is only to determine the degree of confirmation of a hypothesis. His reply is that this only moves his point to different territory. "For the determination that the degree of confirmation is, say, *p* . . . is clearly nothing more than *the acceptance by the scientist of the hypothesis that the degree of confidence is p*" (153, p. 3).

These arguments have been widely discussed and often criticized. Isaac Levi has evaluated them in detail and discussed the central issues in a number of places. It will be useful to begin by considering his initial response to Rudner. He first of all attacks Premise 3. His central criticism is that choosing to accept a hypothesis does not entail choosing to *act* on it in relation to any specific objective. For "a person can meaningfully and consistently be said to accept a hypothesis without having a practical objective," (Levi, 1960, p. 351). Thus, a person can accept a hypothesis in an open-ended situation and hence need not thereby choose to act on it relative to any particular objective.

Levi also considers another line of reply to Rudner, which, at the time, he drew from Carnap, (1950, pp. 205–207) Hempel, (1949, p. 560 and 1966, Ch. 3 & Ch. 12) and Jeffrey (1956, p. 245).[3] On this view, scientists should be content to assign degrees of confirmation to hypotheses relative to the available evidence; and "Anyone who is confronted with a practical decision problem can go to the scientist to ascertain the degrees of confirmation of the relevant hypotheses" (1960, p. 353). Levi rejects this as "like crashing into Scylla to avoid Charybdis," and he attempts to reconcile his view that scientists do accept and reject hypotheses, with the value-neutrality of science. He bases his reconciliation on two contentions. The first is that

The necessity of assigning minimum probabilities for accepting or rejecting hypotheses does not imply that the values, preferences, temperament, etc. of the investigator, or of the group whose interests he serves, determine the assignment of these minima (1960, p. 351).

The second contention is that the value neutrality thesis does not preclude the scientist *qua* scientist's making *any* value judgments.

What it requires is that "given his commitment to the canons of inference he need make no further value judgments." (1960, p. 356).

Regarding this last point, Levi does not commit himself on the
crucial question "whether the canons of scientific inference dictate
assignments of minimum probabilities in such a way as to permit no
differences in the assignments made by different investigators to the
same set of alternative hypotheses (1960, p. 357). He has treated this
and similar questions at length in more recent writings.[4] A discussion
of his views on the issue will not be possible here, but it will not be
necessary since our concerns are essentially neutral with respect to
specific criteria of confirmation or acceptance and should apply to any
plausible set of such criteria. For our purposes we need to take up
criteria of confirmation and acceptance only in relation to the question
whether scientists as such must make moral or other non-epistemic
normative judgments in evaluating scientific hypotheses. In discussing
this, I want to consider two distinct though related issues: whether
moral or other non-epistemic normative judgments must be made in
the scientific acceptance of a hypothesis, and whether they must be
made in the scientific assessment of the degree of confirmation of a
hypothesis. Before proceeding, however, we need to consider what
constitutes acceptance. As often as this notion has been used in recent
literature, it remains very much in need of further clarification.[5]

At times 'acceptance' is so used that one might suppose accepting a
proposition is equivalent to believing it. But as the term figures in discussions of accepting and rejecting hypotheses, belief is surely only a
necessary condition for acceptance. Consider cases in which a scientist
decides whether a hypothesis is acceptable and as a result of his reflection accepts it. Here acceptance is surely an event (though not necessarily an action). Accepting in this sense entails assenting to the
proposition, forming the belief that it is true, and, for a time, at least,
believing it. There is also a dispositional use of 'accept', on which to
say that someone accepts a hypothesis is, roughly, to say that he
believes it and to suggest that he has accepted it in the above sense.
One might argue that there is a quite different dispositional use; that,
e.g. a person who walks into an ordinary well-lighted lecture room
accepts the proposition that there are seats in it, even though he has
not *assented* to this but has merely seen that it is so. This seems to me
at best loose parlance, in which accepting is assimilated to believing.
Perhaps what gives this conception of accepting plausibility is the

thought that if the person did for some reason entertain the proposition he *would* accept it.

These considerations suggest that a person is properly said to accept a hypothesis, *h*, only if he has considered its credibility, however unselfconsciously. Perhaps this does not hold in general, but it certainly seems to hold for scientific acceptance. Indeed, it seems typical of the scientific acceptance of a hypothesis that the person not only consider the credibility of it, but form a belief about its credibility – e.g., that *h* is highly confirmed by the evidence – and come to believe it in part on the basis of that further belief about its credibility. Often, moreover, scientific acceptance of *h* will involve not only a belief of *h* and a further one regarding its credibility, but a second-order belief about the warrant (evidence, confirmation, grounds) one has for one's belief that *h*.

If scientific acceptance does have this twofold character and thus involves, typically, both the belief that *h* and at least one other belief, usually one about the credibility of *h* or one about one's warrant for believing *h*, then we have to be careful to avoid ambiguity. If we talk, e.g., of the strength of acceptance, we must distinguish between the strength of the belief that *h* and the strength (or associated degree of probability) of the quite different belief that *h* is confirmed by the evidence. Degrees of (subjective) probability can also be applied to either belief. Evaluation can of course occur in arriving at either belief and may be different in each case. For instance, in arriving at the belief that *h* is well confirmed by the data, the scientist may simply judge intuitively or may explicitly use certain principles of confirmation. Doing the latter may, depending on the case, be rather straightforward. In arriving at the belief that *h*, however, he may rely on special epistemic principles, e.g. 'Do not accept a hypothesis with a probability on the evidence less than .99'. A non-scientist or a scientist employing extra-scientific criteria of acceptance might instead (or in addition) use an ethical principle of acceptance, drawn from, say, an "ethics of belief." Such a principle might prohibit letting oneself believe a hypothesis whose probability on the evidence is less than .99; it might prohibit this only where the matter in question is in some specified way important; or it might require a minimum difference in probability between an acceptable hypothesis and any competing one.

Given these distinctions, we are in a good position to evaluate Rudner's argument. Consider his crucial Premise 3: The scientist's decision whether the evidence is strong enough to warrant accepting the hypothesis is a function of the *importance*, in the typically ethical sense, of making a mistake in accepting or rejecting the hypothesis. This wording – particularly the phrase 'decision whether the evidence is strong enough to warrant accepting the hypothesis' – runs together the two questions we have been distinguishing: (1) What is the degree of confirmation of *h* on the evidence? and (2) Is *h* acceptable?, where this may be simply a matter of having a certain minimum level of confirmation on the evidence. Granted, adopting an epistemic standard for answering questions of the second kind involves evaluation. But Rudner does nothing to show that it need be moral evaluation, or even need be done with possible application to morally significant cases in mind. He talks, however, as if a *separate* evaluative question – 'Is this evidence strong enough to warrant accepting this hypothesis?' – must come up in typical cases of acceptance. Certainly such questions might come up and might lead to reassessment of the relevant epistemic standard; but that need not happen. Moreover, either of the questions – (1) and (2) – that do come up in deciding the scientific acceptability of *h* could lead to such reassessment. But answering them does not require it, and Rudner does not show that scientific acceptance entails the application of standards beyond those a scientist has already adopted beforehand, quite possibly on purely epistemic grounds.

The above assessment of Rudner's crucial premise is a good background for evaluating some key elements in Levi's reply, as outlined earlier, to Rudner's argument. It seems quite true that, as Levi and others have maintained, one can accept a hypothesis without having in mind any specific practical objective. This point may be sufficient to block Rudner's argument. One might go further, however, in assessing the general issue. Suppose that a scientist does accept a hypothesis with the idea of using it to solve practical problems. (a) He need not and perhaps should not be prepared to stake anything on it – which is the most important case of acting on it. For he may believe that the relevant actions would not be warranted unless he had further evidence, or that until he can expect *others* to accept it he should take no action on it. Thus, (b) he can and should still distinguish two quite different questions: (i) Is the hypothesis scientifically acceptable? and (ii) Given how probable it is on the scientific evidence for the hypothesis, is it reasonable to act on it?[6]

Question (ii) will have different reasonable answers as applied to different actions. Moreover, it is not a question for the scientist as such, though competently answering it may require scientific training or some conceptual sophistication. The plausibility of both (a) and (b) is readily seen in the light of the twofold character of scientific acceptance, stressed above. For instance, if one accepts *h*, while having the second-order belief that one is just barely warranted, by one's evidence, in believing it, it is easy to see how one might be reluctant to act on it.[7] One may embrace before one is ready to trust.

All this can be illustrated by the example regarding a drug's toxicity. The scientist may accept the hypothesis of its non-toxicity with the hope of using it widely as medication and with the intention to test it further, yet – as Levi would agree – justifiably decide not to support its general use. This illustrates the distinction just made: the scientist, as scientist, answers the question of scientific acceptability positively, but, as responsible moral agent, negatively answers the question whether, *given* the scientific evidence, it is reasonable to put the drug in general use. This should not, however, be described as a refusal to "act on" the hypothesis (Levi, 1960, p. 351). The refusal is less general than that, and it is relative to a context. Thus, the scientist's acceptance of the hypothesis might carry a willingness to act on it in some situations, e.g. a situation of forced choice involving using or not using the drug in a life-or-death situation. Presumably here the scientist would, for himself at least, use the drug; and surely he might act on the hypothesis as least to the extent of arguing that it is worth the effort of further testing.

The sort of relation between acceptance and action just illustrated may be part of what motivates Rudner's argument: surely it is plausible to hold that accepting a hypothesis implies a disposition to act on it in *some* possible circumstances. But this disposition is consistent with a second-order belief that one's evidence is just barely sufficient to warrant believing *h* and does not justify staking anything on *h*. Thus, the toxicity example and similar ones do not undermine the distinction between questions (i) and (ii), nor show that (i) cannot be non-morally and objectively answered. As my earlier discussion suggests, one reason why this distinction is missed may be that acceptance is often mistakenly thought of as simply a matter of belief. But in fact acceptance carries varying degrees of conviction that *h*, and varying beliefs about the credibility of *h*. A rational person neither acquires conviction whose strength is disproportionate to his assessment of the evidence, nor, in his behavior, stakes on a hypothesis he accepts more than is warranted by his degree of conviction in accepting it.

These points bear on a quite recent attempt by James Gaa (1977, esp. pp. 551–523) to undermine the view that science is morally autonomous. We are asked to imagine a situation in which a scientist is studying a fresh water lake system and arrives at what he takes to be a number of scientifically acceptable hypotheses about relations among the constituents in the system. Now suppose that a policy-maker needs to decide what, if anything, to do about "nuisance algal bloom" in bodies of water of the relevant kind.

Since a probability value ... on the relevant hypothesis is also needed in making the decisions, a hypothesis concerning what that probability value is, must be accepted – and the costs associated with doing so are, in general, different. Presumably, the costs of error in the policy-maker's case are much higher than those of the scientist ... Now, the autonomy thesis requires in such a situation that the scientist should ignore the needs of the policy-maker – since the needs of the former come first (and, indeed, alone), there is no reason to gather more evidence (Gaa, 1977, p. 529).

The upshot is that "in the kinds of situations just delineated, the scientist qua scientist should act unethically" (1977, p. 530).

Whereas Rudner had argued that scientists *qua* scientists make value judgments, Gaa is arguing in addition that if they do not, the moral autonomy of science thesis may require them to act unethically. I have two points in reply.

First, the phrase "the costs associated with doing so," i.e., with accepting such a value, suggests willingness to act on the hypothesis, if only by leaving the nuisance algal bloom alone. But in fact it does not by itself imply this; it implies it only given (among other things) a further judgment of how well confirmed the hypothesis must be to justify acting on it. That judgment will be in part based on moral considerations, but it is not one the scientist as such should make. One might reply that even apart from such a judgment, the policy-maker must act on the hypothesis, since he must either put a chemical in the lake or not. It is true that he must do this or not do it, but neither action need be *based on* the hypothesis. One might, e.g., leave the lake alone simply on the ground that one has no good reason to do otherwise.

My second point is that nothing in the objectivity thesis (or in the moral autonomy thesis, as I understand it) entails that "the scientist should ignore the needs of the policy-maker." Surely Gaa is forgetting here that the objectivity and autonomy theses concern the scientist *qua* scientist. Such a scientist is, of course, an abstraction, a convenient but unfortunately misleading device for talking about the logic of the scientific enterprise. One cannot instantiate the concept of a scientist *qua* scientist without also being a *person*. A person should be ethically responsible, and, of course, the person who is a scientist studying lakes should, if possible given his other moral obligations and his resources, provide the policy-maker more information. But this is consistent with the scientist *qua* scientist being motivated wholly by a desire to pursue a purely scientific quest for truth. One would hope, moreover, that the figures the policy-maker gets from the scientist are based on just such a quest. It would be most unfortunate if the only scientific assessments of hypotheses the former could get were filtered through the scientist's moral judgments.

So far, we have not discussed the view that scientists as such do *not* accept hypotheses. I agree with Levi that adoption of this view is not necessary to defending the view that scientific evaluation of hypotheses does not require making moral judgments. But why should Levi have said that adopting it is like fleeing from Charybdis into the hands of Scylla? Granted, scientists often accept and sometimes even argue vehemently for hypotheses. But this could be regarded as extra-scientific; one could account even for the scientific search for truth by saying that scientists seek to articulate the best confirmed hypotheses and theories they can discover in the relevant scientific domain (perhaps allowing simplicity to figure as a subsidiary ideal). If, as human beings, they cannot help accepting certain apparently true hypotheses or theories, this only shows that they operate in two roles: the scientific and the pragmatic.

It seems preferable, however, to conceive the scientific enterprise more broadly. If we do suppose that scientists as such accept hypotheses, then I agree with the view, expressed by Levi and others, that the value-neutrality thesis – and scientific objectivity in general – does allow those value judgments which are implicit in the canons of scientific inference. These are plausibly considered purely epistemic, and it is reasonable to suppose that they could be made by a purely epistemic agent, in the sense of one whose only aims are to believe (certain sorts of) truths and avoid erroneous beliefs. Similarly, scientific objectivity allows scientists to make what Nagel calls "characterizing value judgments," (1961, pp. 104–107). These too may be plausibly argued to be neither in any sense moral nor necessarily open to bias by subjective influences. The question of when a hypothesis is acceptable relative to the evidence is one on which rational persons may disagree. But enough has been said above to indicate why it is not a moral question. The values it involves are epistemic: acceptability here is analogous to the notion of a good *argument*, not to that of a right act. Rudner is thus mistaken in claiming that determining the degree of confirmation of a hypothesis is itself accepting a hypothesis whose assessment involves value considerations of a moral or at least nonepistemic kind.

This way of replying to Rudner's claim contrasts with the view taken by Levi, at least in his initial response to Rudner. Levi there suggested that the claim can be refuted only by establishing a positive answer to the question whether (as he puts it) "the canons of scientific inference dictate the assignment of minimum probabilities in such a way as to permit no differences in the assignments made by different investigators to the same ... hypotheses." This is an important question whose answer seems to me not clear. I shall not try to answer it here. What I want to suggest is that the objectivity thesis does not require a positive answer. Let me explain.

It is essential, in discussing the notion of acceptability, to bear in mind an adequate distinction between the vagueness of concepts or claims and their subjectivity. Consider the claim that something is blue. It is vague. Is it also subjective? Granted, vagueness often allows subjective judgments to generate disagreements. The vagueness of 'intelligent', e.g., may lead to disagreements about someone's intelligence, based on subjective judgments of "brightness." But notice that it is possible to be clearly right or clearly wrong about colors (or intelligence), in a way in which it does not seem possible to be clearly right or clearly wrong in many matters of taste (those plausibly considered subjective). Moreover, when people do disagree over whether something is blue, the disagreement can very often be resolved by attention to terminology. These and related points show that vagueness does not entail subjectivity, and surely that applies to 'acceptability' as well as to many other terms.

Notice also that most people will agree that certain specimens are paradigms of blue, and these can be appealed to in resolving some disagreements about non-paradigm cases. A similar point holds for scientific hypotheses in respect to acceptability; and just as the epistemically cautious will not apply 'blue' where their less strict colleagues do –

without this implying either subjectivity or any (non-epistemic) value judgment – scientists may differ in the application of 'acceptable' without this implying either subjectivity or any (non-epistemic) value judgment.

At this point it might be argued that the only reason the judgment that something is blue is not subjective is that we can have recourse to a quantitative terminology, and the issue about acceptability of hypotheses is precisely whether comparably objective criteria are available. But this is too strong. If the conception of objectivity developed in Section I is approximately correct, then it seems sufficient for the objectivity of terms like 'blue', at least in certain uses, that (a) there be a range of cases in which there is very high intersubjective agreement on the application of the term in its non-comparative uses, and (b) if the term is comparative, there is similar concurrence on which, if either, of two things agreed to come under the term, has the relevant property to a higher degree.

Applying this to questions of the confirmation and acceptability of hypotheses, suppose for the sake of argument that we cannot find a precise method, which commands the assent of all rational persons who understand it, for assigning inductive probabilities to hypotheses given the scientific evidence for them, and suppose that even if we can, scientists still disagree on the minimum probability required for acceptability. If there is high intersubjective agreement among scientists on (a) whether purported evidence confirms a hypothesis and (b) which of two competing hypotheses, if either, is better confirmed by the purported scientific evidence, the notions of confirmation and acceptability might still be sufficiently objective to sustain, in the assessment of hypotheses, the value-neutrality thesis and scientific objectivity. I am not sure that such intersubjective agreement does exist among scientists; but I do not believe it has been shown to be an unrealistic ideal, and it is certainly more readily defended than the quantitative ideal to which some people apparently want to tie scientific objectivity.

The conclusion to which we come in this section, then, is that neither Rudner's argument nor similar ones show that the scientific evaluation of hypotheses requires making moral judgments or any other kind of non-epistemic normative judgments. It may be true that even if scientists *qua* scientists do not accept scientific hypotheses they do accept propositions about degrees of confirmation which might be called hypotheses; and there certainly appear to be alternative rational sets of criteria of acceptance and of confirmation. Perhaps the selection of one or another set of either kind can be shown to require moral considerations; but this does not appear to have been shown. Moreover, if it turns out that, on any plausible criteria, 'degree of confirmation' and 'acceptability' remain vague, it may not be inferred either that they are not objective or that differences in their application must be attributed to moral or other non-epistemic normative judgments. When Rudner's arguments – and the many similar ones proposed since his article – are rightly understood in the light of these points and the distinctions developed above, they cease to appear to give significant support to the thesis that scientific evaluation requires making moral judgments.

III. SOME QUESTIONS OF OBJECTIVITY IN THE SOCIAL SCIENCES

The possibility of scientific objectivity has been doubted or denied for reasons quite different from those we have considered in connection with Rudner's argument. It has been especially common for the social sciences to be thought incapable of objectivity. Max Weber has been a powerful influence underlying that position, and in this section I want to examine some of his major views that are relevant to the question whether scientific objectivity is possible and have been taken (even if not, or not unqualifiedly, by Weber himself) to imply that it is not.

Two of Weber's contentions are of particular interest here. They do not appear to have been fully answered in the literature, and they represent recurring themes in discussions of the methodology of the social sciences. In one place he argues that

The *significance* of a configuration of cultural phenomena ... cannot ... be derived and rendered intelligible by a system of analytical laws (*Gesetsbegriffen*), however perfect it may be, since the significance of cultural events presupposes a *value-orientation* towards these events. The concept of culture is a *value-concept* (1949, reprinted in part Brodbeck (1968) p. 88).

One implication seems to be that unless a social scientist adopts a value-orientation towards a culture, he cannot understand its distinctive events. Weber also posed a related, but more general, objection to the possibility of scientific objectivity. He said, e.g., that "All knowledge of cultural reality" is "from *particular points of view*" (in Brodbeck, p. 92) and that

If the notion that those standpoints can be derived from the "facts themselves" continually recurs, it is due to the naive self-deception of the specialist who is unaware that it is
due to the evaluative ideas with which he unconsciously approaches his subject matter . . .
without the investigator's evaluative ideas there would be no principle of selection of
subject matter and no meaningful knowledge of the concrete reality (in Brodbeck, pp.
92–93).

There are at least two important claims in this passage: that scientists'
knowledge is somehow affected by their points of view, which are in
turn affected by their evaluative ideas, and that these ideas heavily influence scientists in choosing what to study.

Weber has been criticized by defenders of scientific objectivity. The
best-known critique is perhaps Ernest Nagel's, and what Nagel says in
replying to Weber will provide a good backdrop for my points about
Weber. A central claim of Nagel's is that

It is not clear . . . why the fact that an investigator selects the materials he studies in the
light of problems which interest him and which seem to bear on matters he regards as
important, is of greater moment for the logic of social inquiry than it is for the logic of
any other branch of inquiry . . . there is no difference between any of the sciences with
respect to the fact that the interests of the scientist determine what he selects for investigation. But this fact, by itself, represents no obstacle to the successful pursuit of objectively controlled inquiry (1961, reprinted in part in Brodbeck (1968) pp. 99–109).

Nagel gives no reason, in the context, for his claim that this selectivity
is no obstacle to objectively controlled inquiry; but one would suppose
he has in mind the point that the way in which a problem is selected
for study entails nothing about the character of the method used in
studying it. He would likely hold that criteria of selection pertain to
issues concerning discovery and that even if objectivity should be un-
obtainable in the selection of problems to be studied or hypotheses to
be tested, it does not follow that scientists cannot be objective regarding confirmation and acceptance.

Nagel is surely right to stress that social scientists are not unique in
selecting their problems in accord with their interests, values, etc.
Moreover, it is clear that this does not necessarily bias their testing and
acceptance of hypotheses once they begin inquiry into a problem. On
the other hand, if scientists as such may be conceived as seeking to
provide a true account of some aspect of the world, it is disturbing to
think that their own values or interests might prevent them from even
exploring rich areas of potential truths. But how likely is this, one
might ask, if scientific inquiry is *free*, as, morally, it must be? Surely

free individuals freely inquiring into the empirical world will differ enough to make it unlikely that major areas will be ignored indefinitely.[9] Unfortunately, there is no guarantee that this will never happen. But its occurrence seems unlikely enough so that whatever obstacle it poses to scientific progress is practical and avoidable. The problem is not intrinsic to scientific method; and it concerns the context of discovery, not that of confirmation.

In distinguishing between the context of discovery and that of confirmation, I do not mean to suggest that the former is dominated by subjectivity and cannot be informed by rational, intersubjective procedures. Whether these procedures should be called a *logic* is another matter. I am inclined to withhold that term, if only because it suggests the use of deductive and inductive reasoning in a way more appropriate to the testing of hypotheses than to their initial formulation. But certainly some of the procedures which have been discussed under the *"logic of empirical enquiry"*[10] are not only capable of objective employment but applicable to the task of selecting appropriate hypotheses.

We have yet to consider two claims of Weber's cited earlier which, so far as I know, Nagel and others have not answered: that all knowledge of cultural reality is from particular points of view, and that the concept of culture is a "value-concept" such that understanding cultural events presupposes a "value-orientation" towards them.

Perhaps the truth which underlies the first point – and indeed makes it plausible – is that every *knower* has a point of view. Let us grant this for the sake of argument, though the notion of a point of view is left rather vague by Weber (and many others). How would the knower's having a point of view affect the quality of his knowledge? It might affect what he *believes* or why he believes it; but that is quite consistent with scientific method's serving as an objective control on the acceptance of hypotheses as representing scientific knowledge. Certainly Weber does nothing to show that knowledge, in the sense of a certain sort of justified (or reliable) true belief, is relative to a point of view. Indeed, his sound theses are consistent with the idea that the use of scientific method is a way to transmit knowledge from a person with one point of view to another with a quite different one.

Regarding Weber's second view, the claim that the concept of culture is a value concept is ambiguous. It might mean (1) that the concept *applies* only to a human phenomena that exhibit values – say in the relevant social practices – or (2) that *applying* the concept requires *making* (or presupposing) a value judgment. (1) seems correct. But it is only on interpretation (2) that one could use the claim against the value-neutrality thesis as applied to social science. On interpretation (1), scientists may *use* the concept of a culture and other concepts that in some sense refer to values, without endorsing those values or making any value judgments. They may need to *have* a concept of a value – say, of the socially accepted values of a culture; but they need not make normative statements involving – say, entailing a commitment to – any (non-epistemic) value. Weber has surely given us no reason, then, to think that social scientists as such must make value judgments.

IV. CONCLUSION

If the main points made in Sections II and III are correct, then some major obstacles to the thesis that scientific method is objective can be overcome. Moreover, while we have not examined all the plausible objections to this thesis, I believe that those we have explored are among the most serious, and that many of the others can be answered by extending the sorts of points made above. It is true that the conception of objectivity in relation to which this paper has defended the thesis is weaker than certain other conceptions of it, such as those on which it is truth-conducive. But the objections we have answered have been criticized here in ways available to proponents of stronger conceptions of objectivity; and if objectivity as we have conceived it is a less inclusive virtue of scientific method than it is sometimes thought to be, it is still of very great importance in science, and by conceiving it as we have we arrive at a more fine-grained view of the characteristics of scientific method than we would reach if we treated such virtues as truth-conduciveness as aspects of objectivity.

Having argued that neither moral judgments nor other non-epistemic normative judgments are required in the proper scientific evaluation of hypotheses, I want to reiterate that even if scientists *qua* scientists did have to make moral judgments, this would not entail that they cannot be objective. For it may not be simply assumed that moral judgments cannot, at least in special cases, be objectively arrived at in the sense specified in Section I; and I believe they can be. It still seems to me unlikely that scientists as such need to make any moral judgments in evaluating scientific hypotheses; but let me emphasize that this is consistent with the view, which I suggested in Section II, that scientists, as moral agents and citizens, should use their relevant scientific knowledge to support their moral judgments and should be willing to take moral stands on at least some pressing issues regarding which this knowledge gives them a good grasp of the relevant facts.

There remain two important and often neglected problems which I have not touched on. First, if *simplicity* is a properly scientific criterion of acceptability, is there an objective way of judging it?[11] Second, even if we can determine precisely the inductive probability of a hypothesis given the scientific evidence for it, how can we determine objectively when we have enough evidence – and enough warrant for believing our evidence statements – to make even a high inductive probability justify actually accepting the hypothesis?

If these problems cannot be resolved, then the defense of scientific objectivity may at least need to move closer to the view that scientists *qua* scientists do not accept or reject hypotheses. I believe that the problems can be resolved and that, for reasons which have emerged in our discussion, the relevant issues are epistemic and prima facie capable of objective resolution. Moreover, it seems likely that the conception of objectivity proposed in this paper can be useful in dealing with them. The conception has the advantage of allowing us to argue that a method is objective, without having to show that it is truth-conducive; and it preserves the distinction – which I have argued is essential in understanding objective methods – between objectivity and quantitative precision, and, correspondingly, between subjectivity and vagueness. In any case, supposing the defense of scientific objectivity does require adopting the view that scientists *qua* scientists do not accept or reject hypotheses, I doubt that taking this narrower view of scientific practice would prevent one from adequately dealing with the central questions in the philosophy of science. But pending a solution of the remaining problems I have mentioned, concerning how the acceptability of a hypothesis is affected by criteria of simplicity and by the credibility and breadth of the evidence statements themselves, the objectivity of scientific method is perhaps best regarded as a reasonable but not fully vindicated hypothesis.[12]

Robert Audi
The University of Nebraska,
Lincoln

NOTES

[1.] For some plausible criticism of Rudner's view of objectivity along lines different from those pursued in this paper, see Martin (1973).

[2.] Notice that in order to regard Rudner's argument as valid we must not take Premise 3 literally and suppose it concerns what in fact determines scientists' decisions regarding the acceptability of a hypothesis; we must take it as intended to express a proper criterion of scientific acceptability. The latter is doubtless Rudner's intent; but the argument's apparent plausibility may partly depend on the initial ambiguity, since the more literal reading yields a far less controversial claim.

[3.] For a more recent statement of Jeffrey's views on this general issue, in which he contrasts his views with Levi's, see Jeffrey (1970).

[4.] See, e.g., Levi (1967), esp. Chs. V and VI; (1980a), Chs. 47; and (1980b).

[5.] R. G. Swinburne (1980, p. 63) is among those who have suggested that 'acceptance' is often left unclear. For a helpful discussion of Levi's views and what Levi and others mean by 'acceptance' see Paul Teller (1980). Acceptance in relation to probabilities is discussed in detail in Burks, (1977), esp. in Ch. 2, Section 5.

[6.] This question must be distinguished from the one Levi raises (and affirmatively answers) of whether one can accept a hypothesis and refuse to act on it. See Levi (1960, p. 351).

[7.] The unclarity of 'acting on a hypothesis' takes on special significance in an interesting recent defense of Rudner's views by Robert Feleppa (1981). Whereas I would hold (as I think Levi and many others would) that *telling* somebody that one accepts a hypothesis is a way of acting on it, Feleppa suggests that the "act of acceptance" is not private, so that, e.g., the scientists in the Manhattan Project did not in the "fuller sense" accept the relevant hypotheses until "their announcing their full and un-mitigated acceptance to their superiors."

[8.] Henry Kyburg discusses roughly what I am calling intersubjective agreement under the heading of "interpersonal uniformity." For discussion of this concept, including some points that tend to support the distinction between subjectivity and vagueness, see Kyburg's (1968), esp. Ch. 3, Section II, and Ch. 4, Section II.

[9.] Cp. Popper's point that "it is the public character of science and of its institutions which imposes a mental discipline upon the individual scientist, and which preserves the objectivity of science and its tradition of critically discussing new ideas." Popper (1960, pp. 155–156).

[10.] See, e.g., Burks (1977). He says, in "listing some of the rules of procedure belonging to this branch of logic," i.e. the logic of inquiry: "given a vague problematic situation, formulate a precise and definite question concerning it. Systematically consider and organize the possible solutions to the problem suggested by what is known in the field at the time. Search related fields for analogies... Take a scientific or mathematical method or technique developed to solve a certain kind of problem and search for other areas in which it may be successfully and usefully applied" (p. 17). This contrasts with Popper's view: regarding the "logic of knowledge," he says, "I shall proceed on the assumption that it consists solely in investigating the methods employed in those systematic tests to which every new idea must be subjected if it is to be seriously entertained." See Popper (1959) p. 21. This and similar views have been critically discussed at length by Maxwell in (1974).

11. For a brief discussion of simplicity as a criterion of scientific acceptability, see W. V. Quine, "On Simple Theories in a Complex World," in (1966). A useful discussion of this relevant to the concerns of this paper is given by Howard L. Rolston; see his (1976).

12. This paper was written for a volume honoring Arthur W. Burks, who, as one of my teachers in the philosophy of science, contributed much to both my knowledge and my interest in the field. He bears no responsibility for any of my errors, however; indeed, I am certain that this paper would have benefited greatly if, before writing it, I had been able to achieve a mastery of his extensive work in the philosophy of science. For helpful comments on earlier persions of the paper I want to thank Albert Casullo, who criticized two versions in great detail, and Wayne A. Davis, Jamees Fetzer, and Merrilee H. Salmon.

REFERENCES

Brodbeck, May: 1968, *Readings in the Philosophy of the Social Sciences*, Macmillan, New York.

Burks, A. W.: 1977, *Chance, Cause, Reason*, University of Chicago Press, Chicago.

Carnap, R.: 1950, *Logical Foundations of Probability*, University of Chicago Press, Chicago.

Cohen, L. J. and M. B. Hesse (eds.): 1980, *Applications of Inductive Logic*, Oxford University Press, Oxford and New York.

Feleppa, R.: 1981, 'Epistemic utility and theory acceptance: comments on Hempel,' *Synthese* 46, pp. 413–420.

Gaa, J. C.: 1977, 'Moral autonomy and the rationality of science', *Philosophy of Science*, 44, pp. 513–541.

Hempel, C. G.: 1966, *Aspects of Scientific Explanation*, The Free Press, New York.

Hempel, C. G.: 1949, Review of C. W. Churchman's *Theory of Experimental Inference*, *Journal of Philosophy* LXVI, pp. 557–561.

Jeffrey, R. C.: 1970, 'Dracula meets Wolfman: acceptance vs. partial belief', in Swain (1970), pp. 157–185.

Jeffrey, R. C.: 1956, 'Valuation and acceptance of scientific theories', *Philosophy of Science*, XXIII, pp. 237–246.

Kyburg, H.: 1968, *Philosophy of Science: A Formal Approach*, Macmillan, New York and London.

Levi, I.: 1960, 'Must the scientist make value judgments?', *Journal of Philosophy*, LVII, p. 345–357.

Levi, I.: 1967, *Gambling with Truth*, Alfred A. Knopf, New York.

Levi, I.: 1980a, *The Enterprise of Knowledge*, M.I.T. Press, Boston.

Levi, I.: 1980b, 'Potential surprise: its role in inference and decision making', in Cohen and Hesse (1980) pp. 1–27.

Martin, M.: 1973, 'The objectivity of a methodology', *Philosophy of Science* 40, pp. 447–450.

Maxwell, N.: 1974, 'The rationality of scientific discovery', *Philosophy of Science*, 41, pp. 123–153, 247–295.

Nagel, E.: 1961, *The Structure of Science*, Harcourt, Brace & World, New York.

Popper, K.: 1959, *The Logic of Scientific Discovery*, Basic Books, New York.

Popper, K. 1960, *The Poverty of Historicism*, Routledge & Kegan Paul, London.

Quine, W. V.: 1966, *The Ways of Paradox and Other Essays*, Random House, New York.

Rolston, H.: 1976, 'Note on simplicity as a principle for evaluating rival scientific theories', *Philosophy of Science* **43**, pp. 438–440.

Rudner, R.: 1966, *Philosophy of the Social Sciences*, Prentice-Hall, Englewood Cliffs, New Jersey.

Rudner, R.: 1953, 'The scientist *qua* scientist makes value judgments', *Philosophy of Science* **20**, pp. 1–6.

Swain, M. (ed.): 1970, *Induction, Acceptance, and Rational Belief*, Reidel, Dordrecht and Boston.

Swinburne, R. G.: 1980, 'Properties, causation, and projectibility: reply to Shoemaker', in Cohen and Hesse (1980), pp. 313–320.

Teller, Paul.: 1980, 'Zealous acceptance', in Cohen and Hesse (1980), pp. 28–53.

Weber, M.: 1949, *The Methodology of the Social Sciences*, The Free Press, New York.

PART II

THE PHILOSOPHY OF LOGICAL MECHANISM

Replies by Arthur W. Burks

347

ARTHUR W. BURKS

THE PHILOSOPHY OF LOGICAL MECHANISM*
Replies by Arthur W. Burks

I. INTRODUCTION

Let me first express my great appreciation to the editor and to the contributors of this volume. I had never expected to be so honored, and it has been a pleasure and a rewarding experience to read and reflect on each of these excellent essays.

My responses are rather general. I do not feel able to reply in degree to the detailed and technical analyses given by some of the contributors, valuable and insightful as they are. My long-range philosophical interests have always been in broad questions of epistemology, logic, metaphysics, and value, such as those treated by Plato, Lucretius, Hume, Kant, and Peirce. Although I have done technical philosophical analyses and developed formal logical systems, such as my logic of causal statements and my calculus of choice, these were developed as tools for specific purposes.

A historical comment on the role of the philosophy of science in the curriculum may be appropriate here. I was a graduate student in philosophy at the University of Michigan from 1936 to 1941. At that time the department offered only one course in the philosophy of science: first, philosophy of relativity, and later, after a change in faculty, inductive logic. When I returned to the University of Michigan as a faculty member in 1946, I initiated a course in the philosophy of science devoted to questions in the foundations of induction, causality, and probability. While teaching this course I developed a line of thought I had begun as a graduate student.

I had become interested in the classic controversy between Hume and Kant concerning the nature of causality and induction, and the related problem of the logical difference between the necessity of causal statements, such as a causal subjunctive (e.g., "If this ring should be placed in aqua regia, it would dissolve"), and the necessity of mathe-

349

Merrilee H. Salmon (ed.), *The Philosophy of Logical Mechanism*, 349–531,

matical statements. This interest led to my logic of causal statements, my work on probability, my formulation of Hume's thesis that there is no noncircular justification of induction in terms of formal inductive logic, and my presupposition theory of induction. *Chance, Cause, Reason* is the culmination of this research.[1]

For many years, my course in the philosophy of science was the Michigan deparrtment's only course in this area. But times have changed. Today there are almost a dozen and a half courses, with about a third of them taught in any given year. While these increases are due in part to a tripling of the size of the university, they are also partly due to the growing importance of science and logic in technology and culture. As a consequence, there is now a very large literature in the philosophy of science, inductive logic and decision theory, and the logic and philosophy of language.

Part of the reason I have not followed recent developments in some of these subjects to which I contributed earlier has been my heavy involvement in computer science. This began mostly by accident, during World War II, when I had the opportunity to work on the first general-purpose electronic computer, the ENIAC, and one of the first stored-program computers, the IAS machine. I find computers as fascinating as philosophy, and much of my research and teaching really belongs to both fields. Computers are both models of nature and models of mind.

My philosophy makes such thoroughgoing use of computer ideas and their relation to logic that I have come to call it "The Philosophy of Logical Mechanism". During most of their history, the doctrines of materialism and mechanism have been bound too closely to the concepts and laws of classical physics, failing to reflect the logical possibilities inherent in mechanisms and in biological systems. "Mechanism" is used here in the broad sense that includes automatic devices, computers, robots, and natural organisms insofar as they function similarly. "Logic" is also used in a broad sense, covering a wide range from computer switches and memory components, at one extreme; through programs (genetic as well as computer) and computer structures; to inductive logic, principles of learning, and the logic of evolution at the other extreme. When a system makes important use of logic and memory, it is appropriately called a *logical mechanism*. Mind is a logical mechanism. The reader will find a summary of my philosophy of logical mechanism in the last section of these replies.

*   *   *

The discussion will be organized under these topics: the logical foundations of science (addressing the articles by Cohen, Suppe, Thompson, Zeigler), the relation of computers to minds (Arbib, Boër, Laing, Lugg, and Nelson), and evolution and induction (Audi, Clendinnen, Moore, Railton, Rosenkrantz, Skyrms, and Uchii). But first, it will be useful to distinguish various points of view from which these topics can be approached.

There are many ways of looking at the world, as many as there are subjects, either in the sense of subject matters or in the sense of cognizers along the evolutionary scale. Consider a rational man, whether a commonsense knower or a scientific inquirer. He is a conscious knowledge system operating in nature, communicating with other similar systems. Any such reflexive knowledge system can be approached from either of two points of view: the inner or the outer.

From the inner perspective, one analyzes the present state of one's corpus of beliefs and one's knowledge of nature and oneself, distinguishing what can be known more or less directly from what can be inferred, and theorizing on the relation between the two. This approach has a long tradition, including Socrates and Plato on innate ideas, continental rationalism and British empiricism, Kant's *Critique of Pure Reason*, introspective psychology, phenomenology, and some forms of Buddhism.

From the outer perspective, one starts with the present state of belief or knowledge (in some one area or in many areas taken as a whole) or with the evolution of a corpus of beliefs and knowledge over some period. The outer approach analyzes the structure of this corpus or the general features of its development. It is pursued by materialists, naturalists, intellectual historians, coherentists, neurophysiologists, and behavior psychologists, among others.

The inner approach is more epistemological, whereas the outer approach is more metaphysical. One can, of course, pursue both approaches and study their interrelations; a fully adequate theory of knowledge must do this. But usually one perspective is emphasized and the other neglected, and their relation left unclear. This has been a common weakness of philosophy: naturalists have started from the outside and never reached the aspects of experience the idealist regards as most basic; idealists have started from the inside and never been able to construct a "full-blown" nature.

I will use the inner-outer distinction in discussing the relation between the evolution and structure of scientific theories and my own presupposition theory of induction.

Another distinction, that between local and global points of view, will also be useful, especially in discussing computers and minds. In studying a large system, a scientist may find that the properties of the whole are very different from the properties of the parts. For example, the density of matter may be fairly uniform, from one large region of space to another, and yet be highly variable locally, as between a solar system and the almost empty space around it. My distinction between a local and a global point of view in philosophy is analogous (*CCR*, sec. 9.2.4). An example is the contrast in science between a local law and a global theory. An ordinary cause-effect law states what will happen in a small or local region of space-time. Because of its inverse-square character, Newton's law of gravity lies in between; when determining the gravity field at a point, one can generally ignore distant objects.

The free will versus determinism issue provides another example of the local versus global distinction. Determinism and its alternatives, indeterminism and probabilism, are global theories. They are metaphysical theories about the nature of the laws governing the whole universe, and also about the completeness of their rule. Are the basic laws causal, and, if so, are there occasional exceptions to them? Or are the basic laws statistical, or partly statistical and partly causal? In contrast, the following questions raise local issues: Can a rational person make free choices and really control his or her own destiny? How can a person be held morally responsible, or justifiably punished, while living in a deterministic universe? These issues are local, for they concern the process of decision-making in an individual, together with the influence of the environment and society on this process. Although the local issues associated with free choice and the global issues associated with determinism are interrelated, they are clearly different and somewhat distant from each other. Philosophers generally pass too quickly from one type of issue to the other.

Corresponding problems arise with respect to society. If physical, chemical, biological, and cultural evolution are all deterministic, does it follow that society is not free to improve itself? If pre-human organisms are much more competitive than cooperative, does it follow that the best human society would be more competitive than cooperative? What difference does it make to these problems if nature is indeterministic or probabilistic rather than deterministic?

Here again local issues are distinct from related global metaphysical issues. Actually, these ethical-social issues concern larger systems than do the corresponding personal issues of free choice, since societies are groups of individuals and evolution encompasses sequences of such groups. This illustrates the fact that the local-global distinction is relative to the context of discourse. The sequence of *individual, society, evolution, nature* is a sequence of ever larger and more inclusive systems.

With these two distinctions in mind (inner versus outer, and local versus global points of view) let us turn to the essays. I begin with the essays treating the nature of our subject, philosophy of science, and address them in terms of my foundational approach to this subject.

2. THE LOGIC OF SCIENCE

2.1 What are the foundations of science?

Frederick Suppe's fine essay "Is Science Really Inductive?" is a good place to start, for it treats very well a fundamental issue in the philosophy and history of empirical science. Everyone agrees that observation and experiment play an essential role in deciding which generalizations and theories to consider, accept, or reject. There is disagreement over whether a probability-based inductive logic plays a fundamental role in this process. My view is that it does. Most contemporary historical analysts of the development of scientific knowledge believe that it does not.

Suppe says I do not really argue for my view, "taking it as given". It may appear so, but I have never taken such a broad, speculative thesis for granted, and in *Chance, Cause, Reason* I tried to argue for it more carefully than did my predecessors. My discussion of the geocentric and heleocentric hypotheses, of medical examples, and of the frequency relation between being guilty and being found guilty, all contain arguments for my thesis (secs. 1.4, 1.5, 2.4, and 3.1). Moreover, Section 10.3 (Causal Models of Standard Inductive Logic) employs inductive probability in a first attempt to explain how non-modal observation statements can confirm causally necessary generalizations.

It is true that my examples come from common sense and early science, rather than from later science. This is a reflection of the present state of formalisms for inductive logic, which are relatively simple and hence serve as better models for simple, early science than for later, complex science (*CCR*, secs. 1.6, 1.7, 3.3.3). Thus there is a gap to be filled before we will have an inductive logic adequate for well-developed science. I have some ideas as to how to fill this gap, both in method and in content (sec. 2.3). Methodologically, the procedure of logical formalization should be extended to computer simulation. In terms of content, I believe that the evolution of scientific theories from common sense knowledge is significantly analogous to evolutionary processes generally, and that the hierarchical structure of well-developed scientific theories reflects this fact (sec. 4.3). But, in any case, I don't think the present gap between formal inductive logic and actual science should count against foundationalism, because the anti-foundationalist school that Suppe is defending in the first part of his paper does not use formal methods at all, and, in fact, does not even employ some important epistemological distinctions, such as that between discovery and evidential proof.

The last part of Suppe's paper brings out very well that my differences with contemporary historical analysts of scientific knowledge are much more a matter of approach and subject matter than of doctrine. Although we are all studying the same system, scientific inference, we are asking different questions about this system rather than giving conflicting answers to the same questions. The writers Suppe discusses are much more descriptive, detailed, and historical than I am, treating science in all its complexity. I am much more analytic, general, and speculative, distinguishing learning from use, distinguishing discovery and history from evidence and organized justification, and distinguishing deductive from inductive considerations. I am looking for simple general principles, rather than for detailed theories of how science operates.

As Suppe points out, one can deny the helpfulness of such distinctions in understanding science. But they are prima facie useful; ignoring them, as so many current writers do, does not show that they are not useful. Here are distinctions that seem to me fundamental. Discovery may take different paths to essentially the same result. Learning a technical language and solving scientific problems are different processes, even though they are interactive and both involve "learning" in a suitably general sense. The value of simplicity in any conceptual structure is understandable on economic grounds, separately from confirmatory issues concerning the applicability of that structure to nature.

Also, normative and evaluative issues are generally separable from descriptive and explanatory questions.

The origin of the foundationalist view of induction – and the strongest argument for it – is to be found in the foundations of deduction and mathematics. The distinction between methods of proof and methods of discovery in mathematics and mathematical logic is now well established. This distinction began with Pythagoras and Euclid and has led, in the past 150 years, to the well-developed subject of the foundations of mathematics. In this subject, the concept of an uninterpreted formal system serves as a model of deductive inference. Of course, the logic of mathematical discovery, the subject that constitutes the other side of the discovery-proof distinction, is still in a rudimentary state; but there now exists a powerful tool for advancing this subject, namely, the modern electronic computer.

Thus I think the following analogy an important one: deductive logic is to mathematics as inductive logic is to empirical science. Deduction is, of course, employed in science, and so is the core of induction. So, also, the distinction between discovery and proof for induction is an extension of that distinction for deduction. It should be emphasized, however, that induction goes beyond deduction and is not reducible to it.

My view, then, is definitely not what Suppe calls "the received view" in his *The Structure of Scientific Theories* (Suppe 1974). Briefly, this is the view that scientific theories should be constructed "as axiomatic theories which are given a partial observational interpretation by means of correspondence rules" (Suppe 1974, pp. 3, 12, 50–56). In my opinion, the received view treats science too much like mathematics, is much too reformatory in its approach (trying to fit theories into the narrow axiomatic mold), and takes a narrow view of the relation of theory to practice. Axiomatic theories are not good tools for actual use in science. Mathematicians do not use axiomatic theories unless they are working in the foundations of mathematics, so why should empirical scientists employ them in their work?

Incidentally, I would make the same criticism of "artificial intelligence", the "received view" in computer science as to how to design intelligent computer systems. Current workers in artificial intelligence rely mainly on deductive methods, to the neglect of inductive methods. All the prominent logic-based program languages are deductive rather than probabilistic. This emphasis is explained by the fact that deductive methods are available – computer scientists can take them ready-made from deductive logicians. But in my opinion the design of really intelligent systems will require the development of new inductive methods.[2]

I am disparaging only the overemphasis on traditional axiomatic systems, and the oversimplified view of their use. Axiomatic systems do clearly have a fundamental place in the philosophy of science. For example, it was not until philosophers symbolized empirical conditionals that the difference between causal implication and material implication emerged. This distinction is basic to the philosophy of science, whether or not a modal formalism such as my logic of causal statements is the best way to elucidate it (see sec. 4.1). Note that once the distinction is explained by deductive means, an inductive problem arises: How can a non-modal observation statement confirm a causal generalization? That is, what is the inductive relation between the actual world and the set of causally possible worlds? (This question is discussed in the next section.)

My general theme is that both philosophers and computer scientists need to take a broader view of what a formal system might be. I have argued that a digital computer is a formal structure, and so a computer simulation of a system is a formal model of that system in an extended but legitimate sense (Burks 1975b, pp. 308–311; 1979a, pp. 408–410). In this extended sense, vagueness, non-formal uses of language (legal, moral, aesthetic), the self-corrective nature of human knowledge, inductive inference and the logic of empirical inquiry, and biological evolution can all be formalized, at least in principle. Of course, this only shifts the problem to that of finding a manageable computer model, but at least it widens the range of possibilities. This topic is discussed further in Section 2.3 (Logical formalization and computer simulation).

We need at this point a positive characterization of inductive logic. Suppose it is granted that inductive logic lies at the foundation of empirical science. The question then arises: What is the nature of induction, and how does it differ from deduction? My own view, of course, is that probability theory, properly interpreted, is the essence of inductive logic, and most of *Chance, Cause, Reason* is devoted to this thesis. As a basis for further comments, let me state briefly the main points of my account.

There is a sense of "probability" that I call "inductive probability"

(P) because it expresses the evidential relation of one empirical statement to another. This concept was explicated at great length in *Chance, Cause, Reason*, both informally and formally, and was contrasted with other concepts of probability, especially frequency probability and empirical probability. The informal specification consisted of examples and comments. The formal specification was done through three axiomatic systems: the calculus of inductive probability, some simple inductive logics, and my calculus of choice. A few words about each of these systems will be useful later.

The calculus of inductive probability treats systems of interrelated probability statements, each an atomic statement of the form $P(c,d) = x$. Here x expresses the extent or degree to which premise or condition d gives evidential support to conclusion c; x may be a real number, or a vague term like "high", "low", "proved beyond a reasonable doubt". By the rule of total evidence, one should use all the information d available in deciding what degree of probability to use in acting on c.

The traditional calculus of probability relates atomic statements to each other only {as in, if $P(c,d) = x$ then $P(\sim c,d) = 1-x$}; it does not determine the truth values of atomic statements except in limiting cases {e.g., $P(c\&\sim c,d) = 0$}. A formal system that also specifies the values of inductive probability statements is called an "inductive logic". Rudolf Carnap showed how to define inductive logics for simple models of reality. *Chance, Cause, Reason* (sec. 3.2) contrasts standard inductive logic, which incorporates a rule of enumerative induction, with two logically possible alternative inductive logics: random inductive logic, which treats successive states of nature as probabilistically independent, and inverse inductive logic, in which instances of a generalization disconfirm that generalization rather than confirm it. These alternative inductive logics show the relation of an inductive logic to prior probabilities and are a means for formulating a probabilistic version of Hume's scepticism.

The axiomatic systems discussed so far formalize what I call the "a priori aspect of inductive probability (P)". There is a second, co-equal aspect of the meaning of "P", which I call the "pragmatic aspect of inductive probability". This aspect involves the application of P to individual choices and actions, and hence its relation to utility (U). The utility of an act depends on the utility (value) of each consequence and the probability of its occurring, relative to the information one has.

Generally speaking, one should choose an act of maximum utility. Thus an atomic probability statement does and should express a disposition to act in certain ways under conditions of uncertainty.

My calculus of choice (*CCR*, ch. 5) is an axiomatic system designed to formalize the pragmatic aspect of inductive probability. It is based on the work of Frank Ramsey, Bruno de Finetti, and James Savage, but it goes beyond their systems in various ways, the most basic being that they considered only choices from among normal form acts, for example, a choice between

> (A) if p is true receive consequence C_1,
>   if p is false receive consequence C_2; and
> (A') if q then receive C_3, if *not–q* then receive C'_4.

Such a choice can be represented by a two-level tree, the root (choice point) leading to chance points A and A', chance point A leading to consequence point C_1 by a line marked with the statement that produces that consequence (namely "p"), etc. In contrast, the calculus of choice concerns choice trees with several levels of choices alternating with chance events until consequences are reached at the lowest level.

These extended choice trees are more like extended strategies in game theory than normal form trees, and so they fit the sequential nature of life (where each choice leads to a new situation that also involves choices) better than do normal form trees. But they are still idealized models of individual decision-making in various respects. For example, it is usually impractical to consider very many alternatives or to make choices many steps ahead, except in a general way. We give a more realistic picture of human choice later (sec. 4.4, Evolution, science, and values).

Many philosophers of science deny the relevance of probabilistic inductive logic to scientific inference. Since I developed the calculus of choice to elucidate this relevance, I will explain the calculus briefly.

The calculus of choice explicates the notion of rational choice, and it does so in a way that distinguishes the "qualitative" or "approximative" essence of using inductive probability (**P**) and maximizing utility (**U**) from the full quantitative application of these concepts. Mathematical theories of probability and utility treat these concepts as quantitative in the sense that probability and utility values are real numbers, or numbers with infinite precision. But in actual practice inductive probability values and utility values are imprecise, rough, qualitative, and approximate, obeying some of the laws of real numbers but not all. In *Chance, Cause, Reason* I called this the distinction between "qualitative" and "quantitative" probabilities and utilities, but I now think the terminology *approximative* versus *quantitative* is better. To bring out the contrast, I will outline the calculus of choice in two stages.

Suppose that you are given a system of extended choice trees and told to mark your preferences on all of them. Suppose further that your set of choices satisfies the following rules. (1) Make your preferences transitive. (2) Treat logically equivalent choice alternatives the same. (3a) If you prefer consequence C to consequence C' in one context, do so in every context. (3b) More generally, suppose two contexts contain the same choice alternatives and have the same background information; then make the same choices in each context. (4) Suppose you are given several choices of whether to bet on statement p or statement q, with the background information the same in each case; if you stake one prize on p rather than q, then stake every prize on p rather than q (i.e., if you bet on p in one case, do so in every case). (5) The final rule is a normal form rule, connecting choices made in extended (multi-level) choice trees to choices made in standard two-level trees (normal form trees), in which one chooses from among normal form acts (e.g., if p then C, while if *not–p* then C'). The normal form rule states that whatever choices one makes in an extended choice tree, he or she makes equivalent choices in the corresponding normal form tree.

The preceding list of rules gives a pragmatic meaning to inductive probability (**P**) and to utility (**U**) of the following kind. If a person – a man, say – follows these rules, then *there is* some utility function U that he is maximizing, even though he has no explicit concept of utility. Moreover, if he follows these rules he is *acting as if* he has assigned an inductive probability (**P**) to each uncertain alternative, even if he has no explicit concept of probability. This pragmatic sense is the sense in which I hold that inductive probability plays a basic role in scientific inference, and the calculus of inductive probability is a useful model of the ordinary use of "probable", although of course one can also be explicit in the use of probability theory and the rule of maximizing utility. Likewise, this pragmatic sense of meaning is what I have in mind when I say that practicing scientists use standard inductive logic, although, again, they can also be explicit in their use of analogy, Mill's methods, and enumerative induction (*CCR*, pp. 97 and 103).

The preceding result about probability and utility is an approximative result, because for any finite number of choices there are many different quantitative probability and utility assignments for which the subject is maximizing utility, so that the agent's choices fix his or her implicit probabilities and utilities only approximately. For there to be unique probability and utility assignment, the subject must make infinitely many choices in a properly structured choice situation. Suppose, for example, that a subject – a woman this time – prefers consequence C_1 to consequence C_2, and she regards a certain coin tossing process $T_1, T_2, T_3, \ldots$ as fair. ("T_n" means "toss n is tails".) Then we can determine her probability for any statement p as follows.

To begin with, we see whether she prefers to stake C_1 on p or on T_1; that is, whether she chooses "if p then C_1, if p then C_2" or "if T_1 then C_1, if T_1 then C_2". This fixes the probability to within 1/2. We next use the sequence of two tosses as the reference frame, comparing p with $T_1 T_2$, with T_1, and with $T_1 T_2 \vee T_1 T_2 \vee T_1 T_2$ in succession; this fixes the probability to within 1/4. In terms of the binary number system, the first two bits of the probability have been determined. This process can be repeated indefinitely, to determine the probability of p bit by bit.

Thus the calculus of choice brings out the basic difference between quantitative and approximative probability, and between quantitative and approximative utility, and shows how these differences are special cases of the difference between the finite and the infinite. Quantitative probabilities and utilities presuppose a structured infinite sequence of choices, and so from a practical point of view they are idealizations. In practice, we always use approximate numbers, which are finite, whereas most real numbers are infinitely complex. On the other hand, the system of real numbers has a simpler structure than a system of approximate numbers. Consequently, although approximate probabilities and utilities fit actual usage more closely than quantitative ones, it is generally better to use quantitative models and treat the discrepancy informally.

Let me now use these results to interpret my claim that the ordinary person and the practicing scientist use standard inductive logic. By my Bayesean analysis of the confirmation process (*CCR*, sec. 2.5), this claim divides into two, one concerning prior probabilities and one concerning posterior probabilities. The first claim is: people generally agree to a significant extent on their prior probability assignments, the extent needed to account for the degree to which they agree on their posterior probability assignments. The second claim is: there are inductive rules whereby one generalization or hypothesis tends to become confirmed and the alternative hypotheses disconfirmed, and by these rules experimentation and data collection gradually bring the members of the scientific community into agreement.

Everybody recognizes that empirical probability assignments can be confirmed or disconfirmed. But one can deny that probability is otherwise relevant to inductive inference, thereby rejecting the conceptual basis of my model. The coherentists do this; I consider their views later (pp. 468–470 below), after developing some ideas about evolution and hierarchy. None of my commentators takes this extreme position. Suppe and Cohen, however, do find my Baysean model inadequate in crucial respects. I will close this section by addressing Suppe's view briefly and then take up Cohen's view in the following section.

Suppe, following Dudley Shapere, believes that scientific inquiry begins with "plausibility assessments" of hypotheses, which are "more basic and fundamental than inductive probability assessments" (Suppe, p. 15). But it seems to me that these plausibility assessments result from applying general rules to the background information and specific knowledge one has about the hypotheses under consideration, and that these plausibility judgments are in fact rough inductive probability estimates. They are, of course, approximative, not quantitative, and the reasoning that produces them is partly intuitive, but they are probability judgments nevertheless.

How might Suppe respond to this? He could grant that rules are employed in making plausibility assessments and that plausibility is a matter of degree, yet still question whether any system of inductive probability can model these plausibility assessments at all well. Furthermore, he could point to a "structure-complexity gap", a gap between the complex structures of modern scientific theories and the simple structures expressible in formal systems of inductive logic. Suppe's major work is his *The Structure of Scientific Theories* (Suppe 1974), which demonstrates very well the complexity of the structure of modern scientific theories.

I agree that there is such a "structure-complexity gap" between real science and current formal models; the conclusions that I have drawn about science from formal models have always been tempered by an awareness of this (*CCR*, sec. 3.3.3). Current formal models of induction do not go beyond first-order quantification theory and modal versions of it. Attention is usually focused on generalities of the form "All A are B" and instances of these. It is an exaggeration, but not too far from the truth, to say that the current state of inductive logic is comparable to the state of deductive logic in Aristotle's time.

Granted all this, it does not constitute an argument against my version of foundationalism, because it fails to make the burden of proof for the inductivist greater than the burden of disproof for the anti-inductivist. It means only that inductive logic is highly speculative and programmatic, a feature it shares with most of philosophy. Indeed, when formal inductive logics for complex theories are developed, as I think they will be, because of their complexity they will no longer be handled by humans with paper and pencil, but by humans interacting with powerful computers.

In the meantime, we must start from where we are. My claim for inductive logic has much the same status as reduction claims' in general, especially the claims that culture can be reduced to evolutionary biology and evolutionary biology to chemistry and physics. It is closely related to such claims, because well-developed science evolved from beginning science, and beginning science from commonsense knowledge. Hence we must at some stage discuss the relation of induction to evolution, and we do so later (sec. 4.2). We will now discuss the nature of inductive logic, including the specific question of whether enumerative induction is more basic than eliminative induction.

2.2 *Rules of induction and their formalization.*

Cohen's essay on Bolzano is a good place to start a discussion of the formalization of inductive rules, because both have raised a fundamental issue concerning the nature of these rules. Bolzano held that "two factors should affect the confidence with which we expect an observed regularity to continue; one is the number of observed cases, the other is the extent of their variety". (Jonathan Cohen, "Bolzano's Theory of Induction", p. 32.) Bolzano had universal generalizations in mind, but Cohen has pointed out that the same two factors apply to statistical generalizations as well.

These two factors of *number of cases* and *variety of cases* may be formulated in terms of two different types of rules, those of *enumerative induction* and those of *eliminative induction*. In enumerative induction, instances are taken to confirm a generalization like "All A are B" or to make it more likely that the next A will be a B. Eliminative induction involves observing or creating situations with different qualitative structures to ascertain if one property is causally related to another, and if so, to determine the circumstances under which one produces the other or occurs simultaneously with the other. Rules of eliminative induction are given by Bacon's tables, Mill's methods, what Cohen calls "the method of relevant variables" (Cohen 1970, 1977), and what I call "the method of varying causally relevant qualities" (*CCR*, p. 102).

As Bolzano states, enumerative and eliminative induction can operate together. It is illuminating to look at the history of science in terms of their relative importance. Science really grew out of commonsense knowledge, and it has developed continuously, sometimes in spurts, but usually gradually. Enumerative induction plays a major role in commonsense knowledge and so was very important in the beginning stages of science. But as science developed, enumerative induction gradually became less important. Advanced science does not pile up quantities of the same kind of instances. On the other hand, since advanced science evolved from commonsense science, we should expect a residue of enumerative induction in it. I think this residue is found in the fact that experiments are repeated often enough to make sure things go as planned, although this involves qualitative and quantitative variation as well as repetition.

There are many intriguing questions to ask about enumerative and eliminative induction. How are they related to each other and to other inductive rules, such as analogy and the method of testing hypotheses by their consequences? Is one of the two, enumerative or eliminative induction, more basic than the other? We should not count all instances or vary every property, but only special ones, such as carefully selected instances and properties we have reason to believe are causally relevant. No doubt these restrictions are related to the nature of causal laws, true theories, and good predictions, but an explanation of these relations is needed, and it is not contained in traditional formulations of the rules of induction and the conditions under which they may be applied, such as those of Mill, Bolzano, and Cohen.

Underlying these questions – and rendering them more complex than they might otherwise be – is the fact that the strength with which an inductive conclusion can properly be asserted is a matter of degree. The concept of degree is explicit in enumerative induction (more instances, more confidence, at least up to a point), and at least implicit in eliminative induction (some experiments are better than others). As Cohen points out, John Stuart Mill tried to formulate his methods so that they would yield conclusive results. But this is hardly a plausible position; it rests on the same kind of error as Mill's claim that arithmetical truths are empirical, namely, the failure to take full account of the conditions of application of a logical structure. It is true, as Mill said, that the membership of the sets to which one is applying an arithmetic formula can be changed without one's knowledge; but this makes the formula irrelevant, not false. Similarly, if one could know all the relevant variables, whether each was absent or present, and were sure of all the presuppositions involved, one could be sure of the inductive conclusion. But as most writers recognize, it is the function of induction to deal with such matters, and that is why inductions are subject to degrees. Francis Bacon, David Hume, Bolzano, J. F. W. Herschel, and Charles S. Peirce all held that an inductive conclusion must be asserted with an appropriate amount of strength.

There are some who hold that deduction is also a matter of degree: coherentists who extend their coherentism so far as to include pure mathematics in its scope are committed to the view that there are various degrees of deductive validity. But if one makes the kinds of distinctions I suggest in the preceding section (learning a language versus using a language already learned, the logic of discovery versus the logic of argument, etc.), there seems to be no basis for degrees of deductive validity. Hence I think that deduction is all-or-none. Once the mathematical language is fixed, the conclusion either follows from the premises, or it does not. The more-or-less feature of induction is a fundamental difference between induction and deduction, a difference that has many consequences. For example, it is reflected in the different role that completeness plays in applications of the two kinds of logic. Both are subject to the rule of total evidence: one should use all the information (evidence, knowledge) available in deciding whether a proposition is true or false, probable or improbable, and in deciding what degree of probability to use in acting on a proposition. However, once the premises of a deductive argument are strong enough to yield the conclusion, strengthening them can have no effect on the validity status of the argument. In contrast, as more information is added to the premises of an inductive argument, the probability of the conclusion may go down *or* up.

Granted that induction is a matter of degree, a basic question arises: What features of the evidence account for the degree to which one can justifiably draw an inductive conclusion? David Hume gave an exceptionally perspicuous answer to this question (Hume 1777, sec. IV, part II).

Nothing so like as eggs; yet no one, on account of this appearing similarity, expects the same taste and relish in all of them. It is only after a long course of uniform experiments in any kind, that we attain a firm reliance and security with regard to a particular event. Now where is that process of reasoning which, from one instance, draws a conclusion, so different from that which it infers from a hundred instances that are no-wise different from that single one? This question I propose as much for the sake of information, as with an intention of raising difficulties. I cannot find, I cannot image any such reasoning.

I think Hume's basic point is correct, even though his formulation is faulty. Substitute "deductive reasoning" for "reasoning", and Hume is saying that because deduction is all-or-none, reasoning in which instances count is not deduction. What Hume elsewhere calls "reasoning from causation", "reasoning concerning matters of fact", and "probable reasoning", is what we now call "inductive reasoning". And in induction the strength of the conclusion *does* depend on the number of experiments. Thus Hume was right: enumeration does distinguish induction from deduction.

But there are still further questions concerning the degrees of inductive arguments. How much does an instance count, and under what circumstances? There are no precise answers to such quantitative-sounding questions, but one can describe general properties of the inductive "counting function"; for instance, later instances typically count less than earlier ones. Besides the number of instances, what else counts? Surely there are features of an eliminative induction that contribute to the strength of the conclusion. What features of arguments by analogy and arguments by the method of hypothesis determine their strengths? There are even more basic philosophical questions about induction. Why should additional instances, or more experimental variations, increase our confidence in a conclusion? How are the different kinds of induction related? Can some of these be reduced to others? Which of the two, enumerative or eliminative induction, is more basic than the other?

Jonathan Cohen and I have argued on opposite sides of this last issue (Cohen 1980, Burks 1980a), and in a general way my present discussion of his paper "Bolzano's Theory of Induction" is a continuation of our debate. I appreciate very much his perceptive comments on my views. He thinks eliminative induction is the more basic. I think that a fully adequate formal and computer model of inductive logic, the logic of discovery and learning, and the logic of evolution shows that enumeration has a basic role in induction. My argument for this thesis continues through this subsection and is renewed in later discussions of evolution (pp. 452, 474).

Answers to basic questions about inductive logic are not found by describing inductive practices, although that is valuable and our answers must conform to and explain these practices. Since induction differs from deduction in being a matter of degree, one promising approach is to develop the notion of inductive degree. Now Hume, Bolzano, Peirce, and many philosophers in our century have noted that probability is a matter of degree and that the term is used in an inductive or confirmatory sense in common and scientific discourse. Peirce said, for example, that probability theory gives a quantitative treatment of logic. So let us link induction to probability in the confirmatory sense.

Consider an inductive argument that claims that the data D strongly confirm the hypothesis H when the background information B is taken into account. This argument corresponds to the statement "the inductive probability of H relative to D and B is high" – more briefly, $P(H, D\&B)$ is high – in the sense that the argument is sound or valid if and only if the probability statement is true. Thus the degree of an inductive argument may be taken to be a conditional probability.

This method of interrelating argument or rule status (valid or invalid) with statement status (true or false) facilitates the comparison of induction and deduction. A deductive argument

$$P_1, P_2, \ldots, P_N; \text{ therefore C}$$

is valid if and only if the conditional

$$\text{If } (P_1 \& P_2 \& \ldots \& P_N)) \text{ then C}$$

is logically true. Thus there is this analogy between induction and deduction: a probability conditional is to an inductive argument as a logical conditional is to a deductive argument.

Both similarities and differences emerge from the comparison. The more-or-less feature of inductive validity is associated with the probability operator, the all-or-none feature of deductive validity with logical implication. One can say more specifically how induction is related to deduction. There is no premise that, when added to a valid inductive argument, will convert it into a valid deductive argument. But atomic inductive probability statements like "$P(H, D\&B)$ is high" have important deductive or a priori aspects. Systems of them are governed by the deductive rules of the calculus of probability. Moreover, on my view, *single* atomic inductive probability statements have an a priori aspect: each such statement is true or false according to whether it conforms to the rules of inductive logic. This view is in the general tradition of John Maynard Keynes and Rudolf Carnap, though our versions of it are very different (*CCR*, sec. 5:7).

With probability (**P**) as a guide to induction, we can ask again about enumerative induction: Why should observed instances of a general statement increase the probability of that statement, and why, after a large number of favorable instances or experiments, do further experiments do little good? I think the answer lies in Bayes' theorem, and more specifically in an application of it that involves repeated testing of a mixture of independent processes. This application yields a generalization of Bayes' theorem that I call "the confirmation theorem" (*CCR*, sec. 2.5).

Although the mathematics of this is a little complicated, the basic idea can be illustrated fairly simply. An "experimenter" starts with boxes of marbles, the marbles being either white or non-white. Suppose there are 101 boxes, with white marbles occurring in the proportions .00, .01, .02, ..., .98, .99, 1.00. The (male) experimenter knows all this, but the boxes are not marked and he is not allowed to look inside them, so that he does not know directly the character of any particular box. He chooses a box at random and makes repeated drawings from it with replacement, the box being carefully shaken each time to randomize the drawings. Using these observations, he can decide probabilistically on the constitution, or approximate constitution, of the box. (This indirect character of inductive verification is analogous to the indirect nature of evolutionary development; see Burks 1988a, sec. 3.)

Suppose, for example, that the experimenter draws an uninterrupted sequence of white marbles. By enumerative induction he eventually concludes that the chosen box is the all-white box; after, say, 500 draws, he stops drawing, on the ground that further drawings will not give him enough information to justify further tests, and infers inductively to the universal generalization "all balls are white". Now Bayes' theorem explains much of this. After 125 draws, the posterior probability of the all-white hypothesis is over 0.7; after 250 trials over 0.9; and after 500 instances, over 0.99. The slope of this curve clearly decreases, so that later instances count less than earlier instances.

Thus Bayes' theorem answers Hume's question as to why, in induction, instances count. It also explains why later instances count less than earlier instances. I know of no better explanation of enumerative induction, indeed, of no other explanation. Descriptivists in the philosophy and history of science, such as the writers discussed by Suppe and Cohen, offer no explanation, nor do coherentists. This shows why Bayseanism has a basic and pervasive appeal in inductive logic. (See also Burks 1988a sec. 10.)

The above model brings out features of induction favorable to the presupposition theory (see sec. 4.2 below). For the confirmation results depend both on the general set-up, which is presupposed, and on the prior probability assignment.

One need not treat the presence and the absence of a monadic property equally. Imagine the following set-up. A second (female) investigator watching the marble drawing experiments believes that the process whereby a box was selected strongly favored the predominantly non-white boxes, so that a predominantly white sequence is extremely unlikely. Imagine further that the box drawn is in fact the all-white box. Still, given the character of her prior probability assignment, this investigator will require more than 500 white draws to be convinced of the truth.

Because this simple model captures the essence of Charles Peirce's attempted justification of induction, and of Hans Reichenbach's attempt as well (*CCR*, sec. 3.4.2), it is worthy of further comment. With enough drawings the second investigator will become convinced of the all-white hypothesis, and the two investigators will finally agree, as Peirce said they must. But there is a finite side to this infinite coin: at any given stage of inquiry there is only a finite amount of evidence, and if two investigators differ too much on the prior probabilities, this evidence will not be sufficient to bring them into agreement on the posterior probabilities. Thus limit theorems alone cannot account for the objective character of science (see further sec. 4.2).

Our marble example is an exceedingly simple model of confirmation. Its universe of discourse contains only one property, and that a monadic property (whiteness, its absence), so that the model does not even have room for eliminative induction. Hence we need to explain how to extend it to the much richer universe of actual inquiry. For this we consider a sequence of systems, each an extension of its predecessor: (0) our simple marble example, (1) many monadic properties, (2) relations, (3) causal modality and space-time structure.

(1) It is easy to extend the marble example to include a finite number of monadic properties; just imagine multi-colored marbles. The calculus of probability then yields arguments by analogy: this marble has properties P_1, P_2, P_3 and Q: that marble has properties P_1, P_2, and P_3; what is the probability it has Q? One can see immediately that the strength of an analogy depends critically on the prior probability weights assigned to the various properties and to their correlations.

That properties should not all be treated the same and that some sets of properties are more likely to be interconnected than other sets, are common-place maxims of inductive logic. Thus in eliminative induction, the properties to be tested should be prima facie relevant to each other. Enumerative induction is not applied to indexical properties nor to complex properties believed to be non-enumerative, such as tosses of a coin made by a random process. Moreover, enumerative induction does not treat negative and positive properties equally, as Carl Hempel's example of instance confirmation shows: a white swan is more confirmatory of "all swans are white" than a non-white non-swan. These facts show, I think, that inductive inferences build on one another, and that the intellectual materials developed by induction include useful concepts as well as verified statements. In other words, there is concept-induction as well as statement-induction. This is a point to which I will return when discussing the relation of inductive reasoning to genetic evolution (p. 452).

(2) Next, we add relations to the model. This can be done by imposing a space-time organization on the exemplifications of monadic properties, a method we will discuss under our third point. But relations can also be added directly to the model, by a method Rudolf Carnap developed. There are two cases to consider according to

whether there are a finite or an infinite number of individuals; in both cases there are only finitely many properties and relations. Carnap used the concept of two universe descriptions being isomorphic to one another and defined the *isomorphism measure* of a universe description as the number of universe descriptions isomorphic to it (including itself). One can then define an inductive logic for a model by assigning probabilities to universe descriptions as a function of their isomorphism measures. Where possible, an inductive logic is defined for a model with infinitely many individuals by taking the limit of the inductive logics of a succession of finite models.

But while Carnap's definitions of confirmation apply to models with relations, his analytic results are mostly limited to models with monadic properties. For these models, he showed how to obtain a rule of enumerative induction by assigning prior probabilities to universe descriptions. For example, in a model with a finite number of individuals (and hence a finite number of universe descriptions), if each universe description is assigned a prior probability that is inversely proportional in its isomorphism measure, the result is a rule of enumerative induction. Moreover, this assignment can be extended to models with more and more individuals, to obtain a rule of enumerative induction for a model with infinitely many individuals.[3] Finally, by varying the weighting with respect to the isomorphism measure, Carnap generated a continuum of enumerative induction rules, varying with respect to how much an instance contributed to confirming a generalization, that is, how rapidly an experimenter came to accept a generalization (Carnap 1950, 1952; Burks *CCR*, ch. 3).

Many interesting results about enumerative induction flow from these formal models of induction, of which we will mention two. The first concerns the relation of enumerative induction to the uniformity of nature. The isomorphism of a universe description is inversely related to its uniformity, in a first-order statistical sense of uniformity. Consider a sequence of five draws from a box of white and non-white marbles, and compare these possible outcomes or universe descriptions:

$$W_1 \, W_2 \, W_3 \, W_4 \, W_5 \qquad \text{isomorphism measure} = 1$$
$$W_1 \, \overline{W}_2 \, W_3 \, W_4 \, W_5 \qquad \text{isomorphism measure} = 5$$
$$W_1 \, \overline{W}_2 \, W_3 \, W_4 \, \overline{W}_5 \qquad \text{isomorphism measure} = 10.$$

The first universe description is the most uniform and the last the most irregular, so that the statistical uniformity of a possible universe varies inversely with its isomorphism measure. Thus, to use enumerative induction is to favor, probabilistically, statistical uniformity over irregularity.

The second interesting result about enumerative induction follows from Bruno de Finetti's theorem about exchangeability in models with finitely many monadic properties but infinitely many individuals (CCR, p. 126). Our boxes-of-marbles examples with multi-colored marbles are such models, as are Carnap's models with monadic properties and infinitely many individuals. For our models, we derived enumerative induction by applying Bayes' theorem to a mixture of independent processes. Carnap derived enumerative induction by assigning prior probabilities to possible universes by a function that favors statistically uniform universes over statistically irregular universes. This function is symmetric in that it treats positive and negative properties equally. De Finetti's theorem shows that Carnap's models are a special case of Bayesian confirmation by repeated confirmation.

(3) These Carnapian models of inductive logic lack structure in two essential respects: they are extensional, equating nature to an actual universe and not treating what would have happened had circumstances been different, and they do not have a space-time framework. I have discussed these topics extensively (CCR, chs. 9 and 10) and will make only a few comments here.

I find spatio-temporal ("indexical") properties (including relational properties) to be very different from properties that are not spatio-temporal ("non-indexical"), mainly because indexical properties violate the identity of indiscernibles. See my discussion in Section 3.1 below of the distinction between singular and general terms and its relation to the distinction between particulars and universals. It is possible that both indexical properties and non-indexical properties can be constructed from some neutral kind of property (this is the relational view of space and time), but I think commonsense science begins with a space-time framework in which non-indexical properties are exemplified. Moreover, I think the distinction between properties and the matrix in which they occur is the metaphysical basis of enumerative induction, which involves repetition, and hence presupposes both a repeatable factor and a matrix or framework in which repetition can occur. A non-indexical property is repeatable, for it may have many spatio-temporal instances or exemplifications. Thus enumerative induction is based on the repetition of non-indexical properties in space-time.

Space-time constitutes one structural or relational constraint satisfied by enumerative induction. The causal modalities are another such constraint. In commonsense inference and in science we verify laws of nature and causal laws, and thus as a matter of practice we distinguish between causally true and contingently true statements. Of course, we also infer to non-modal generalizations, but often this is done via modal statements. Similarly, causal uniformities underlie statistical (extensional) uniformities.

Because the present section is devoted to the formalization of inductive rules, I will postpone consideration of questions about the nature and meaning of causal necessity and its role in the presupposition theory of induction to Section 4.1 (Causal laws) and Section 4.2 (Evolution and the presuppositions of induction). We now consider how to extend the results for enumerative induction already established for the extensional, unstructured formal models of (0), (1), and (2) to modal space-time formal models. There are two issues here: the general nature of the extension, and whether or not traditional methods of formalization are adequate for the task.

My suggestion is to obtain a formal model of induction by combining three logical systems and making an appropriate prior probability assignment to its universes. The three logical systems are: a logic of causal statements (to give the causal modalities), a cellular automaton structure (to give discrete space and time), and a system of constraints on causally possible universes (involving the existence of local laws and their uniformity over space and time). Enumerative induction to causal laws results from making a suitable assignment of prior probabilities to the possible universes of this combined system.

My concept of a "causal model of standard inductive logic" is a first attempt at such a logical system (*CCR*, sec. 10.3). A causal model contains many possible causal systems, each corresponding to a set of possible causal laws and theories. In turn, each causal system contains a set of causally possible universes. Every system and universe has a cellular space structure and operates in discrete time steps. The laws of a system satisfy three principles: causal uniformity, causal existence, and limited variety. One causal system is actual and one possible universe within it is actual; the former corresponds to the true set of laws and theories governing nature, the latter to the extensional facts of nature.

A causal model of standard inductive logic has two parts: the doubly modal structure just described, and a probability assignment to this structure. Unconditional inductive probabilities are distributed equally across possible causal systems and, within each system, equally across its possible universes. The causal uniformity and causal existence requirements eliminate possible universes that are "too" statistically random, leaving only those with a minimal amount of statistical uniformity. As a consequence, an equal assignment of prior probabilities to all causally possible universes leads to standard inductive rules of confirmation, including enumerative induction from non-modal instance statements to causally necessary generalizations. Thus these causal models of standard inductive logic extend enumerative induction from the earlier models (0), (1), and (2) to a model with a space-time structure and the causal modalities.

These causal models were suggested in *Chance, Cause, Reason* (sec. 10), and further results have been obtained by Soshichi Uchii (1972, 1973, 1977; see also sec. 4.1 below). But because these models are extremely simple in comparison with actual inductive reasoning, they are only preliminary. More generally, all current formal models of induction are relatively simple and theorems for those that embody relations are quite limited in scope. Thus while there are good indications that enumerative induction does hold when the extension is made, for the present this is only a philosophical speculation.

This state of affairs could be taken as good reason for not attempting to formalize induction; the philosopher of science who pursues a descriptive and historical approach might so take it. My conclusion is the reverse: we should not abandon formalization, but should improve its methodology by using computers, as I argue in the next subsection.

2.3 Logical formalization and computer simulation.

It is clear from the foregoing that the philosophical activity of logical formalization plays an essential role in my foundational approach to the philosophy of science. Some remarks about formalization are therefore in order. The modern subject of mathematical logic arose in large part from the construction of axiomatic systems intended to model the basic steps of deductive reasoning. In my own time this method has been applied to inductive reasoning and probability theory, rational decision theory, and grammar. The method is an extension of the traditional philosophical activity of definition or analysis: a formal axiomatic system designed to model some systematic aspect of language use is a holistic, precisely defined structure that stands to its subject matter as a definiens stands to its definiendum or an analysan stands to its analysand. I suggest that the next natural step in this evolution of philosophic method is to extend logical formalization by employing computer simulation in the modeling of reasoning and other uses of language (Burks 1975b, 1979a).

I begin my justification of this proposal with further historical remarks. An adequate account of enumerative induction must be based on a relational structure (space-time), and the logic of relations is much more complicated than the logic of monadic predicates and unstructured classes. That is why relational logic was the last branch of logic to be developed. It was fewer than 150 years ago when Augustus de Morgan and Charles Peirce discovered two key features distinguishing relations from monadic properties: relational arguments are not always covered by syllogistic rules (e.g., a horse is an animal, therefore the head of a horse is the head of an animal), and relational arguments valid in every finite model may be invalid in an infinite model.

Moreover, as Peirce emphasized, reasoning is only one aspect of language use; applicability is equally essential. At present we are far from having good models of language use in general, so that, despite its weakness, our formal understanding of induction is perhaps as good as our formal understanding of language in general. An effective criterion of success for logical studies in language and reasoning is the computerizability of a model, and present models of language are not even adequate for high-quality translation of informative discourse between natural languages. Computerizable knowledge of the psychology of reasoning and learning is equally limited.

For both inductive logic and language use generally, formalization and other studies by traditional methods are nearing a barrier defined by the capacity of the human mind. While it is not literally true that man can do only one thing at a time, there is a rather small limit on what he can hold and manipulate in consciousness, and a not very large limit on what he can store in memory. This limit is best characterized by tracing how man's idea of what constitutes a solution to a problem about nature has changed over time. The traditional three-body problem provides a good example.

Isaac Newton's theory of gravity yields differential equations for the motions of the bodies of a planetary system. People wanted analytic solutions to these differential equations, simple formulas that would describe the paths of the planetary bodies in a way that was directly understandable and would facilitate computation of these paths. The problem of describing the path of a planet around the sun, ignoring all other bodies, is called the two-body problem. Johannes Kepler's three laws constitute the solution to this problem, for they are simple enough to remember easily and they convey their generality directly. Hence they give an intuitively understandable description of the motion, and constitute a neat and simple algorithm for making planetary predictions.

However, this method of human understanding fails for complex phenomena, where no simple laws or formulas exist. This is the case with the real problem of planetary motion, which is many-bodied, since each body (sun, planet, moon) acts on all the others. Once the two-bodied problem was solved, scientists attempted to solve the three-body problem, but they found no formulas analogous to Kepler's laws and concluded that the three-body problem was unsolvable. They could still predict planetary motions, but only by taking the basic differential equations case by case and calculating the solution step by step, which had to be done by hand. Originally they used logarithm tables as aids, and later small mechanical calculators to do the elementary arithmetic operations. In either case, the calculation was laborious and did not contribute much to understanding the phenomena.

With the development of electronic computers, this situation changed drastically. Now it is possible to calculate planetary trajectories rapidly enough to control rockets, satellites, and interplanetary spacecraft. Of even greater importance in the present context is the use of computers to understand the phenomena. Planetary and other trajectories can be calculated very rapidly as compared to the human attention span, and can be computed under human control for many different starting states and on-line modifications. By working interactively with a powerful computer that has good graphic displays, a human can both predict specific trajectories accurately and obtain a good general understanding of how a many-bodied system behaves in a wide variety of circumstances. Of course, analytic formulas are not abandoned, but only play a different role. The differential equations of motion are analytic formulas, and they guide the computation. The nature of the solutions has changed: instead of obtaining simple formulas, as

Kepler did, we obtain representative cases and *experiment with these cases*.

Thus with modern computing technology the many-bodied problem can be solved in both its aspects: prediction and understanding. We will apply this historical lesson to the problem of formulating rules of induction, in two stages. In this section I will discuss computer simulation and its relation to embedding and reduction, using Bernard Zeigler's interesting paper "Cellular Space Models: New Foundations for Simulation and Science" as a basis. Later I will describe a computer simulator for learning and discovery that is related to evolution and induction (pp. 449–455).

Zeigler's pioneering book *Theory of Modelling and Simulation* suggests a new formal framework for simulation. Simulation theory is closely related to the philosophy of scientific theories, so that Zeigler's paper leads directly into a host of issues on which I wish to comment.

Evolutionary considerations suggest the following practical criterion for evaluating conceptual structures: How many worthwhile cases does each handle in comparison with its cost? Thus in computation we seek systems (hard, soft, or both) that are general (have many applications) and that are efficient for calculation. To define a modeling formalism appropriate for systems operating under the direction of natural laws, Zeigler combined some earlier models used in simulation, called "discrete event simulation models", with the cellular automaton models I suggest in *Chance, Cause, Reason* (ch. 9). The result is a *next event cell space* (NEVS) model. Roughly speaking, a NEVS model combines the uniform spatial structure of a cellular automaton with the asynchronous temporal structure of a discrete event simulation model. In his paper Zeigler shows how a NEVS model could be used to represent such diverse phenomena as the motion of a single body, traffic congestion, plant growth with gravity effect, and the behavior of ecological and chemical reaction systems.

The design of models for simulating law-governed systems is a special case of the general problem of applying mathematics to nature. One is studying a very complex system; metaphysically speaking, it might even be infinitely complex, but in any case it is orders of magnitude more complex than any possible conceptualization or simulation. To deal with the system, one must cut it down to size or, equivalently, find an embedded subsystem of it (*CCR*, p. 588) that is intellectually manageable. The semiotic relation here can be viewed as triadic: one system (a theory or model) represents a second system that is an embedded subsystem of some third system (the full, possibly infinitely complex, system). Questions about this relation are basically questions about reductions. They are very important for both computer science and philosophy, and this is an opportune place to pursue them.

Zeigler's NEVS models involve two main modifications from my cellular automaton models. (I) The first involves a very strong kind of intantaneous action-at-a-distance, and it leads to some interesting questions about the role of contradictions in logic and inquiry. (II) The second employs a notion of event that is defined in terms of property changes rather than in terms of a temporal framework, and it raises questions about the economy of terms and concepts. Both modifications are made to simplify the computation. We discuss them in turn.

(I) The usual cellular automaton is defined so that causal action is local. Space is divided into cells; for example, a two-dimensional cellular automaton might have the structure of an infinite checkerboard. Time is divided into moments $t = 0, 1, 2, 3, \ldots$, and at every moment each cell is in one of a finite number of cell states. A finite neighborhood is defined for each cell; for example, the neighbors of a square might be the four squares contiguous to it. There is a law governing the transitions from state to state: it gives, for each assignment of states to a cell and its neighbors, a unique next state for that cell. Thus time and space are linked by the constraint that information or causal influence can travel no further than to a neighbor in one time step. The resultant maximum transmission rate is sometimes imaginatively called "the speed of light". (There are also cellular automata with indeterministic or probabilistic laws; they are subject to the same restriction.) The usual cellular automaton is uniform in all these respects (e.g., all cells have the same shape), but such uniformity is not necessary. The essential point is that these space-time restrictions are a simple way to guarantee determinateness or computability, so that the cellular automaton functions similarly to a computer in the ordinary sense.

Consider now the restriction that causal influence can travel no further in one time step than to a neighbor. This almost requires that the effect be spatially and temporally contiguous to its cause, but not quite, since the neighborhood of a cell can be defined to be any finite area encompassing the cell. Thus the restriction of direct causal influence to a neighborhood makes all laws of the cellular automaton *local*, where locality is a slight generalization of contiguity. The essential point about both locality and contiguity is that they are incompatible with action-at-a-distance. This local-action versus action-at-a-distance issue has two aspects, the theoretical and the practical.

On the theory side, action-at-a-distance has always been a problematic concept. Although Newtonian gravity involves it, Newton himself felt uncomfortable with it. The concept of action-at-a-distance is a pivotal notion in current discussions of whether determinism can be made compatible with quantum mechanics by embedding an indeterministic quantum mechanics in a lower-level deterministic system. It is easier to do this if action-at-a-distance is allowed, but most current physicists reject action-at-a-distance. The ultimate outcome of this controversy seems to me uncertain. Action-at-a-distance violates our natural intuitions about causality, but so does the relativity of simultaneity, which everyone accepts.

On the practical side, because of its quantum structure, a cellular automaton model is limited in its ability to represent rates of propagation. There is a minimum time step from cell to neighbor, and information can travel faster in some directions than in others. Consider, for example, a cellular automaton with square cells in which the neighbors of a cell are limited to the contiguous cells. In this type of cellular automaton, causality can travel no faster than one cell per time step in the x and y directions, and at correspondingly slower speeds in other directions. This feature of cellular automata is inconvenient when one is simulating a system with widely different propagation rates, as in Zeigler's example of plant growth (example 3). Plant growth is relatively slow, but for all practical purposes the gravitational load of new growth travels instantaneously. Because of the propagation restrictions of the neighborhood relation, a cellular automaton cannot handle both speeds efficiently.

To simplify and speed up the computation in a NEVS simulation, Zeigler modifies the cellular automaton model by allowing action-at-a-distance. He thereby gives up locality and allows ad hoc deviations from the original uniform neighborhood relation. Moreover, as a consequence, his NEVS models are not locally deterministic. The state of a cell and its neighborhood does not determine a unique next state for that cell, because an instantaneous propagation from afar may affect this next state. This much was true of Newtonian physics. But Zeigler allows a NEVS model to have two logically powerful additional features. First, a NEVS allows the instantaneous propagation of influence or information without diminution, in contrast to Newtonian gravitation which falls off rapidly with distance. Second, communication paths can be constructed from within the model by rules that allow cyclic paths with no time delay. The combined consequence of these two features is that the computation of the NEVS may become indeterminate or even contradictory. The indeterminate case is like that of the question "What is the truth-value of 'This statement is true'?" which can be answered by *either* "true" or "false". The contradictory case is like that of the question "What is the truth-value of 'This statement is false'?" which cannot be answered by *either* "true" or "false".

The possibility of indeterminacy or contradiction is a strange consequence for a computation system designed to help a user reason about some natural system. As Zeigler points out, the formalism could be modified to avoid this (Zeigler's note 10). The standard concept of a cellular automaton can be altered to allow the step-by-step construction of instantaneous paths under a rule that prevents cycles. It might seem obvious that NEVS should be modified in this way, but Zeigler does not do so, saying instead that it is "the responsibility of the user to prevent" such instantaneous cyclic paths. What shall we say about all this?

Consider first Zeigler's claim that a NEVS model, though not locally deterministic, *is* globally deterministic. This is not true as it stands but only with a significant qualification. A NEVS system is globally deterministic *only if* it contains no indeterministic or contradictory paths. If a system is indeterministic it has at least one complete state that leads to two or more successor states, whereas if a system is contradictory it has at least one complete state that has no successor state.

Second, it seems counter-intuitive to choose for practical use a computing system that is potentially contradictory, and so Zeigler's choice needs justification. This also raises the general issue as to whether contradictions play a positive or a negative role in the evolution of ideas and theories. This general issue is important, but in the present case we need only note that in computer simulation, the issue is a practical one, depending on the particular problems being solved, and the criterion of choice is utility. In computing, one wants a useful answer at an acceptable cost. The NEVS simulation system could be restricted so as to guarantee consistency and also indeterminacy, but at the price of making it harder to use. It is probably better to leave it as it is and

rely on the user to avoid contradictions and indeterminacies. Hence
Zeigler's choice is reasonable. It also fits historical practice. For example, Oliver Heaviside's operational calculus for solving differential
equations contained contradictions, but electrical engineers found it
useful and avoided the contradictions in the applications they made of
it.

(II) Zeigler's second modification of my cellular automaton model is
a shift in the nature of the temporal reference frame, from a
synchronous discrete frame to an asynchronous continuous frame. This
difference corresponds to two alternative philosophies of computer
design, the synchronous and the asynchronous, and is also somewhat
like the timing difference between two classical theories of mind-body
interrelations, Leibniz's pre-established harmony and Descartes' inter-
actionism.

Cellular automata behave synchronously in a pre-established way,
for all cells act on the same schedule $t = 0,1,2,3,\ldots$, no matter how far
one cell may be from another. But cellular automata are idealized
abstract constructions. When it comes to actual man-made computers,
we do not know of any perfectly timed, intrinsically synchronous computer components. Left alone, each component or subsystem of a computer would operate on its own, somewhat unpredictable, speed. To
make the different parts of a computer cooperate and work together,
we must add a layer of timing control. There are two basic ways of ac-
complishing this, the synchronous and the asynchronous. We will explain them separately, though they could be mixed.

In a synchronous machine all local actions are regulated by a central
electronic clock, which produces a repeated sequence of timing signals
that is broadcast throughout the machine. Computer clocks exert a
form of central control, just as clocks have done throughout history.
Asynchronous computers embody the alternative philosophy of
decentralized control; each circuit operates at its own pace. But most
computing tasks require the cooperation of many circuits, working in
parallel and in sequence. The required synchronization is accomplished
at the local level. Each input stream of signals contains a flag,
semaphore, or other mark to indicate when it is complete, and the
receiving circuit has an interlock or other coordinating means to determine when all needed inputs have been received. Thus in an
asynchronous computer the local interactions time themselves.

We have explained the distinction between synchronous (centralized) and asynchronous (decentralized) timing in computers at the basic level of hardware. Now the fully programmed modern computer is the most complex, well-defined hierarchical structure ever built by man. The bottom level or levels are hardware, the higher levels software. The same synchronous-asynchronous distinction applies at each software level, for when many programs operate in parallel and in series on the same task, their activities must be coordinated. This can be done with central control (synchrony) or local control (asynchrony).

To evaluate the asynchrony of Zeigler's NEVS simulation models, we need to imagine an appropriate software package in which any NEVS model could be specified easily and computed (simulated) efficiently. Such software would operate at a high level in a computer. The question then becomes: Is the simulation more efficient because the model-theoretic framework is asynchronous rather than synchronous? It might seem that this question is a technical question of computer science and so not of philosophical interest. But Zeigler wants his NEVS structures to be generally useful in modeling natural systems; thus the structure of his framework should match the structure of nature in basic respects. Hence in this case the theory of computer modeling interfaces with the foundations of induction. The key concept is "event", and this concept belongs as much to philosophy as to computer theory.

The name "next event cell space (NEVS) model" contains the subname "next event". In ordinary and scientific language an event is "a more or less noteworthy occurrence". Thus an event involves change, usually a change in some underlying substance or continuant. An event is noteworthy when it influences other things or events around it. This use of "event" contrasts with that in technical metaphysics, as when a continuant is construed to be a succession of events satisfying certain conditions, such as space-time contiguity and causal connectedness (Burks 1967). Zeigler is using "event" in the first sense. An "event" in a NEVS model is a change in a thing or subsystem that affects neighboring things or subsystems.

The distinction between the two senses of event can be explained with a synchronous cellular automaton. View each cell and its successor of cell states as a continuant. Assume that, statistically speaking, a cell does not often change state in a way that influences its neighbors. Let us call such a system a *slow-changing system*. In the ordinary sense of "event" events occur infrequently in a slow-changing system, whereas in the technical metaphysical (and automata theoretic) sense of "event", an event occurs at every time step in every cell.

Consider next the simulation of a slow-changing cellular automaton. One could have the simulating program communicate the state of a cell to each of its neighbors at every time step. But it is more efficient to communicate only the effective changes, the "events" in the ordinary sense, and program each cell to treat the absence of communication as the absence of change. This is the essence of Zeigler's *next event* simulation (NEVS) models: whenever an event occurs in a cell as a consequence of activity in that cell, the fact of that event is communicated to the cell's neighbors.

Once this procedure is adopted there is no reason to retain the same clock sequence for each cell (pre-established harmony) and so each cell is allowed to operate at its own speed. Thus a NEVS simulation model is asynchronous. In mathematical terms, the derivatives of the functions being computed change much more slowly than the functions. This aspect is what makes asynchronous parallelism computationally more efficient than synchronous parallelism.

For us the basic question is an applications question: In modeling natural systems, is an asynchronous simulation system more efficient than a synchronous simulation system? In general the answer is "yes", because most natural systems are slow-changing systems in the sense characterized above. The philosopher's perennial example of a table illustrates the point. There are frequent changes in the table at the molecular level, but they make no difference to the philosopher's perception and practical use of the table. These changes are events for the molecular physicist, but not for the ordinary person.

The preceding analysis shows that our ordinary and scientific concepts of event and substance reflect a principle of economy in information handling. When an object changes slowly a list of changes is much shorter than a list of states at each moment; in ordinary life, then, the common-sense meaning of "event" is informationally more economic. On the other hand, the metaphysical sense of "event" is more useful in general scientific and global speech about systems and their histories.

This completes my discussion of Zeigler's cellular automaton simulation models. These are asynchronous and even somewhat non-uniform with respect to the neighborhood relation. Since most on my work on cellular automata has been on the uniform case, and since this is the standard case, some comments on the reason for my interest may be worthwhile.

Uniform cellular automata are simpler than non-uniform, and so are better for defining general concepts and establishing theoretical results. Thus they were appropriate for von Neumann's definition of universal construction and his proof of automaton self-reproduction (von Neumann 1966, Burks 1970). For the same reason, uniform cellular automata are suitable for my causal models of induction. Uniform cellular automata are also useful for studying the relation of computation theory to the basic characteristics of mechanical and thermodynamic systems. In recent years many physicists, biologists, and computer scientists have studied and simulated cellular automata as models of complex non-linear dynamic systems, including systems with chaotic behavior, systems evolving to complex limits (such as strong attractors), and systems in which organization emerges with disorganization.[4]

The relation of cellular automata to basic types of physical systems can be illustrated with an interesting theorem concerning backward-determinism. In a forward-deterministic system each present complete state determines the future history, while in a backward-deterministic system a present complete state determines a unique past (*CCR*, p. 579). These two kinds of determinism are not usually distinguished, but they are logically separable. The classical concept of determinism, as formulated by Pierre Laplace, was generalized from the mechanical systems of classical Newtonian physics, and hence included both backward- and forward-determinism. These classical mechanical systems are reversible (e.g., the equations governing planetary motion could be computed backward), and also conserve entropy (are frictionless, and do not dissipate energy). In these respects, classical mechanical systems are very different from thermodynamic or probabilistic systems that are not reversible and do not conserve energy. Charles Peirce thought this contrast was basic, and his rejection of determinism included the claim, as I would put it, that irreversible probabilistic systems cannot be embedded in systems that are both forward-deterministic and backward-deterministic.

It has seemed to me for some time that this traditional classification of systems mixes levels of categories. The logical distinction between unique laws or rules and probabilistic laws or rules is basic, and for each type of law or rule we can distinguish whether it applies forward in time, backward in time, or both.

When I considered the matter a while ago, I noted that all universal computing machines, whether actually constructed or theoretically defined, were forward-deterministic *but not* backward-deterministic. That is, they all had complete temporal states that might have come from two or more prior states. I then wondered if this did not have to be true for realistic automata, that is, automata that take account of the spatial arrangement of logical switches and memory elements. Thus I was led to the conjecture that a universal computing machine cannot be embedded in a backward-deterministic cellular automaton.

This conjecture has been proved false. My student Tommaso Toffoli first showed in his doctoral thesis how to embed any given n-dimensional deterministic cellular automaton in a backward-deterministic automaton of $n+1$ dimensions (Toffoli 1977). Then he and others showed how to embed any deterministic cellular automaton in a backward-deterministic cellular automaton of the same dimensionality (Fredkin and Toffoli 1982). Furthermore, each such backward-deterministic cellular automaton is equivalent to a perfect, classical billiard-ball mechanical system (Margolus 1984). And this shows that dissipationless computing is in principle possible! I find this last connection very interesting, because a perfect billiard-ball mechanical system is an infinitely precise analog computer and so in an important sense is an order of infinity more complex than a digital computer!

To summarize my brief review of the relative merits of uniform and non-uniform cellular automata: uniform automata are more valuable for theoretical studies of a quite general kind, while non-uniform automata are better for simulating more specific and detailed theories. For similar reasons, non-uniform cellular structures are better as computer architectures. For example, a hypercube interconnection network superimposed on the regular cellular structure allows much faster communication between remote cells.

Let me say in closing this subsection that the issues discussed herein have been a prime topic of research in my Logic of Computers Group, since its beginning in 1957. We have studied the relation of computational systems (both abstract and practical) to natural systems (both biological and physical). We have established theoretical results in this area and developed computational methods for studying complex control systems, especially biological systems. Two contributors to the present volume were long-term members of the Group, Bernard Zeigler and Richard Laing. My colleague John Holland has been a member from the beginning. The reference items of Codd, Langton, and Toffoli were also generated in the Group.

More recently, the Logic of Computers Group has been interested in the theory and simulation of evolutionary, adaptive, learning, and reasoning processes in both natural and artificial systems. This interest is grounded in a Peircean-type belief that there is a significant commonality in the laws or logical rules governing the following phenomena and other phenomena similar to them: chemical and biological evolution and the adaptation of species; competitive economic market and exchange mechanisms; learning of all kinds and at all levels, including the growth of concepts; inductive discovery and verification; and intentional planning and decision-making. The laws and rules of each type of phenomenon are unique, but underlying all of them are some basic common principles, which we think are mechanizable.

A new type of computer system, called a "classifier system", has been developed in the Logic of Computers Group, primarily by Holland. A classifier system employs market economy principles for learning (induction) and genetic principles for discovery. I will discuss these computing systems later, in connection with Peirce (p. 489). Now I need to consider logical truth and its relation to empirical truth.

2.4 Modes of truth.

The papers by Manley Thompson and Robert Audi raise separate but related questions about the nature of logical truth and empirical truth.

In *Chance, Cause, Reason* I distinguish fairly sharply between logically true-or-false statements and empirically true-or-false statements (ch. 1). Thompson thinks my account of this distinction is incompatible with my view that man is functionally or behaviorally equivalent to a computer robot. In the present subsection, I will argue that this is not so. This answer will lead naturally into a discussion of the evolutionary origin of man, and it will be continued in Section 4 (Evolution and Induction).

Audi argues in his paper that one can make a similarly sharp distinction between empirically true-or-false statements and normative or value statements. I agree with Audi on the separability of values from science at the foundational level. Scientists make many value judgements, especially in the process of inquiry, but these do not undermine scientific objectivity. Audi's arguments for this position are good. However, because my own arguments for this position derive from the theory of biological evolution, I will postpone discussion of this issue to Section 4.4 (Evolution, science, and values).

I classify non-normative judgments or propositions into three basic types:

> Empirically true-or-false statements,
> Inductive statements, e.g., causal uniformity principle, causal existence principle, limited variety principle,
> Logically true-or-false statements.

Actually, the classification applies to quite complicated structures of statements, such as axiomatic systems and scientific theories. Moreover, the classification is hierarchical, in that a statement on a higher level may involve statements on lower levels, as when mathematics is employed in the formulation and verification of empirical theories. In *Chance, Cause, Reason* I began with the distinction between logical and empirical statements (sec. 1.2.1) and argued at length that the verification of empirical statements presupposes inductive statements (sec. 10.4.2).

In his challenging paper "Some Reflections on Logical Truth as A Priori", Thompson discusses whether logical truth should be explained epistemologically (How do we know logical truth?) or metaphysically (What are logical truths about?). I think a complete account of logical truth should do both, and should also relate the epistemological account to the metaphysical account.

The same is true for other modes of truth. Each mode of truth should be explained epistemologically in terms of what goes on inside the learning system; it should also be explained metaphysically in terms of the relation between what goes on inside the system and what is external to it. An issue of the latter kind is the question of what concepts and statements are about. Some suggestions will be made below in terms of correspondence and evolutionary fitness (pp. 399, 454–5).

At the moment we are concerned with logical truth. Since *Chance, Cause, Reason* was primarily about induction and its relation to deduction, the book emphasized the epistemology of logical truth. I will say a little about the metaphysical side now.

My short answer to the metaphysical question is that logical statements are about the logical-mathematical structure of reality. This answer is partly Platonic, since it asserts the existence of a logical mathematical structure existing independently of any knower. It is not fully Platonic, however, for this structure is conceived modally, as an organization of possible universes that includes nature as an essential constituent, and so the realm of mathematics does not exist or subsist independently of nature. This broad answer is indicated in my development of modal logic (*CCR*, ch. 6).

Chapter 10 establishes metaphysical-epistemological correlations between various modal structures, on the one side, and inductive and empirical statements on the other (see especially p. 624). When this classification is extended to encompass logical truth, we obtain the following classification of basic types of metaphysical systems and correlated types of epistemological judgments.

Kind of statement	Kind of system
(3) Empirical	
(3b) Non-modal, e.g., an observation statement	Actual universe
(3a) Causal law	Actual causal system
(2) Inductive	Possible causal systems
(1) Logical	Logically possible universes

The levels are numbered from bottom to top to symbolize the dependence of the higher levels on the lower.

The statements I call "inductive" resemble some of Immanuel Kant's synthetic a priori statements of transcendental logic, especially the doctrine of determinism (*CCR*, 10.5.2). In *Chance, Cause, Reason* I used the modern tools of symbolic logic and probability theory to redo the classic controversy between Hume and Kant concerning the nature of causality and induction. My solution was the presupposition theory of induction, discussed in Section 4 of these replies. I concluded with Hume that there is no non-circular justification of induction, and with Kant that the concept of causal necessity is a priori rather than empirical.

The war between the empiricists and the rationalists was over the status of concepts as well as the status of propositions. Compare the following two questions. Are some very basic concepts, such as *cause-effect*, *substance*, and *God* a priori or empirical? Are some very basic principles such as "every event has a cause", "the soul is immortal", "God exists", "one should obey the moral law" logically true or empirically true? Concept empiricists were not always statement-empiricists, John Locke being an example. Locke was a thoroughgoing concept-empiricist, but he thought that God's existence could be proved and that morality, like mathematics, is capable of demonstration.

Consequently, it is important to distinguish a priori from empirical concepts as well as to distinguish a priori from empirical statements. This should also be done hierarchically, and for me the hierarchy is:

> (3) Empirical concepts – common sense and scientific con-
> cepts of properties of natural objects, of systems, and of
> laws,
> (2) A priori concepts of induction – causal modalities, in-
> ductive probability,
> (1) A priori concepts of logic and mathematics – *not, or,
> and, statement variable, individual variable, some, all,
> eleven, integer, plus*, etc.

The a priori concepts are constituents of "man's *innate* structure-program complex", his inborn physiological abilities to process infor-mation and interact with his environment. The empirical concepts are acquired from experience and contribute to his "*acquired* structure-program complex" (*CCR*, p. 612).

As Thompson points out, I actually employ two quite different definitions of the distinction between a priori and empirical concepts. The one just mentioned, expressed in terms of man's structure-program complex, is a computer science definition. The other defini-tion is mental, for it is expressed in terms of one's conscious introspective and reflective powers, as in "if one can directly ex-perience instances of a concept and abstract the concept from these ex-periences, the concept is empirical" (*CCR*, p. 609). These two definitions belong, respectively, to the outer and inner points of view I distinguished at the beginning of these replies. To explain their com-patibility, I must say something about my theory of mind, even though this topic belongs mainly in the next section (sec. 3, Computers and Minds).

The internal distinction between a priori and empirical concepts in-volves the reflective operations of conscious mind. The external dis-tinction between a priori and empirical concepts involves the human nervous system viewed as an information processing computer. This computer carries out the processes we call mental perception, language use, reasoning, learning, imagination, visual intuition, dreaming, etc., and so it constitutes the computational basis of mind. Consciousness involves the higher-level functioning of the nervous system, so consciousness is a part of mind. That is, functional consciousness is reducible in principle to mind, or in my technical terminology, consciousness is an *embedded subsystem* of mind (*CCR*, p. 588; see also sec. 3.4 below). This means that the properties and states of consciousness are sets of the properties and states of mind, and that the laws of consciousness are simplified versions of the laws of mind.

The size or complexity of an information processing system can be measured in terms of the amount of information it can hold and its rate of processing. By this criterion, consciousness is a very small embedded subsystem of mind. What we are aware of at any given moment is a very small part of the total information in our minds, and conscious reasoning is a very small part of total mental information processing.

This reductionist view of mind has implications concerning the relation of my two definitions of the distinction between a priori and empirical concepts, that in terms of conscious experience and that in terms of man's structure-program complex. On the one hand, since the inner and outer points of view concern different systems, the two definitions should be different. On the other hand, since functional consciousness is reducible in principle to mind in the broad sense, the first definition should be reducible in principle to the second definition. I think this is the case, even though we are a long way from a very good understanding of the form of the reduction. It is to be expected that computer studies and simulations interacting with physiological and psychological studies will appreciably advance our understanding of the relation of conscious mind to total mind by the end of this century.

Thompson argues that because physiology and neurology are empirical disciplines, we cannot "discover which concepts are a priori and which are empirical by observing the nervous system" since "knowledge acquired by observation is always a posteriori" (his paper, p. 75). I do not accept this. One can in principle establish by observation how a brain or a computer works, what basic language or machine language it uses, and what concepts and rules are ingredients of this language.[5] Furthermore, by simulations one could in principle establish that a certain core of concepts and rules are needed for that brain or computer to adjust to a certain general type of environment in a limited amount of time. I think that adequate empirical knowledge of how the human develops from genome through the infant's nervous system to the mature nervous system of the adult would show that humans do have a priori concepts.

Moreover, one can acquire knowledge of the truth-value of a logically true-or-false statement by empirical means. One can approximate the value of pi by throwing match sticks at random on a uniform parallel grid and counting the frequency with which the sticks intersect a grid line. Electrical engineers use network analyzers to study and predict the behavior of electric utility distribution systems. If the laws of electricity were not known, a network analyzer could be used for the experimental investigation of these laws. Actually, these laws are known, and, in fact, the engineers use the network analyzer to obtain the solutions of mathematical problems empirically. Similarly, aerodynamicists sometimes use wind tunnels to solve mathematical equations.

There is a fundamental epistemological difference between physics and computer science that makes the relation between each and the philosophy of mind very different. A computer is the physical realization of some logical system. By observing the design of a computer and how it operates, one can certify that it is doing logic or arithmetic correctly. One is then making empirical statements about logical systems that reason and use language. A computer is a logical system, and a sufficiently complicated logical system can know itself. Hence a mechanistic philosophy based on computers is very different from a mechanistic philosophy based on physics. This is why I call my form of philosophical mechanism "*logical* mechanism".

It is for these reasons and for similar reasons advanced later in this section that I do not accept Thompson's position that there is no empirical knowledge of man's innate structure-program complex. But I understand why he holds it: the opposite position sounds paradoxical. I return to this issue in Section 4.2, Evolution and the presuppositions of induction (see especially pp. 470–474).

My pursuit of the inner point of view has been influenced by Peirce and Kant. Peirce's phenomenology of mind was, in a broad sense, an empirical discipline. Reflective observation played a basic role in Kant's transcendental logic, and yet Kant was doing logic in the broad sense of that term. In using the inner approach to the philosophy of deduction, the philosophy of induction, or the philosophy of mind generally, one meets the problem of whether any classification of concepts or statements applies to itself. I do not agree with Thompson's suggestion that this problem disappears on a coherence theory of truth (his paper, p. 80), though coherentists in fact ignore the problem. A hierarchy of languages is a much better model for dealing with self-reference. A structure at one level need not apply to itself, and a corresponding structure on a higher level is significantly different, at least contextually.

It is appropriate to emphasize that my treatment of the a priori in *Chance, Cause, Reason* was directed primarily at induction, not deduction. The discussion of level (1), especially the definition of "logically true-or-false statement" was provided as necessary background. However, since Thompson has questioned the consistency of my account of deduction with my thesis that a machine can perform all natural human functions, I need to say more about the nature of logical-mathematical truth to show that my views on this subject are in fact consistent.

"The truth-value of a logically true-or-false statement *can be* established, *in principle*, by deductive reasoning alone, without the use of observation or experiment. Included within the scope of deductive reasoning are such activities as defining, intuiting, reflecting, and calculating" (pp. 636, 6). This definition claims only the existence of some kind of proof or disproof, leaving open the possibility that one cannot find it. But it leads naturally to the question: What is a proof?

An answer that seemed plausible at one time is given by what I call Leibniz's conjecture: There is an interpreted formal language L of complete expressive power such that (1) every theorem of L is logically true and every formula that is logically true is a theorem, and (2) there is a computer program for calculating whether or not a formula of L is logically true (*CCR*, pp. 336–367). This conjecture yields an explanation of what a proof is: a proof is simply a calculation in a suitable formal language. Leibniz's conjecture also yields a simple, precise criterion for a priori logical truth: If a statement can be formulated in language L, a properly programmed computer will decide whether it is logically true or logically false.

This neat and simple analysis of logical truth and falsity turned out to be inadequate, as a consequence of Gödel's work. Nevertheless, Leibniz's conjecture captures a substantial part of the notion of logical truth, and it is the only precise general definition of logical truth I know of. Let us pursue it a bit further. When combined with my thesis that a finite automaton can perform all of man's natural functions, Leibniz's conjecture yields a good approximate elucidation of my definition of a logically true-or-false statement. There are two key points. First, the truth-value of a logically true-or-false statement *can be* established by an internal process of deduction and calculation, without any outward appeal to observation and experiment. Second, this holds only *in principle*, since man's calculating powers are limited, while the computational definition postulates a computer with an unlimited amount of tape storage working for an unbounded amount of time.

This computer account of deduction makes clear a sense in which logical truth is a priori; in principle, the calculation can be done by the organism without collecting data about or learning the laws of nature. This is a weak sense in which mathematical truth is a priori, but it is the only sense I advocate in *Chance, Cause, Reason*; I do not advocate a stronger Kantian sense, as Thompson seems to think. Nevertheless, it does seem to me an important philosophical claim that mathematics can be done internally in a way that empirical science cannot. But it is only a foundational thesis, not a thesis about how mathematics is learned or developed. In practice, most of us learn mathematics from others, and historically mathematics developed interactively with business, science, and technology. Also, man relies on his ability to adjust to his environment in order to confirm that his internal computer is functioning correctly. Insofar as one's practical and intuitive mathematics is valid, however, it is in principle reducible to calculation.

Let us turn now to the refutation of Leibniz's conjecture and see what that implies concerning our account of logical truth. The refutation is a consequence of Gödel's incompleteness theorem for arithmetic, interpreted in terms of Alan Turing's concept of a universal computer. Gödel's theorem states that there is no interpreted formal language L_A that expresses all of arithmetic and is such that the class of theorems of L_A equals the class of logical truths of L_A. It follows by Turing's result that there is no machine that can decide whether an arithmetic statement is true or is false. An even stronger result actually follows: there is no machine that can enumerate the elements of the set of arithmetic truths. Since arithmetic is only part of logic and mathematics, there can be no mechanical definition of logical truth. Actually, these results and their proofs can be understood fairly easily by anyone who has an intermediate understanding of logic and knows how to program a computer (e.g., see *CCR*, sec. 6.4).

Even though it is false, Leibniz's conjecture is of value in dealing with the a priori character of logical truth, because the computational model fits much deductive reasoning and gives us a precisely defined base on which to build. Since the answer to any really finite problem is calculable in principle, the incompleteness of mathematics concerns the infinite. The philosophical problem raised by Gödel's result can be expressed very simply in terms of our thesis about the mechanical nature of man: How can a finite automaton understand the mathematical infinite?

If logical truth were as simple as Leibniz's conjecture implies, there would be no problem here, for we understand in principle how a finite automaton with an indefinitely long tape computes. But there is no satisfactory answer to the question of how a finite automaton can understand the infinite. Constructionist theories of mathematics, such as Peirce's iconic theory and L. E. J. Brouwer's intuitionism are attempts to provide a finite basis for the infinite. Gödel's result implies that mathematics is open ended in the sense that one can always add new axioms and get a stronger system, and this fact accords well with a constructionist account of mathematics. But there are important and substantial areas of mathematics that constructionist theories have proved unable to capture. It would be easy for a mathematical Platonist to say that mathematical man is more than a finite automaton, but no Platonist has ever given a plausible account of how such a non-finite or transcendent ability to deal with the infinite would work.

In any case, my foundational claim about our inner access to logical truth and falsity is not affected by Gödel's incompleteness result. This is the claim that logical truth and falsity can, in principle, be established by internal processes of reasoning, without outward appeal to observation and experiment. Leibniz's calculation model of deduction fails only for the mathematics of the infinite, whereas all the information the organism receives from the environment is finite in character; therefore, Gödel's result is no argument against the inner assessability of logic and mathematics. It also seems to me that a rejection of the distinction between logical and empirical truth by an appeal to coherence or pragmatism involves bringing in the environment, and for the same reason is irrelevant to the foundations of mathematics.

It is time to leave the level of deductive logic and pure mathematics (level 1, above) and move up to the level of induction and causal laws (level 3a, above). This is the level of most concern in *Chance, Cause, Reason*. My definitions of a priori and empirical concepts were given as criteria for deciding whether causal necessity ($\square^c$) was a priori or empirical, and I concluded after an argument conducted from the inner point of view that it was a priori (*CCR*, sec. 10.2). From the outer point of view, this means that the concept of causal necessity occurs in man's innate structure-program complex, rather than being learned and appearing later in his acquired structure-program complex. I made a similar claim concerning the concept of inductive probability (*CCR*, pp. 635–636).

Such claims are, I think, verifiable in principle. Human innate structure-program complexes are a product of Mendelian-Darwinian evolution, and each person's innate structure-program complex develops into his acquired structure-program complex under the influence of environment and culture. If we really understood the actual workings of evolution, we would know in a constructive way the sequence of steps whereby the genome, interacting with the environment and competing organisms, produces a mature organism. Interpreting the genome as a computer program, and comparing it with the computer program operating in human inductive inference, we could see whether categories like causal necessity and inductive probability occur in man's *innate* structure-program complex, or only appear in his *acquired* structure-program complex.

We noted a few pages back that the issue of empirical-versus-rational with respect to concepts is somewhat separable from the corresponding issue concerning statements. Indeed, since the role of concepts in knowledge is different from the role of statements and beliefs, the two issues are different issues, even though related. This means that my somewhat a priori account of the concepts of causal necessity and inductive probability (*CCR*, secs. 10.2 and 10.4.1) do not entail my partially a priori accounts of the presuppositions of induction and applied inductive probability statements (*CCR*, secs. 3.3, 5.6, 10.4), and I have given separate arguments for the two accounts.

This completes what I have to say for the present about modes of truth. I have found Thompson's criticisms of my views highly stimulating and useful in rethinking my position. This subsection has emphasized the topic of logical truth. Section 4 (Induction and Evolution) will discuss the evolutionary basis for inductive truth and value judgments.

3. COMPUTERS AND MINDS

3.1 Language and the environment.

The papers by Arbib and Boër on the philosophy of language allow me to return to the branch of philosophy in which I first published. My four papers "Empiricism and Vagueness" (1946a), "Icon, Index, and Symbol" (1949). "A Theory of Proper Names" (1951a), and "Ontological Categories of Language" (written in 1953 but not published until 1967) were influenced by Peirce's pragmatic semiotic, Plato's *Timaeus*, the analytic method taught me by C. H. Langford, and my work in designing computers and computer language.

My goal in these papers was to achieve a general understanding of how terms in language (as contrasted to statements) are employed by a language user in dealing with his environment. In working on this problem, I found that it helped to think of a computer interacting with its environment. The central question about the nature of concepts then became: How might we program a robot to use general terms and proper names?

My development of the logic of causal statements involved the same emphasis on the applicability of language, and I regarded it as part of the philosophy of language. People use causally modal statements in making judgments of regret (Burks 1946a), as well as in expressing laws of nature (Burks 1951b; *CCR*, ch. 6). My interest was primarily in the *concrete interpretation* of a formal language that would connect that interpretation to ordinary and scientific language and reasoning; it was only secondarily in an *abstract interpretation* of the kind that logicians use informal semantics (*CCR*, p. 352). Thus in *Chance, Cause, Reason* a whole chapter is devoted to the former (ch. 7, The Logic of Causal Statements as a Model of Natural Language), but only a section is devoted to the latter (sec. 6.3, An Abstract Interpretation).

With this background, I can answer one of Michael Arbib's criticisms in his provocative paper "Semantics and Ontology". On pp. 92–93 he expresses great surprise that I have not written about Richard Montague's modal grammar. Let me explain why. The meaning of a compound symbol is derived from the meanings of its atomic constituents in accordance with the syntactical rules governing their arrangement. It is, of course, an oversimplification to view language constructions in this two-level way of atoms and compounds, but it is a useful first approximation. (There is a similar complexity in genetics between the gene and the genome.) Using this approximation, we can say that statements are grammatical compounds of atomic terms and that by virtue of their meanings these compounds obey rules of deductive logic.

Language is tremendously complex. For my studies, I chose two basic distinctions that needed clarification: the difference between indices and symbols, which is closely related to the difference between particulars and universals; and the difference between causally modal statements and extensional statements (such as a material implication), a difference relevant to the analysis of causality and determinism. To analyze the latter distinction, it seemed best to extend already existing modal logics to cover the causal modalities as well as the logical modalities. This procedure involved using the standard kind of definition for well-formed formulas. For anyone who has studied English grammar, such definitions are obviously too simple to be useful models of natural language grammars. Montague's formal grammar is undoubtedly a better model, as are various models proposed earlier by Noam Chomsky. But these grammars are much more complicated than the standard definition of well-formed formulas, which was adequate for my purposes.

Arbib raises a number of interesting questions about my philosophy of language, most of which can be answered by discussing the problem of universals from the pragmatic viewpoint of an organism adjusting to its environment. The traditional problem of universals was actually a bundle of interrelated epistemological and metaphysical problems, of which the following are perhaps the most important. *What* are universals? Metaphysically and metaphorically speaking, *where* are they? Are universals "in" the environment, or are they "in" the organisms adjusting to the environment? If they are "in" the environment, are they properties, or are they laws of nature? In either case, how are universals related to the corresponding analogs in the organism?

Let us begin with general concepts at the sensory recognition level, for example, greenness, squareness, chairness. Peirce said that a concept is a conscious habit. I once quoted this to G. E. Moore, who dismissed it as obviously wrong. Of course, in the dictionary sense of "habit" as "an inclination to perform an act, acquired through frequent repetition", Peirce's statement is unduly restrictive. But Peirce needed a general term to cover the whole metaphysical spectrum of actualized conditionals or relations that connect a stimulus or cause in a context or circumstance to a response or effect. He viewed this spectrum as ranging over a continuum from the laws of physics and chemistry at one end, through the reactions of simple organisms to environmental stimuli, and on up the evolutionary scale to behavioral patterns of complex organisms and societies, including the practical conditionals of his pragmatism. "Habit" was the term he chose to cover this wide range of relations. (See further Burks 1980a, pp. 177–183; Burks 1988a, sec. 4.)

Peirce's characterization of a concept as "a conscious habit" was derived from Kant's dictum that "a concept is always, as regards its form, something universal which serves as a rule". The point is better put in computer terminology. Suppose a system receives specific reports of sensory inputs in binary form. Then a general concept will be a computer program that classifies these inputs into categories. For example, the general concept *green* will be a computer program that distinguishes green stimuli from non-green stimuli, and similarly for the concepts *square* and *chair* (Burks 1967, pp. 34–38; 1979a, pp. 404–407; 1988a, sec. 5). The development of programs for pattern recognition is now well advanced, although the problem of recognizing significant patterns by computers has turned out to be extraordinarily difficult, attesting to the intelligence of animals as well as humans (cf. pp. 404–449).

With this computer model of a concept we can understand better what Peirce meant by "a concept is a conscious habit". In his use of "habit" he was clearly generalizing from the ordinary sense of that word; thus he would apply it to newly compounded or recently learned concepts not yet fixed by repetition. When Peirce added the qualifier "conscious" to "habit" he was distinguishing the more or less conscious use of conditionals in language and reasoning from the unconscious role of stimulus-response conditionals in lower organisms. One implication of Peirce's theory that a concept is a rule is worthy of note. Terms and statements play very different roles in language; e.g., terms apply or not, declarative statements are true or false. But in both cases, the correlates in a language user are rules or computer programs. An instance of this underlying similarity occurs in Peirce's famous formulation of his pragmatic maxim – he analyzes a concept like *hard* into a set of conditional statements that state various empirical tests for hardness. Also, Peirce's belief that concepts are mental rules is reflected in his view that concept knowledge evolves along with propositional knowledge.

We are now ready to take up some of Arbib's criticisms and questions about reality and universals. Are properties real? Are they relative to the observer? It is worthwhile to look at these problems from the evolutionary perspective of organisms adapting to their environment. There are the properties of the environment that physicists and chemists study – what John Locke called the primary qualities – and these are quite independent of the observer. There are properties that normal humans recognize and respond to, such as colors and visual shapes, sounds and pressures in certain ranges, some smells, etc.; these are Locke's secondary qualities. Secondary quality spectra vary from species to species, e.g., dogs can respond to higher frequency sounds than humans, and thus are species relative. (We will discuss primary and secondary properties further at pp. 423f. and 441f.)

So far we have been discussing the problem of universals with respect to terms. Similar distinctions can be made about universal statements. For example, a psychological conditional connecting secondary qualities may correspond to a physical fact about primary qualities. Moreover, under what are called "standard" conditions the former is a reliable indicator of the latter. As an example, consider a heavy green object, a lighter red object, and a balance scale in good condition. The practical conditional "if one places these two objects on the balance, the red side will appear to go down" will be true under normal circumstances. Similarly, an object that seems heavier than air will normally be observed to fall when released.

It is relevant to note here what "standard conditions" or "normal circumstances" mean. These are the cases when secondary properties are good indicators of primary properties, e.g., the appearance of redness is a reliable sign of physical redness. For these correlations of secondary and primary properties to hold, both the organism and the environment must be "normal". Statistically speaking, the normal cases are the usual or common cases, and hence the concept of "normal conditions" is clearly linked to evolutionary success.

Although there are important similarities between universals in the sense of properties and universals in the sense of laws, especially at the conceptual level (concepts of both are rules), there are of course significant differences between the two, corresponding to the different grammatical roles of terms and statements. Terms fit (apply) or do not fit, while empirical statements are true or false. There is a corresponding difference between the reality of a property and the reality of a law of nature.

Let us now examine the issue of realism and the correspondence theory of truth from the evolutionary adaptive, pragmatic point of view. Ontologically speaking, an organism, its environment, and the other organisms with which it is competing or cooperating are systems of events and objects having properties that function according to laws. An organism that adapts successfully does so because, as compared to competing organisms, it takes good account of a subset of properties and laws that are relevant to it. In biological language, this subset of properties and laws defines its niche.

This evolutionary view of organisms adjusting to their environment and to other organisms establishes a sense in which universals are real: there is a reality to which organisms must adjust in order to survive. This reality involves properties of the environment and also laws connecting the occurrence of these properties. These considerations show that the fact of evolution establishes a minimal requirement for realism. From the pragmatic point of view, this is all there is to realism (see also sec. 4.1).

This evolutionary form of realism is closely related to the biological concept of fitness, and thereby to reductionism as conceived by the philosophy of logical mechanism. Let us see how this is so.

Darwinian evolution has two essential mechanisms: chance variation and natural selection. Chance variation operates mainly on genetic programs; it is accomplished primarily by the statistical recombination (including crossover) of the genes of parental chromosomes to make new chromosomes for offspring, and secondarily by such random factors as mutation and inversion. Natural selection operates mainly on the organisms produced by genetic programs. Some of these survive and are successful in producing offspring that also survive and are successful in producing offspring that also survive . . . , while others do not survive or are less successful in being represented in subsequent generations.

Natural selection is often stated as "survival of the fittest", where

the fitness of an organism is defined in terms of the survival of its descendants, more technically as the proportionate survival of the
organism's genes in the whole population, including the genes of all
relatives properly weighted. (This concept is generally called "inclusive
fitness".) But so stated this formulation of Darwinian evolution sounds
circular: "natural selection is the survival of the fittest" and "the fitness
of an organism is measured by the long-run survival of its genes".

The way out of this circle is to consider the properties and habits of
organisms in relation to the properties and laws of the environment.
An organism is fit to the degree that its physical, physiological, and
computational abilities enable it to survive, procreate, and successfully
raise kin in an environment that has certain characteristics, in competition and cooperation with other organisms having their own physical,
physiological, and computational abilities. Thus the fitness of an organism ultimately reduces to the chemistry and physics of the organism
in relation to its context. For example, the contribution of a particular
kind of food to sustaining life and growth depends on the chemistry of
the food in relation to the powers of the digestive system. The digested
food in turn contributes to the strength of the organism, which must be
assessed relatively to the physical strengths of other organisms and the
physical properties of the environment.

Thus to investigate fitness in a non-circular manner, one must study
a complete ecological system consisting of organisms of several species
competing and cooperating in an environment. Such systems are so
complex that the only way to study them is with the help of computer
models.

I have suggested the general framework of such a model of evolution in my "Computers, Control, and Intentionality" (Burks 1984).
John von Neumann's kinematic or robot model of self-reproduction
was based mainly on computing and input-output atoms, although his
girders were used for structural support as well as for making a computer storage tape. To extend this model to an evolutionary model,
one must provide atoms for genetic programs and noncomputing atoms
for the environment, such as energy sources and parts for constructing
shelters.

This robotic computer model involves a biological cycle and a con-
current environmental cycle. In the biological cycle, robots mate to
produce robotlets with genetic programs in them, and the robotlets
grow into robots and take the place of their parents. In the environ-

mental cycle, the environment changes according to the impact of the robots and robotlets in it and according to the laws of nature.

For brevity and simplicity of exposition, the evolutionary model sketched in Burks 1984 was limited to a single species. Hence it explicitly covered only the fitness of individuals of a single species operating in a single environmental niche, and it would encompass competition and cooperation only between members of the same species. But biological evolution is a process involving a historical sequence of competing and cooperating species, as Darwin's title *On the Origin of Species by Means of Natural Selection* implies. Hence we should think of my evolutionary model as extended to include many species.

This completes our outline of a robotic computer model of evolutionary ecology, formulated to show how biological fitness could be reduced to complex physical and chemical relations between organisms and the environment. Although our discussion of this model may seem like a digression from the topic of universals and realism, it really is not. For the evolutionary process of organisms adjusting to their environment and to other organisms establishes a pragmatic sense in which universals are real, both universals qua properties and universals qua laws of nature.

This is the kind of realism Peirce believed in. The following doctrines of his, listed in historical order, were all aimed at characterizing aspects of an evolutionary, adaptive sense of realism: his conception of truth as the limit of inquiry by the community of investigators, his pragmatic theory of meaning, his emphasis on the reality of Thirdness, his thesis that causal and probabilistic laws are modal "would be's", and his Aristotelian teleology of Thirds as final causes (Peirce 1931, 1982; see also sec. 4.1 below).

This evolutionary perspective also throws light on the pragmatic philosophies of John Dewey and George Mead. They described the problem-solving and communication processes whereby individuals and societies adjusted to one another and to reality. Dewey and Mead were more specific and far-reaching in this area than Peirce was, and their work was very influential in establishing modern social psychology and sociology. But their adaptive accounts were mainly limited to contemporary man and society, that is, to the single species that is the last product of the evolutionary process.

In contrast, Peirce was interested in the whole evolutionary process, starting with the primeval soup of physics and proceeding through chemistry and biological evolution to human cultural evolution and science. His doctrine of synechism was that this evolutionary process was continuous. As we explained near the beginning of this section, he extended the term "habit" to cover hypotheticals over the span of this continuum, starting with the conditional character of the laws of physics, covering the stimulus-response reaction of simple plants and animals, and proceeding on up the evolutionary scale to the conscious use of conditional rules in human reasoning.

In Burks 1984, I called the biological and cultural portion of this continuum "the teleological continuum", and I characterized it as proceeding from direct-response goal-seeking systems to intentional goal-seeking systems, namely, humans. It is of interest to apply the idea of the teleological continuum to our explication of the reality of universals in terms of their role in the evolutionary process. The idea that the properties and laws of nature are the realities to which organisms must adapt applies all along the teleological continuum.

In the case of humans, the relation of knowledge to reality is explicit: our concepts do or do not fit properties found in nature; our judgments and assertions do or do not correspond to the facts and laws of nature. Such knowledge and error occur at the conscious, intentional level, where language is well developed and explicitly used. And so traditional epistemologies apply only to the high end of the teleological continuum.

These considerations lead directly to a generalized form of the problem of universals. The concepts of successful and unsuccessful adaptation apply over the whole span of the teleological continuum. Hence the human contrasts of knowledge and error and of reality and non-reality developed gradually out of earlier organic forms that existed in the stages of the evolutionary continuum prior to the explicit use of language. The generalized form of the problem of universals is to give an evolutionary account of this whole process of how the explicit and conscious use of language and knowledge of universals developed in this gradual fashion from pre-explicit forms. Standard epistemologies would be contenders for the last or present stage of this account.

Peirce's epistemology comes the closest of any to solving this generalized problem of universals. But in a kind of Hegelian way, Peirce's definitions of reality, truth, and knowledge in terms of infinite limits contain the seeds of their own destruction, for they presuppose, and therefore cannot supply, accounts of adaptation, local fact, and partial knowledge for each stage of the evolutionary process.

Different minds may set out with the most antagonistic views, but the progress of investigation carries them by a force outside of themselves to one and the same conclusion.... This great law is embodied in the conception of truth and reality. The opinion that is fated to be ultimately agreed to by all who investigate, is what we mean by the truth, and the object represented in this opinion is the real. (Peirce 1931, vol. 5, par. 407, as in the original 1878 paper "How to Make Our Ideas Clear". In a footnote Peirce said, "Fate means merely that which is sure to come true, and can no how be avoided.")

Here the whole process under consideration consists of an unlimited series of inquiries on the same problem. Assuming that these inquiries converge on a single answer, Peirce defined reality, truth, and knowledge in terms of this limit point.

Peirce intended this process to embrace what I have called the teleological continuum, and he made attempts to show how it could do so. But for such limit definitions of reality, truth, and knowledge to work, they must be based on accounts of adaptation to reality that apply to each particular stage of the teleological continuum. Now every stage of the teleological continuum is finite, so that these stages cannot be defined in terms of the limits of the evolutionary process without circularity; and, if they are left undefined, limit definitions of reality, truth, and knowledge lack foundations, as is the case with the definitions Peirce gave (Burks 1980c).

Let us now turn to Steven Boër's paper "Names and Attitudes", a very carefully worked out criticism of description theories of proper names and a defense of causal-historical theories. It is a valuable contribution to the philosophy of names.

Boër classifies my descriptive-indexical theory of proper names as an early form of description theory. But it is also an indexical theory, and the combined indexical-descriptive meaning of a proper name can involve historical and causal connections. Thus a philosopher learns the meaning of "Socrates" as the result of a causal chain running from Socrates to Plato on down to the present. In using "Socrates" a philosopher picks out the teacher of Plato rather than another person of the same name, because the indexical-descriptive reference goes back down this causal chain.

The best way to explain and defend my theory of proper names is to put it in the context of my philosophical goals and presuppositions.

Singular empirical sentences (of the form "b is ø" or "c is R to d") are the building blocks of truth-functional and quantified empirical statements. Thus an understanding of how singular empirical statements are connected to reality is basic to understanding how empirical sentences in general are connected to reality. Prima facie there are two interrelated but distinguishable problems here, for the semantics of general terms like ø and R seem very different from the semantics of singular terms like b, c, and d. This view has been challenged, however. Frank Ramsey, in a paper on particulars and universals, said there is "no essential distinction between the subject of a proposition and its predicate, and no fundamental classification of objects can be based on such a distinction" (Ramsey 1931, p. 116), yet one of my objectives in the philosophy of language was to see if I could find a good semantic way of distinguishing subjects from predicates, and particulars from universals.

Consider first general terms that are directly applicable in experience. Their meanings are best understood by modeling them with the rules or programs a robot would use to apply such terms (Burks 1967, sec. 5). The traditional philosophical distinction between specific universals and generic universals is useful here. The input sensors of a robot detect quite specific universals, and it is easy to write programs that connect these to linguistic symbols. A generic universal is a class of specific universals. If it is a class that is natural to a detector, such as a continuous range of colors, it is easy to write a program connecting any color of this range to the appropriate general term.

Most general terms are not of this simple kind, however. Rather, the cluster of specific properties they represent can be characterized only by complex rules. Often these rules must take into account a considerable amount of the context of the term. Most of the words in a dictionary are general terms of this kind. Even words that have strict formal definitions become members of this class when they are applied. For example, begin with a near-perfect square and gradually distort it; when precisely does it cease to be a square?

Computer scientists working on pattern recognition have discovered how difficult it is to program or design a machine to recognize general patterns. Though much effort has been expended, no programs yet exist to recognize handwriting or to make good translations from one language to another. Despite optimistic predictions, it took over thirty years of substantial effort to achieve a machine that could play master-level chess, and no progress has been made on a machine to play GO. These experiences demonstrate that basic human skills involve a great deal of program complexity and intelligence. Furthermore, since such skills are very close to common sense, I think the great difficulties encountered in robotizing them validate Peirce's claim that commonsense beliefs contain a great deal of wisdom.

Before leaving the topic of general terms, let us note two consequences of the theory that a concept is a rule or computer program. First, the relative complexity of concept rules and the corresponding universals in the environment to which they apply are a basic source of vagueness in terms (Burks 1979a, p. 408; 1980a, p. 174). Second, general concepts evolve along with general statements, and this is important in scientific inquiry (pp. 369, 385, 452, 474; Burks 1980a, sec. 4).

Let us turn now to the semantics of singular terms, such as proper names or indexical references like "now" and "this table". Since my theory of proper names is the most controversial part of my philosophy of language, I will concentrate on it.

It has often been held that proper names have no meaning, but are somehow directly connected to their designata. In contrast, I hold that a proper name does have meaning. But to see how much difference there is between the two views, we need to discuss the meaning of "meaning". My use of "meaning" was inspired by Peirce's pragmatic theory of meaning, which linked meaning closely to verification and application (*CCR*, sec. 4.2). Accordingly, in "Icon, Index, and Symbol" I said that "the meaning of a sentence is whatever must be understood in order to be able to verify that sentence" (Burks 1949, p. 685). In my semiotic papers I used that principle to analyze how a language user connects terms (proper names and indexical terms as well as general terms) to the environment.

Let us apply this procedure for the analysis of meaning to singular empirical statements. A language user who understands a singular judgment of the form Rcd must be able, in standard circumstances, to apply the singular terms to the objects c and d as well as to decide whether or not they stand in the relation R. Moreover, the essential mental mechanisms whereby this is accomplished constitute the meaning of the terms involved and of their syntactical interrelations.

Thus my problem of the meaning of singular terms was derived from Peirce. My main tool for solving it was also Peircean, his sign trichotomy of icon-index-symbol, which I sought to refine. The most important refinement was to distinguish the process of learning a language from the process of using a language. (Compare the distinction between an evolutionary process and its product.) Although learning and use are continually intertwined, with respect to each symbol or semiotic aspect of language the learning process must lead the using process.

Let us apply this distinction to our problem concerning proper names and general terms. Both may be learned through experience, which involves the indexical operations of pointing and ostensive definition. But even when terms have been learned by means of indices, they may or may not function indexically. "This", however learned, is indexical when used in a pointing context. But an unambiguous use of "red" to refer to the color red is not indexical, even when it was learned by generalizing from examples designated by means of indices. One's pattern recognition program or Peircean "habit" of applying "red" to red colors (and not other colors) does not need to incorporate these earlier examples.

Indices are normally used in assigning a proper name to an object, and most uses of a name inherit this indexicality. Consider Boër's example of the famous Roman orator "Marcus Tullius Cicero" (106–43 B.C.). Presumably this name was assigned to Cicero indexically, and that assignment was the initial event of a long and complicated causal chain that, over time, branched to many references to Cicero made by means of his name. Some of these references were made in Cicero's presence, but most were connected to him only by various representational models. Later, for example, "Tullius" was anglicized to "Tully". Thus a use of "Cicero" or of "Tully" today is part of a continuing referential process that began more than 2000 years ago.

The Cicero-Tully problem arises in connection with two branches of this causal-indexical chain. Suppose John knows about Cicero *only* that

(α) Cicero was a Roman orator,

and hence not that

(β) Cicero = Tully.

It follows logically from (α) and (β) that

(γ) Tully was a Roman orator,

but John does not know this either. Consequently, the argument

(α') John believes that Cicero was a Roman orator
(β) Cicero = Tully.
∴ (γ') John believes that Tully was a Roman orator

is invalid, and thus the rule of interchanging equals fails in this belief context.

Let us contrast John with Johanna, who identifies Cicero as a Roman philosopher who lived in the first century B.C , who also calls Cicero "Tully", but who does not know that Cicero was an orator. For her, (α) and (γ) have the same meaning, but she does not know the truth-value of either of them. Suppose she later learns that Cicero was in fact an orator and says to John, "Did you know that Cicero was an orator?" John would be puzzled, for his method of identifying Cicero makes essential use of the fact that he was an orator. Thus, for John, the two sentences (α) and (γ) differ in meaning.

On my view, the meaning of a proper name has two abstract components or constituents working together. One component is descriptive, the concept of some identifying property. The other component is indexical, an indicated reverse causal route from speaker to designated object. The need for an indexical component is shown by my mirror-image or symmetrical universe model, in which for every object there is a mirror-image object having exactly the same non-indexical properties, relational as well as monadic. Twin speakers of the same words can refer to different objects because the two indexical reference routes start from different space-time points (Burks 1949, p. 683). Imagine that both I and my doppelganger utter "my brother". The descriptive components apply to both my brother and his doppelganger. But because it is indexical, my utterance refers only to my brother. Similarly, when I say "Socrates" I refer to Socrates, not his doppelganger.

Still, the two sentences

(α) Cicero was a Roman orator
(γ) Tully was a Roman orator

are close in meaning, for they concern the same individual and say the same thing about that individual. John believes *of* Tully that he was a Roman orator, but without knowing *that* Tully was called "Tully". In the sense of "information" introduced in Burks 1949 (p. 685), sentences (α) and (γ) convey the same *information*, for their subject terms refer to the same object and their predicates attribute the same property to it.

Thus the property used to identify the individual named plays a different role from that of the property predicated of that individual. The user's goal in employing a proper name (or other subject term) is to identify an object as the subject of discourse so that he can attribute some property or relation to that object. This goal is achieved even if both speaker and listener use different properties to identify the subject of discourse, provided that the property attributed to the object is different from both. The goal is not achieved when the property used for indexical reference is the same as the property attributed to the object, as when Johanna informs John (who identifies Cicero as a Roman orator) that Cicero was an orator.

One dictionary definition of "meaning" is "that which one wishes to convey", and this definition is perhaps closer to my definition of "information" than to my definition of "meaning". But ordinary accounts of meaning do not address the problem of how words are connected to the environment.

Let us now return to Ramsey's argument, which we cited at the beginning of this discussion of singular terms. Ramsey said, concerning singular statements about the environment, that there is no basic epistemological distinction between subject terms and predicate terms, from which he concluded that particulars and universals cannot be distinguished on this basis. I think the foregoing analysis establishes the opposite, showing at the same time that there is a partial symmetry between subjects and predicates. A subject term is like a predicate term in having a symbolic component, but a predicate term can be purely symbolic while a subject term must be indexical.

There is a metaphysical difference between particulars and universals corresponding to this epistemological distinction between subjects and predicates. Particular things and events have a specific space-time locus, whereas genuine universals or properties do not. The qualifier "genuine" is meant to limit the statement to *non-indexical properties* (e.g., being near a desk) and to exclude indexical properties (e.g., being near *this* desk). If Leibniz's identity of indiscernibles principle is restricted to non-indexical properties, it is not logically true. There could be two objects that have all their non-indexical properties (relational as well as monadic) in common (*CCR*, sec. 9.2.2). This metaphysical fact illustrates the profound difference between particulars and universals.

3.2 Persons and robots.

My philosophy of logical mechanism is a form of mechanism that makes computers basic. In this section and the next two, I apply this philosophy to a complex of problems that have been difficult for traditional materialisms and mechanisms: how to account for goal-directed intentionality, consciousness, free choice, and morality, and for the self-reproduction of systems that have these capabilities. My discussions of these issues are stimulated by the challenges of Laing, Lugg, and Nelson. My answers make crucial uses of the powers of logic and memory in computers, powers not easily understood in terms of traditional materialisms and mechanisms.

Let us start with the "man=machine", or "man=robot", or "person=robot" thesis: a finite deterministic automaton can perform all natural human functions (Burks 1973). The qualification "natural" excludes such phenomena as supernatural and mystical experiences, extrasensory perception, and telekinesis, if these phenomena are bona fide and have no natural explanation. In automata theory terms, this is a thesis about the relation of input sequences to output sequences. In psychological terms, it is a theory about behavior: for each person there is a behaviorally equivalent robot.

The person=robot thesis is quite naturally extended both inward and outward. For the former, I have offered a computer account of human intentionality and functional consciousness; I have argued that evolution produced the intentional mode of operation and the computer organization involved in functional consciousness, because these are relatively efficient computer designs (Burks 1984, 1986a). For the latter, I have suggested a robot-like model of evolution based on von Neumann's theory of self-reproducing automata (von Neumann 1966, Burks 1970, 1984, 1986b). Thus, overall, I am using the digital computer or finite automaton as a universal kind of model.

The philosophy of logical mechanism is an extension of Greek atomism and modern philosophical mechanism, taking into account the stored-program computer and developments in modern evolutionary biology. The term "logic" seems appropriate, because computer and evolutionary systems employ logical principles in ways not envisaged by traditional materialistic philosophies. Other terms closely related to logical mechanisms and their study are "automata theory" (von Neumann 1966, 1986), "cybernetics" (Wiener 1948), artificial intelligence, self-organizing systems, adaptive systems (Holland 1975), computer and communication sciences, logic of computers, intelligent machines, and cognitive science.

In his "Finite Automata and Human Beings", Andrew Lugg argues that the use of computer models for treating problems of the philosophy of mind significantly transforms these problems in the process of clarifying them, especially when purposes are taken into account (pp. 145, 147–148). This is a very interesting thesis, with implications for the application of mathematics to social as well as to mental phenomena.

Lugg's view of the effect of computer modeling in these areas is much more relativistic than mine. I agree that the use of computer concepts brings significant changes, but I think these are more in the methods than in the problems, and I think there are important continuities with the past.

Compare the Greek atomists with modern atomic and nuclear physicists, on the one hand, and those of us who want to develop computer models of evolutionary and mental processes, on the other. The metaphysical doctrines of the Greek atomists were very general, and they were intimately connected to normative statements about how one should live. Modern physics yields detailed knowledge, with many engineering applications; but though that knowledge has much greater social impact than Greek atomism ever had, it is remote from the philosophy of life. Similar comments will be true of the theory of logical mechanisms when it is better developed. Indeed, this theory will probably have greater social impact than physics because it deals with the mind, and because computers extend the power of the mind.

The history from Greek atomism to modern physics and the history from Greek atomism to recent computer theories of mind are both more or less continuous. Lucretius believed that physical objects were composed of invisible particles, as did nineteenth century materialists; by the twentieth century, physics had extended the view of an "atom" or building block as hard and indestructible to a more fluid view that encompasses the transient character of matter and the interchangeability of matter and energy. Lucretius believed in a mechanical theory of mind, and many today believe in a computational theory of mind.

In both the evolution from Greek atomism to modern physics and the evolution from Greek atomism to recent computer theories of mind, the problems and methods gradually became more mathematical, formal, and computational. Because of computers, the philosophy of mind is breaking off from philosophy and combining with psychology and neurophysiology. In this process the problems are being transformed, as Lugg says. But I think the transformations are similar to those that occur whenever mathematical and logical formalization is applied to an empirical subject-matter (cf. sec. 2.3 above).

Von Neumann's models of self-reproduction constitute another example of this phenomenon of change. Richard Laing has contributed significantly to their development (Laing 1975, 1977; Cliff 1982). In von Neumann's models and in all the standard variants of them (e.g., Codd 1968), the automaton contains a description of itself from the start. The automaton then uses that description as a guide and builds what is described, namely, a copy of itself. Laing developed an automaton system in which the self-reproducing machine had no description of itself, but generated a description by inspecting its own structure or form. This is certainly an extreme form of consistent self-reference!

In his imaginative paper "Machines and Behavior", Laing tests my theory of consciousness with an interesting thought-experiment about robot consciousness. His gedanken-experiment proceeds as follows. Start with a normal human being. Replace the physiological parts of this person by artificial parts, one after another, until he or she is completely artificial, that is, has become a robot. Will this robot be conscious? An affirmative answer conflicts with our usual idea of a machine. A negative answer generates the problem of finding the transition point between humans and robots.

Laing's thought experiment is an interesting extension of an old paradox about the identity of complex substances. Start with an automobile A and gradually replace all its parts until it becomes automobile A'. Are A and A' the same? In Laing's example, when did the human cease to be a human and become a robot? Suppose further that all the original parts of automobile A had been removed while they were still in good condition, and that they were later reassembled to constitute automobile A^*. Is not A^*, rather than A', the same as A?

To extend Laing's example similarly, suppose that as the parts of the human body were removed they were placed in vitro and later reassembled so that the part-by-part replacement process led to two substances, the original human organism and its robot counterpart or Doppelganger.

I think these thought experiments raise two similar philosophical questions, one general and one more specific. The general question concerns the concept of substance or thinghood: To what extent does thinghood depend on materials and to what extent does it depend on the organization and arrangement of materials? The specific question concerns the nature of consciousness, intentionality, and goal-directedness: Is each of these a "thing", or an organizational or structural aspect of a complex of "things"?

Consider first the notion of thinghood or substance as it applies to a wide variety of things: tables, chairs, buildings; chemical atoms and compounds; plants and animals, institutions and societies; habits and computer programs. The traditional dichotomy of atom and compound should be replaced by a level-dependent notion of what I call a *natural thing functioning on some level of a hierarchical system* (Burks 1986b, p. 48). This is a relatively stable structure composed of natural things from levels below and operating in certain ways. It may be part of one or more natural things on levels above.

This notion of "thing" is very general. It encompasses habits of action and fixed patterns of behavior of organisms and organizations. Computer programs and genomes are stable in appropriate contexts, and thus are natural things operating at a certain level. Humans function in social groups and institutional structures of various kinds, each containing or overlapping other institutions. These in turn are combined into societies, whole nations, and even international organizations. All these are natural things functioning on levels of a hierarchical system.

This concept of thinghood is subject to degrees and is vague. For example, the important concept of a gene is not sharply defined and is essentially statistical (Dawkins 1976, ch. 3). Consider a person who has changed radically as the result of extreme brain surgery or the psychological equivalent of it (brain washing); is he or she really the same person as before? A similar question can be asked of Laing's example: When does the human cease to be human and become a robot? There are no clear answers to such questions.

On the other hand, while it is not clear which automobile is which, there is no problem in distinguishing the robot duplicate from the original human, for the materials of the two are radically different. This brings us to the main points of Laing's question: Will the robot be conscious? Will the robot have the other mental characteristics of humans? His thought experiment is a good way of formalizing the question: Are consciousness and other mental abilities and characteristics fundamentally organizational?

My answers are in general affirmative. Assuming that a part-by-part replacement is really possible, I think the robot would have drive and will, would be capable of intentionally goal-directed action, and would be conscious. But these claims are not so radical as they may seem. For my analysis of the concept of complex substance – the notion of a natural thing operating on some level of a hierarchy – is organizational and functional. You are the same person despite changes of the materials of your body; indeed, your structure and mode of operation have changed considerably, though continuously, since your birth. And so, to see how mind is different from body, we must look at the particular forms of organization involved in mental abilities. Actually, we need to start "below" the level of mind, and consider desires, drive, and will first. Then we can discuss intentional goal-directedness, and after that consciousness.

The goals of a natural system are rooted in an underlying desire structure that is ultimately grounded on survival and on the fundamental replication mechanisms of evolution. In human societies, the desires of individuals reflect the interaction of this underlying desire structure with cultural influences, the relative influence of culture and heredity varying with the circumstances. Basic desires are directed primarily to the individual's welfare and that of what he takes to be his kin, but actual goals also show the impact of culture, including morality and law.

How might we install desires and will power in a robot? Insofar as a robot is instructed to respond in definite ways to definite stimuli, the concept of desire is not really needed: the robot just reacts to each stimulus with its appropriate response. Bona fide desires enter the picture only when the value structure is complex and the environment restrictive, that is, when there are competing goals and difficulties in achieving goals. The designer can handle these more complicated cases in a robot by assigning weights or relative priorities to various goals and by introducing into the machine competitive mechanisms and criteria for resolving conflicts among its goals. These weights will in turn control how much energy the robot will devote to satisfying its different desires. This system can be established so that it will evolve as the robot interacts with the environment. Note that a desire has two components: a propositional content, characterizing the goal; and a strength. There is a hierarchical system of desires, and their strengths compete for the role that each will play in governing conduct. Thus desire involves both information and control. (This topic is discussed further in sec. 4.4.)

As Charles Peirce emphasized, man's highest ability is that of self-controlled thought and action. Included is the ability to formulate various alternative goals, to devise and weigh the consequences of each possible goal, to choose a goal and then pursue it intelligently. The first part of this process is a free-choice process, which I discuss in the next two sections. The second part of the process is that of working towards a goal already chosen. I call this "intentional goal-seeking".

Intentional goal-seeking has a relatively static part and a dynamic part. The static part consists of a goal representation and a plan for attaining the goal. A goal is some possible future state of the environment, of the goal-seeking system, or a relation between the two. Often a future goal-state is represented in relation to the present state of the system and its environment, perhaps as a consequence of intermediate steps or means to the end sought. This representation merges with the sequential plan or strategy for achieving the desired end. There are alternative routes for reaching a goal, each with subgoals. Which route is best depends on the circumstances at each step, circumstances that in turn may depend on the actions taken at earlier steps. The plan may include a procedure for modifying the goal or terminating the intention under certain conditions. The cost of the effort to attain a goal can be compared with the probable reward, and the goal modified or replaced if the price of continued efforts becomes excessive (Burks 1984).

The dynamic part of intentional goal-directedness is a repeated feedback cycle of sensed input, internal information processing, and action output. A goal-directed system receives information from its environment and possibly from itself. It updates the representation of itself in relation to its goal, evaluates that relation, makes predictions, consults the strategy (and perhaps modifies it), decides what to do, and does it. This cycle repeats until the goal is reached, modified, replaced, or withdrawn.

The description just given is of the most explicit and complete form of goal-directedness. Most actual forms of goal-directedness are clearly less explicit and less complete.

A natural intentional system, such as a human, has a dynamic hierarchy of goals. Basic inborn drives occur at the lowest levels. Acquired habits dominate intermediate levels. Explicit goals, and possibly a life-plan, occupy the highest levels. Moreover, the goals of this structure are only partly unified, being partly conflictive, and they change over time. As John Dewey emphasized, we change not only our means but our ends as we learn from experience what we want and how to get it. This is especially the case with creative work. Also, there may be a higher-level goal of modifying and harmonizing the goals on the lower levels of the system.

We have now explained computationally how desires, interests, will, and also intentional goal-directedness might operate in a robot, and so by analogy in the robot's human equivalent. At the lowest levels these depend on the organization of the parts, hardware or physiological. At higher levels they depend on the symbolisms and languages employed in the systems. We turn next to the problem of consciousness, which is the most difficult problem in all philosophy.

As a first step I should like to distinguish two aspects of consciousness, *functional consciousness* and *immediate experience*. Consider an instance of pain. Suppose one's toe is injured. This is on the physiological side. On the experiential side, one feels a sharp pain in the toe, sees that the toe is bleeding, and puts a bandage on it. It seems to me that this experience of pain has two aspects: the felt pain as such, an immediate feeling of pain; and the experienced functional connection from an immediate feeling of pain as stimulus to the immediate experience of repair action as response.

It is convenient to postpone the problem of immediate experience to the next section and sketch my theory of functional consciousness here. I do so in two stages. The first stage envisions a robot that performs the functions associated with sense experiences and such internal experiences as pain; the second, a robot with conscious unity.

The functional aspect of conscious pain may be illustrated by the experience of lepers. Leprosy damages the nerves that carry signals from the periphery to the central nervous system. Lepers injure their extremities, but because they feel no pain they are unaware of the injuries. Consequently, they do nothing either to repair the damage or to avoid further damage, and ultimately the members deteriorate and fall off. But we know how in principle to make computers that detect their own mistakes and correct them, and we could do the same for robots.

The functional aspect of color experience is to respond to objects in terms of their colors. This can be done with suitable programs (Burks 1967, sec. 5). The general problem of pattern recognition and appropriate response is more difficult, but considerable progress is being made and the problem seems solvable. The design procedures that result can be applied to other types of conscious experience as well. On that basis, I advance the following strong claim: *A robot could be built that could perform the functional aspect of every type of conscious human experience.*

This robot need not be fully conscious, however. A mere collection of entities is not ipso facto a unified system, and so a robot with a collection of specific conscious abilities need not have the unity of consciousness. Intentionality is not sufficient for consciousness, nor is the ability to reflect on conscious experiences, for these do not require conscious unity.

Consciousness is closely associated with having a high level of awareness and being awake, whereas unconsciousness is closely associated with being asleep or being under a total anaesthetic. A survey of examples of functional consciousness shows that there are generally ways in which the organism controls itself and how it acts. Functional pain and color experiences often involve short-term control. Intentionality is a procedure for long-term control. From these observations, I conclude: *Functional consciousness is a real-time control system of relatively small capacity that exerts short-term control of the person and is capable of long-term, intentional control* (Burks 1986a).

My theory of intentional goal-directedness and my theory of human consciousness are both computational, involving notions of organization and control. They sound as if they belong to the subject of computer architecture rather than to the philosophy of mind. I think they belong to both, because I arrived at my views by comparing the architectures and modes of operation of man-made computers with those of natural computers.

These are testable theories, for in principle we can design robots along these lines and see if they are intentional and conscious. Actually, I hold the stronger thesis, that someday it will actually be practical to build a robot capable of performing *all* natural human functions and to organize the control system of that robot in such a manner that the robot will be conscious. I think that artificial materials can be found having adequate speed, reliability, and repairability or replaceability. These characteristics are essential for constructing beings that can compete in nature. But "artificial materials" are not the only possibility. Engineers are already talking about "biological computers", and as biotechnology advances there will be fewer differences between "artificial" and "natural materials".

It is interesting to test this organizational view of consciousness against one of Leibniz's thought experiments, designed to show that consciousness is an essential property of simple substance.

Moreover, it must be avowed that *perception* and what depends upon it *cannot possibly be explained by mechanical reasons*, that is, by figure and movement. Suppose that there be a machine, the structure of which produces thinking, feeling, and perceiving; imagine this machine enlarged but preserving the same proportions, so that you could enter it as if it were a mill. This being supposed, you might visit its inside; but what would you observe there? Nothing but parts which push and move each other, and never anything that could explain perception. This explanation must therefore be sought in the simple substance, not in the composite, that is, in the machine. However, there is nothing else to be found in the simple substance but perceptions and their changes. In this alone can consist all the *internal actions* of simple substances. (Leibniz 1714, p. 150.)

Leibniz wrote this in 1714. It is not hard to imagine Laing updating Leibniz's thought experiment about the mill as follows. For any computer it is possible in principle (though not in practice) to build an entirely mechanical computer equivalent to it. Take Charles Babbage's Mill and Store, and place a modern computer program in the Store. Now imagine an electronic robot equivalent to a person. Convert this robot into a mechanical computer, expand it in physical size, as Leibniz suggests, and wander through the Mill and Store. You will not find consciousness.

My answer to this Leibniz-Laing thought experiment is simply that in wandering through this mechanical computer one gets only a local view, not a global view. One sees only its computing, storage, and communication components. One does not see how these are organized, either their static, hierarchical organization or their dynamic organization (the lines of control). One does not see the language of the computer or how the computer is instructed. This thought experiment, then, provides no method for distinguishing between an inefficient direct-response robot and an efficient intentional and conscious robot.

There are still Laing-like questions to consider. Consciousness emerged gradually in an evolutionary stream of organisms, each with will and motivation and each operating fast enough to adjust to its environment. Hence these characteristics are causally and historically essential to consciousness. But how tightly are they connected logically and functionally? Suppose someone constructed a person-robot and experimented with it in an artificially controlled environment. What effect would the gradual removal of the robot's will and motivation have on its consciousness? What effect would a gradual slowdown of the whole system have on the robot's consciousness?

Let us pause now to consider how far this analysis has taken us toward answering the question: Are the mental abilities and characteristics of a person fundamentally organizational? We will discuss the status of immediate experiences in the next section, which approaches mind from the inner point of view. Desires and drives are matters of strength; although they often compete with each other, they are generally organized into some kind of unity. Intentional goal-directedness and functional consciousness are highly organizational.

This last conclusion should be viewed against the background of our earlier conclusion about thinghood: the individuality of complex entities of everyday life and technology is more a matter of structure and organization than of materials. Furthermore, the simple all-or-none identity criterion implicit in the traditional notion of substance should be replaced by a more complicated identity criterion involving degrees and permitting vague cases. Hence my claim that functional consciousness is organizational does not mark it as so different from other metaphysical entities as it might at first seem.

3.3 Mind from the inner point of view.

Because humans are reflexively conscious or self-conscious systems, we learn about ourselves by both the internal approach and the external approach (sec. 1). Thus one may learn about a bodily injury by feeling pain or by examining the injured part. The accounts of intentionality and functional consciousness of the last section were based on both internal and external considerations; the latter concerned mainly the design of functionally conscious and intentional robots. A human knows of his or her immediate experiences directly by the inner approach and indirectly by analogical inferences about others. This seeming primacy of the inner approach to immediate experience is the foundation of almost all idealist philosophies.

In the last section, we distinguished functional consciousness from immediate experiences and gave our theory of the first of these. To complete our theory of consciousness, we will study the nature of immediate experience, doing it from the inner point of view. Using this background we will address Nelson's questions about the difference between rules that fit and rules that guide. Next we will take up Lugg's question about free will. The topic of free will is appropriate here because free choices usually involve conscious rule applications. However, since the free will question involves both the outer point of view and the inner point of view, we will conclude our discussion of it in the next section (Mind and body).

Consider now the conscious intentional mind of some (female) person. This person has immediate experiences, which are events in her mind. Indeed, her functional consciousness consists of immediate experiences linked and sequenced in various ways. Some immediate experiences are directed towards objects or events (an experience of an actual table, an imagined table, a memory, a pain resulting from a cut on a finger), while others are not (a generalized pain, a feeling of exhilaration, a feeling of determination or compulsion). When objects and events are involved in immediate experiences, the distinction between veridical (true) and non-veridical (false) may apply (Did she see a table as she thought she did? Was her finger really cut? Does she remember it the way it happened?), or this distinction may not apply (The cube was only imagined, The character is fictional).

Immediate experiences of objects and events to which the categories of true and false are relevant are best understood via their symbolizations. I call a statement or judgment that describes an immediate experience or expresses its content an "appearance statement" and distinguish it from the corresponding "object statement".

Take as an example a simple monadic perceptual judgment, such as "I *see* a round table over there". This object statement may be false; there may be no such table (it may be an illusion) or the table may not be over there (I have taken a mirror image of the table to be a table). Nevertheless, my faulty perceptual judgment was based on an immediate experience of a round table over there; I can truly say "I *seemed to see* a round table over there" and "it *appeared to me* there was a round table over there". Moreover, this appearance statement is technically correct even when there is a round table over there, although in that case the statement is superfluous and to assert it would be misleading. This example is monadic, but it could have been relational, e.g., "I see a vase on the table" or "I saw the vase fall off the table onto the floor and break".

There are appearance statements about one's body. Suppose a person with amputated legs says "my toe hurts". There is no toe; the object statement is false. But the subject feels as if his or her toe hurts, and so the appearance statement "It seems to me that my toe hurts" is true. Again, if a (male) subject chooses to act and does act, he can make an object statement about the act and a corresponding appearance statement, both true. But if the subject judges that he is acting when he is not – for example, in a dream – then his appearance judgment is true while the corresponding object judgment is false.

Likewise, there are object statements about the mind and appearance statements corresponding to them. When the subject thinks he is remembering a past experience, the appearance statement is true but the object statement is true if and only if the memory is veridical. He can have a succession of thoughts, each directed to its predecessor, so that the immediate experience of one stage becomes the object of the immediate experience of the next stage. He can follow a rule consciously, as when applying modus ponens or using a recipe, and then have a memory (immediate experience) of the process.[6]

To generalize, we can divide elementary statements made by a judging subject on the basis of his or her immediate experiences into two closely related kinds. *Object statements* are about objects and events of the subject's environment, body, or mind. *Appearance statements* report on the corresponding *immediate experiences* of the subject.

Although appearance statements report on immediate experiences, it is important that both these statements and these experiences are *about* objects and events in a certain technical sense of "about". Suppose "I seem to see a table" is true, and compare the cases where "I (actually) see a table" is true with that where it is false. There is no intrinsic difference in my immediate experiences in the two cases: in both, I am having an experience OF a table. Let us call the object in my experience an *appearance object* to distinguish it from the corresponding *actual object*. I intend "object" here to include events, as when one perceives an explosion or experiences a flash of pain. I intend also that the distinction apply to the mental realm as well as to the physical realm.

An appearance of an object is an immediate experience and so is usually fleeting. Nevertheless, it is an appearance OF an OBJECT, even when the corresponding actual object does not exist. In semantic terminology, the term "table" occurs *intensionally* in "I seem to see a table" rather than *extensionally*.

It is appropriate to note here the relation of intenSion to intenTion. We explained intenTional goal-directedness in the previous section. It involves thought as well as commitment, and it arose in evolution along with consciousness and the explicit use of language as an efficient way of exploiting a niche in the environment. Now, thought models reality only partially, so that distinctions and connections existing in reality often get lost in the model. Hence intensional logic, a logic of thought, is more restrictive than extensional logic, a logic of actuality. For example, extensional equivalents are not in general exchangeable in intensional contexts (ct. pp. 406–407).

Let us focus next on physical object appearances. Much of traditional epistemology has been devoted to this case, and much is to be learned from it because it concerns the relation of the organism to its environment.

A subject does not consciously infer from an appearance statement to the corresponding object statement; rather, there are unconscious processes starting from stimuli that cause the subject to make the object judgment. In a normally functioning organism these processes usually produce approximately true object statements, but they do not always do so. Since an appearance statement claims only an appearance object and not a physical object, its claim is weaker than that of the corresponding object statement. Some philosophers have said that appearance judgments are known with certainty (e.g., Descartes' "I think") and that in the realm of appearances "seeming is being". While there is some truth in this doctrine, a subject can make verbal mistakes or be confused. Moreover, appearance statements are indexical and fleeting, and so cannot constitute knowledge.

The evidence that a physical object statement is false, or further evidence that it is true, normally comes from further experiences. For example, one gives up the claim that *there is a round table over there* when one runs into a mirror. Let us say that *an appearance object is veridical* if the corresponding physical object exists. As we noted a moment ago, there is no intrinsic mark or characteristic of an appearance object to indicate whether or not it is veridical. It follows that one cannot infer an object statement from the corresponding statement with certainty, in the sense of a probability of one, even though in a practical sense one can be certain (*CCR*, p. 85). This is the point of Descartes' dream argument.

I once heard G. E. Moore maintain that although we cannot find an intrinsic difference between the veridical case and the non-veridical case, there may still be one. This seemed to me wrong, on the ground that in the realm of appearance "everything is on the surface". Moore believed he could be certain about some object statements. In a lecture at the University of Michigan, he pointed to what looked like a skylight and said, "I am certain there is a window overhead". But there was no window, the ceiling having been designed to make it appear so!

Actually, Moore was talking about sense data, rather than what I have called "appearance objects". Sense data were supposed to be items of consciousness very close to what the senses received from external objects. They were quite specific color patches, shapes, sounds, etc., that were directly given or intuited. The traditional realist view of our knowledge of the external world was something like this: in the veridical case the mind receives a sense datum from an object and by inference judges that the object exists and has certain general properties. For example, the sense datum caused by a table gives a two-dimensional perspective, while the mind judges that a solid table is out there. The latter type of experience is sometimes said to involve a percept; on this view the knower *senses* part of the surface of the table and by inference *perceives* a solid table.

My idea of an appearance object is close to the traditional notion of a percept, provided we add in the case of immediate experiences of one's body and mind. To explicate my view further, I will contrast it to Peirce's extreme view of sense data. For Peirce the given is a psychological First, a quite specific experience or bare feeling. Since conception involves Thirdness, the given is intrinsically inexpressible. George Santayana and C. I. Lewis held similar views (see p. 440 below and Lewis 1946, ch. 7). Both views are derived from Kant, who said that a perceptual judgment involves both intuition and conceptualization. Sensations are the intuitions of objects given in sensibility, whereas concepts are contributed by the faculty of thought. Although "thoughts without content are empty, intuitions without concepts are blind" (Kant 1781, B75).

Undoubtedly, one's sensory organs receive messages from the environment, and one's central nervous system makes inferences from these inputs that normally result in appearance objects. But these sensory messages and inferential steps do not appear in consciousness. Using the phenomenological method, I find only immediate experiences of various kinds, nothing like sense data.

There are cases in which one can see on reflection that an immediate experience has more content than is expressed by the corresponding appearance judgment, as when one looks at the stars in the sky and says "I (seem to) see lots and lots of stars". But I do not think this fact implies the existence of a (perhaps inexpressible) sense datum that is the core of an immediate experience. Rather, vagueness is a characteristic of immediate experience that is to be expected on my theory of vagueness. According to this theory, a concept is vague if its complexity is less than the complexity of the set of its instances (Burks 1979a, p. 408; 1980a, p. 174). In the case of the appearance of a starry sky, the image of the sky is much more complex than the conceptual apparatus of immediate consciousness.

I think an adequate version of the distinction between primary and secondary qualities helps explain the relation of physical objects to their appearances. Consider a round red table and a veridical appearance of it. Because the appearance is veridical, it is also round and red, that is, the table appears to be round and red. Moreover, the roundness of the actual table is distinguishable from the roundness of the appearance table, and similarly for the redness of the actual object and the redness of the appearance object. Thus we need to distinguish primary properties from secondary properties both for actual objects and for appearance objects (immediate experiences).

In this way, the primary-secondary property distinction leads to four kinds of properties, as shown in the following table:

	PROPERTIES OF ACTUAL OBJECTS	PROPERTIES OF APPEARANCE OBJECTS
PRIMARY PROPERTIES	Actual primary property	Appearance primary property
SECONDARY PROPERTIES	Actual secondary property	Appearance secondary property

In veridical perception there is a close correlation both between the two kinds of primary properties and between the two kinds of secondary properties, but the correlations are different in the two cases. Consider the primary properties first. The shape of the actual table and others of its primary properties constitute its causal boundaries to certain forces. For example, the table supports other objects, and light reflected from the table can cause the appearance of a round table in the mind of an observer. Now the roundness of the appearance table is quite similar to the roundness of the actual table, for the mapping of the latter to the former is relatively simple.

In contrast, appearance redness is much less similar to objective redness. The primary properties of the microstructure of the table explain how the table responds to white light, reflecting red light. Under normal conditions this red light affects the organism so as to produce the appearance of a red table. Thus the transformation of objective redness into appearance redness is much more complicated than the transformation of objective roundness into appearance roundness. (This claim is supported by the historical fact that the theory of static shape vision was developed long before the theory of color vision; indeed, color vision is still somewhat of a puzzle.)

Originally the distinction between primary and secondary qualities was intended to differentiate the measurable properties of matter from the immeasurable properties of mind. I think there is merit in this. The shape and color of the physical table can be measured with considerable precision, while the shape and color of the appearance table can be assessed only very roughly. Concepts of object properties, both primary and secondary, are only quantitatively vague, while concepts of appearance properties are qualitatively vague (Burks 1946b).

I think this four-fold distinction of primary and secondary properties undermines the force of many of Bishop Berkeley's arguments for subjective idealism. This distinction will also be useful at the end of the next section, where we use it to analyze the reversibility argument for the irreducibility of immediate experiences.

The preceding discussion of immediate experiences constitutes an adequate background for my answers to Nelson's questions about rule following and Lugg's questions about free decisions. Their questions are about the conscious application of rules to cases, as in logical reasoning and decision-making.

Raymond Nelson's "On Guiding Rules" analyzes a basic and important distinction, the one between rules that guide and rules that describe. He uses computer concepts to say many interesting things about these two roles for rules. In the same vein, I will elaborate on the difference by means of examples of automaton functioning, taken from various levels, software and language levels as well as hardware and organic levels.

A disjunctive switch with inputs p, q and output r satisfies the conditional

IF p *THEN* r.

A truth-functional hardware switch gives a pure behavioral or input-output response to each of its input states. These connections can be expressed by a set of rules of the form

IF input state I, *THEN* output state θ.

A well-formed combination of switches and memory elements is a finite automaton (or if infinitely many components are allowed under suitable restrictions, a Turing machine). An automaton's behavioral responses to external stimuli depend on its internal states; its functioning is described by a set of rules of the form

$$IF \left\{ \begin{array}{c} \text{input state I} \\ \text{\& internal state S} \end{array} \right\} THEN \left\{ \begin{array}{c} \text{next internal state S'} \\ \text{\& output state } \theta \end{array} \right\}$$

Similar rules describe the behavior of "fleshware" such as neural nets.

We pause here to note a fundamental limitation of behaviorism. The internal state of an automaton usually plays an essential role in determining the response of the automaton to a stimulus. Hence an investigator can generally learn only a small amount about the behavior of a complex automaton such as a human by observing only its responses to stimuli. This is the Achilles' heel of behaviorism, both as method and as philosophy.

Rules and statements of the forms just illustrated, which define or fit the functioning of switches and automata, are clear-cut examples of rules that describe and explain. Notice that statements and rules are often directly interchangeable; thus we can say either that the switch satisfies "IF p THEN r" or that the switch infers r from p. We could use more complex examples of rules that describe repetitive phenomena: rules with quantifiers, modal and probability operators, algebraic or differential equations, etc. But for our purposes computer examples are sufficient for distinguishing rules that describe from rules that guide or control.

The examples given above are symbolic expressions corresponding to certain facts or behaviors that exist independently of the symbolism; switches and automata transform their input signals independently of our so describing them. Contrast the execution of an instruction or program by a computer, for example, the use of a disjunctive command to transform data. In this case the symbol is interpreted and executed by a control, and so the symbol plays an essential role in the disjunctive transformation. A computer instruction expresses a rule or simple algorithm, in order to make a certain transformation on data.

Thus the contrast of hardware with software illustrates the essential difference between rules that describe and rules that control. Roughly speaking, the difference is that between a declarative and the corresponding imperative. "If . . . then . . ." can describe the operation of a hardware switch, or it can guide (control) branching in software.

Note that either kind of rule can be probabilistic. Hardware is often characterized by probabilistic rules, and a computer instruction could consult a random number source and do one thing part of the time and something else the rest of the time.

For a complete elucidation of the distinction between rules that describe and those that control, we need to make a further distinction between conscious and unconscious rule control. The execution of programs in present-day computers illustrates non-conscious rule control or guidance. But everyone has had experience with conscious rule control or guidance, as well as with failed attempts to follow rules or resolutions. For example, logic students are asked to draw conclusions from premises by modus ponens and other rules of inference. To learn a procedure like binary addition, one may repeatedly practice it. Cooks memorize recipes and follow them in preparing dishes. When watching the luge sled races in the Winter Olympics, the observer could see the racers sitting at the top of the course weaving back and forth; they were rehearsing the sequence of bodily moves they would make as they shot down the curves of the race course.

Thus we can distinguish the control aspect of rule-following per se from the conscious exercise of this control. Over the history of biological evolution there has been a gradual development from mechanisms whose functioning is described by rules, through mechanisms that control themselves by employing symbolic expressions of rules (e.g., von Frisch's language of the bees), to mechanisms like humans that apply rules consciously. These stages developed gradually over time and each stage has included its predecessors; see my earlier mention of the teleological continuum (p. 402). Peirce held that the capacity for self-controlled thought is the essence of man. This capacity is integral to consciousness, intentionality, rational decision-making, and freedom.

Before discussing free choice, we will use the distinction between rules that control unconsciously and rules that are used consciously in order to clarify Peirce's dicta that a concept is a conscious habit and that the meaning of an empirical statement consists of its practical consequences. Humans consciously apply concepts, as when one counts the sides of a geometrical figure to see if it is a decagon, or deliberates carefully before deciding a vague case. But humans also apply concepts automatically and unconsciously, as we saw in our discussion of universals (sec. 3.1) and our analysis of immediate experiences (earlier in this section). I think these unconscious rules employ some kind of symbolism.

Peirce analyzed the meaning of a singular judgment like "that is a table" into a set of practical conditionals, such as "if an object is placed on it, the object will be supported" and "if I move directly forward, the table will block my passage" (*CCR*, sec. 4.2). But there is a more immediate sense of "meaning", according to which the meaning of a term is the pattern-recognition rule for applying the term (p. 405). In this stricter sense of meaning, Peirce's practical conditionals do not express the meaning of "that is a table", but express additional properties and laws about tables. Consider, for example, a robot that knows how to identify a table visually and that correctly judges "There is a table directly in front of me". This robot may not yet know enough about tables to infer "If I move straight forward I will be blocked by the table".

Let us turn now to the freedom aspect of conscious, intentional goal-directedness. One can choose freely a course of action, a goal, an object, the making of a statement, a rule to be followed, a tactic, a strategy, an ideology, a mate, one's friends, etc. The traditional free-will problem concerns the nature of these free choices and whether or not they are compatible with a deterministic account of nature. To understand and solve this problem, we need to employ both the inner and the outer points of view and to reconcile the two accounts thus obtained. Since our most direct knowledge of free choice comes from within, we begin with the inner perspective.

The first question to ask is: How does a free choice look from the inner point of view, that is, what is the phenomenology of the free choice process? The exercise of rational freedom, like consciousness and intentionality, is a matter of degree. Moreover, the ease with which freedom can be exercised – even freedom of thought – varies widely from case to case and country to country. Let us consider first a fully explicit rational choice process, while recognizing that many (if not most) choices are not made so systematically and easily.

The choice process usually starts from a stimulus, such as the appearance of an opportunity. One then formulates the alternatives and perhaps a strategy, considers the evidence and arguments, estimates the chances that various consequences will occur, and evaluates the consequences (*CCR*, chs. 4 and 5). Next, one chooses from among alternatives. Finally, one carries out the course of thought or action chosen. In continuing or complicated cases, the whole procedure may be reviewed and revised periodically. (The similarity of rational decision procedures to scientific inquiry and to good problem-solving methods was pointed out by both Peirce and Dewey.)

Focus now on the act of choosing among alternatives. This is a spontaneous event, not determined by its phenomenologically observable background and context. It is a conscious choice, influenced by reasons and other factors, but not determined by them. The subject could have chosen otherwise. Some advocates of free will hold that an act of free choice is a "first cause" leading to a voluntary action or thought. If we restrict ourselves to the inner point of view, this seems to me close to the truth, though I prefer the term "spontaneous" because it does not have the theological connotations of "first cause".

As we noted at the beginning of this section, however, for an adequate discussion of the free will issue one must employ both the inner and the outer points of view. The way in which freedom appears inwardly must be integrated with the role of freedom in society, and both must be made to fit with the nature of the laws governing the environment. For these issues we shift to the next section.

3.4 Mind and body.

The internal account of freedom just given is an account of certain activities of conscious mind. Our next task is to relate conscious mind to the rest of the person, to social functioning, and to the nature of the environment.

A human society is a goal-directed system composed of goal-directed systems. Its goals are based on those of its members, though its goals can take on a life of their own. Most people have strong self-interests and strong kin-interests, together with the capacity to pursue those interests. Because the interests of different individuals generally conflict, a society can function successfully only if it has mechanisms or institutions that induce individuals to pursue common goals. Laing and Lugg both introduce this problem.

Organizations use many different methods to control the behavior of their members: rules of etiquette, morality, ideology and religion, economic institutions, political administrative systems, civil and criminal legal systems, and in many societies the secret police. Thus a society is a somewhat decentralized system composed of partially free individuals, which it controls to some degree by various kinds of social forces. These range from the extreme of symbolic rewards and punishments through social and material rewards and punishments to sheer physical force.

Normally these various incentives are taken into account by a free person during the choice-making process; acts arising from extreme passion are exceptions. Most adults are generally in control of their choices, although there are many bothersome borderline questions as to whether individuals should be held responsible for their choices: cases of insanity, juveniles, and people who have been brain-washed. Of those in control of themselves, some are highly rule-abiding while others frequently break the laws.

The matter of rule-obedience is complicated by the fact that almost everyone belongs to many groups, which may have competing rules. Thus a person may be a citizen of the United States and a member of the Mafia; what is forbidden by one may be ordered by the other. To keep our analysis of free will simple, we will proceed as if there were a single society with rules of morality, law, etc., and enforcing mechanisms for them.

Humans by nature have certain interests and abilities, including the capacity for free choice. Morality, law, and other control institutions are means for directing people's choices so as to obtain the individual and cooperative behavior needed for social functioning. These means do not, of course, achieve perfect, or even nearly perfect, norm-following. Most societies have substantial amounts of deceit, immorality, crime, etc., and all large societies do. This seems inevitable, given the desire structure of human nature and the limited resources of most environments.

We will discuss briefly the role of physical force and threats thereof in controlling a society. There are very few stable societies that do not rely on force. Some religious communities do not, but are still parts of states that do. One basic question of social philosophy is whether the use of force is required for a stable society. The extreme anarchist theory of the state is that if people were properly trained, morality alone would suffice for social control, so that ultimately "the state would wither away". This seems to me wrong. I think that human nature is statistically such that some use of force is required for a stable society.

An important feature of my theory of free choice and morality is that morality does not need the universality of Kant's categorical imperative to perform its function. The concept of an *evolutionarily stable strategy* as used in recent evolutionary biology and game theory is useful here.[7] The level of norm-following in a society is an important factor in the health of that society, and the breakdown of a society is usually accompanied by a breakdown in norm-following. Suppose a society responds to increased norm-breaking by more police activity and to decreased norm-breaking by less police activity, and similarly for other control mechanisms such as morality. If this interplay of the degree of norm-breaking and the amount of enforcement holds the degree of norm-keeping at an approximately constant level, then the control mechanisms of morality, law, etc., constitute an evolutionarily stable strategy. Nevertheless, in actual societies with an evolutionarily stable strategy or stable level of norm-following, the amount of norm-breaking is usually quite substantial (see further p. 504).

These are factual statements about morality. They are very different in kind from such moral judgments as "We should all follow the same moral rules all the time" and "Everyone ought to choose so as to maximize expected social utility". Many attempts have been made to justify or prove these general normative statements and others like them. Kant's putative proof of universalizability (e.g., society would collapse if no promises were kept) and Mill's attempted proof of utilitarianism are examples. None of these attempts rests on a clear sense of the meaning of normative terms such as "ought" and their relation to factual terms such as "is". I think the account given above of specific moral statements like "Keep your promises" and "Obey the law" also applies to general moral principles like universalizability and utilitarianism. All these normative statements derive their meaning from their actual and potential control role in social organizations (Burks 1986b, Lecture 6).

The preceding theory of how freedom is related to morality and social control is, I think, correct in principle, though obviously much more can be said about the details. Furthermore, this theory provides a basis for explaining how morality and freedom arose gradually in the evolutionary process, along with intentional goal-directedness and consciousness (Burks 1984, 1986a, 1986b, 1988b). Most accounts of morality and freedom do not do this.

This completes my analysis of free choice at the social level, though I will return to the evolutionary origin and nature of morality later (sec. 4.4). I turn now to the question of how free choices are related to the general character of the laws governing the environment. The laws human organisms use in their everyday actions are high-level laws, generally partial, approximate, and statistical. But philosophers and theologians have long held that the nature of the underlying microlevel laws is relevant to the nature of free choice, and so we must examine this issue. As we said in Section 1, the free will and determinism issue involves the interrelation of global and local points of view, for determinism is a global theory whereas free choices are local phenomena.

There are several different metaphysical questions to be asked about the laws of nature, questions that are often not distinguished. The first two concern the character of the laws themselves: Does an antecedent cause or state have a unique consequent or more than one consequent? If there are two or more consequents, is there a probabilistic distribution over their occurrence? A third question concerns the completeness of the rule of law.

Determinism is the doctrine that the rule of law is complete; in other words, that the causal laws of nature and the present state of nature determine a unique future history of nature. (The traditional doctrine of determinism contains additional features, but we can ignore them here.) Indeterminism is the doctrine that some events are not predetermined; acts of spontaneous choice are usually cited as examples, but probabilistic laws are also examples. I prefer the term "near-determinism" to "indeterminism" because responsibility of any kind presupposes that nature is for the most part governed by causal laws, that is, that nature is nearly deterministic. (See *CCR*, secs. 9.3 and 9.4, for a further discussion of these doctrines and their relation to choice, chance, and embedding. See also sec. 4.3 below on the alternatives to determinism.)

There are two traditional views about the relation of free choice to determinism and near-determinism. The *free-will thesis* is that a free choice is partly uncaused, and that this lack of complete causality is essential to freedom and responsibility. The doctrine of *compatibilism* is that free choices require only inner conscious control, and that such control is compatible with determinism. I am a compatibilist and will defend this position here both constructively and by criticizing the free-will thesis (see also Burks 1986b, Lecture 6).

Free choices occur on the personal and social levels, whereas determinism and near-determinism are theories of the microlevel. My criticism of the free-will thesis is that it does not take adequate account of this difference in level. Thus the advocates of the free-will thesis jump back and forth between phenomenological spontaneity and microlevel spontaneity, ignoring their difference in level, and hence fail to show why the latter is required by the former. The use of pseudo-randomness to show the irrelevance of the determinism issue to the freedom issue points up this weakness. As Lugg points out, computer simulations cannot distinguish between spontaneous choices that occur by real chance and spontaneous choices that are pseudorandom. He takes this to be an objection to my embedding theory of freedom, but I think it is an objection to the free-will thesis. For the difference between real chance and pseudorandomness is a matter of computational complexity, and this seems irrelevant to the freedom question.

One can see that the issue of chance versus determinism is irrelevant to the free-will issue by considering a traditional argument to show that determinism is incompatible with freedom. The argument runs as follows. Determinism implies that a person's acts are predetermined by each state of the universe that occurs prior to that person's birth. One cannot control those states, and one should be held responsible only for the states that one can control. Hence if determinism is true, a person cannot properly be held responsible for his or her acts. – My rejoinder is that this argument confuses local states with global states, and hence local control with global control. Furthermore, if it is unreasonable to punish or reward people for their free acts because these are predetermined, it is unreasonable to punish or reward them because their free acts are manifestations of indeterminate or chance events.

I turn now to my constructive argument for compatibilism. This is an application of the idea of one system being an *embedded subsystem* of another, *underlying, system*. Let us review this idea briefly (*CCR*, p. 588). It is closely related to our earlier discussion of computer modeling (sec. 2.3, Logical formalization and computer simulation), for a good model of a system is isomorphic to an embedded subsystem of that system.

Consider a gas that has the global properties of pressure, volume, temperature, and weight, and is governed by the gas law: pressure × volume = constant × temperature. This system is an embedded subsystem of another system, which operates according to the laws of mechanics. This other system is the gas as constituted of trillions of small particles, each particle having local or microproperties of mass, velocity, and acceleration. These particles interact with one another and the walls of the container according to the laws of mechanics.

The gas system described by the gas law is reducible in principle to the gas system of interacting particles. Reducibility is demonstrated by the following procedure. First, the set of allowable initial states is limited, any initial state leading to a gas behavior that does not fit the gas law being excluded. The remaining microstates are lumped together (grouped into sets) to produce the microstates of the embedded subsystem. For example, all the energy distributions over particles that have the same average kinetic energy are grouped together, because each of these microstates has the same temperature. Under the two operations of initial state restriction and lumping, the mechanical laws of the underlying system reduce to the gas law.

The application of the notion of an embedded subsystem to the free-will problem is fairly direct. Consider an example of a person making a free choice intentionally and consciously. We gave a phenomenological description of this process at the end of the preceding section, making clear that it is an embedded subsystem of the whole person, the unconscious mind as well as the body. It is the whole person, the underlying system, who made the decision and who is held responsible for it by society. We also saw earlier that the embedded subsystem is indeterministic, for the actual act of choice has a spontaneous quality.

It is noteworthy that my phenomenological account of free choice is very similar to the total account of free choice given by the free-will theory referred to in the preceding section. This theory is that choices are made by a willing and rational mental substance. A free choice arises spontaneously in this substance. The event of choosing is not causally determined by prior events or states of the choosing substance, though of course the event occurs against a background of conditions that give relevant information and limit the alternatives. This free choice (event) then brings about (causes) a free action. Since the free choice is spontaneous it is a kind of first cause in this causal chain of events.

My internal or phenomenological account of free choice holds that the embedded, conscious, and intentionally goal-directed self makes a free choice with apparent spontaneity, for introspection reveals only conditioning factors, not a determining complex of factors. Because of this spontaneity, the embedded subsystem depicted by my phenomenological account is clearly not deterministic. However, the complete system of conscious mind, unconscious mind, and body might well be a deterministic system.

Thus the compatibilist holds that the underlying system could be deterministic without the indeterminism of the embedded subsystem (and hence the free character of the choice) being destroyed. The crucial point is that the system of laws governing the embedded subsystem need not have the same deterministic-indeterministic status as the system of laws governing the underlying system. For when certain starting states of the underlying system are excluded and other states lumped together to make the states of the embedded subsystem, the nature of the laws connecting the remaining states may change. There are many examples of such shifts (CCR, sec. 9.4). We will analyze the case of fair tosses of a coin to demonstrate how the deterministic-indeterministic status of an underlying system and its embedded subsystems can differ. (Note that a probabilistic system is an indeterministic system that has a probabilistic assignment to alternative states.)

Consider the whole dynamic system of a sequence of coin tosses as made, for example, by some coin-tossing machine that produces highly random sequences of heads (H) and tails (T). Focus on the embedded subsystem that consists of a sequence of tosses, each starting from the state of the coin (H or T) as it leaves the machine and ending with the state of the coin (H or T) as it comes to rest on the table. Since one cannot predict how the coin will come up from how it starts out, the embedded subsystem is indeterministic. Yet it is reasonable to hold that the underlying system is a deterministic dynamical system; for example, there is no reason to believe that the indeterminism of the sequences of tosses depends on quantum effects. A positive case for this thesis is provided by some recent studies of the deterministic differential equations governing a system of coin tosses.

The following story will provide a useful starting point.

I once knew a student who had the remarkable ability to toss a fair coin in the air so that the toss appeared random, but it almost always fell into his hand on the side an observer called out before he tossed it. I do not know how he exerted such control, but let us assume that he imparted a spin to the coin so that it precessed like a top and never actually turned over, even as it seemed to do so. Of course, he could not control the starting state of the coin precisely – beyond its heads-or-tails status – but in his mode of tossing the coin a small variation in starting state did not alter the outcome, for the coin processed stably and always kept the same side up. In the ordinary mode of tossing a coin, a small difference between two starting states becomes amplified (this is sometimes called "the multiplier effect") and the number of times the coin will turn over is not predictable from an approximate knowledge of the starting state.

This difference between two modes of tossing a coin has been studied recently by mathematicians. They have analyzed the solutions of the complex non-linear partial differential equations governing the dynamical behavior of a coin tossed in a gas (the air). They assume that the precise starting state of the coin is not known, but only an approximation to it; that is, the observer knows that the starting state is one of a closely related (neighboring) set of starting states. They then consider two cases, analogous to the two tossing modes of the preceding paragraphs. In the first case the system is stable, and small variations in starting state do not affect how the coin falls; here the outcome of the toss is predictable. In the second case, the behavior of the coin is unstable, small variations in starting state being amplified to produce large fluctuations in behavior; here the outcome of the toss is not predictable (Zeng-yuan and Bin 1985; Kolata 1986).

The underlying dynamic coin-tossing system is deterministic. Given a precisely specified starting state, the partial differential equations of the system yield a unique outcome, head or tail. The studies show that embedded subsystems whose initial states are neighboring sets of specific initial states can be either deterministic *or* indeterministic and random. Also, there is a mathematical account of how random (probabilistic) behavior evolves from deterministic behavior, for example, how a coin-tossing process that is deterministic evolves gradually into a probabilistic coin tossing process. Furthermore, these studies show that the transition from a deterministic subsystem to a random subsystem is very similar to the transition from laminar flow to turbulent (chaotic) flow in hydrodynamics.

Let us look at the situation in terms of a simple example. Suppose that in a deterministic system property P leads after a period of time to property X, property Q leads to Z, R leads to Y, and S to W. Let the properties of the embedded subsystem include the groups (P,Q), (R,S), (X,Y), and (Z,W). If the embedded subsystem starts in state (P,Q), it may end up in either state (X,Y) or state (Z,W) and so is indeterministic. Put in technical terms, the question of whether an embedded subsystem of an underlying deterministic system is deterministic or not reduces to the question of whether the state trajectories of the underlying system lie close together or diverge.

One of Peirce's strongest arguments for indeterminism was based on the observed variety in the universe. He said that this variety could be accounted for in only two ways: as entering the universe in a single dose (the deterministic way), or as arising gradually over time (the indeterministic or tychistic way). But the foregoing analysis shows how variety can arise gradually from uniformity in a deterministic system. Even the great variety of a chaotic system can arise in this way. This refutes Peirce's variety argument for indeterminism.

Since observable properties are groups of underlying properties, there is a close parallelism between metaphysical and epistemological questions here. The question "Do the elements of the starting states of the embedded subsystem have histories that are close enough to make the embedded subsystem deterministic?" is a metaphysical question. The corresponding epistemological question is "Does the statement that the embedded subsystem is in a given starting state contain enough information to enable one to predict the outcome?" Such questions are sometimes of practical importance, as in long-range weather forecasting and earthquake prediction.

A human making free choices is many orders of magnitude more complex than a coin-tossing system. Nevertheless, there is an important similarity between the two. What happens to the coin depends on what happens below the level of observability. Similarly, what happens to a freely choosing individual depends on what happens below the level of consciousness. Consider the problem of explaining why apparently similar people vary in the degree to which they violate ethical and legal norms. I think an important part of the explanation will be found in the interaction of their desire structure with those of others and the conditions of the environment. It is even possible that small variations in heredity and environment may produce large variations in norm-breaking. Such a result would set a theoretical limit on the efficacy of moral and legal control systems.

Let me now summarize my positive case for compatibilism. I agree with the free willer that the conscious process of free choice is indeterministic. But this process occurs on a high level in the hierarchy of a person, for it occurs in the person qua conscious intentional subsystem. This subsystem is embedded in an underlying system, the person qua microscopic biological system. It is in general possible for the deterministic-indeterministic status of an embedded subsystem and its underlying system to be different. And in this particular case there are good reasons for holding that the indeterminism of the subsystem is compatible with the underlying system's being deterministic.

Lugg makes an important point about the analysis of free choice by means of the concept of embedding. Embedding is a triadic relation: when an appropriate aggregating function is applied to the states and laws of a basic system, the result is a simpler, embedded subsystem. This makes embedding relative to a choice, and hence relative to some interest. In the case of the free-will problem the interest is in reconciling the inner and outer points of view. A self-conscious person may be interested in relating his or her powers of control to the countervailing forces of other organisms and the environment.

* * *

This completes our discussion of the free-will problem. Since conscious mind exerts control over the body, the free-will problem is a special case of the mind-body problem. From this discussion of free will and our earlier theories of intentionality and consciousness, we are in a good position to attack the most general mind-body problem, the question of whether mind is reducible to body. For the philosophy of logical mechanism, this question becomes one of whether mind is completely reducible to logical atoms, such as truth-functional switches, simple stores, communication lines, and sensing and acting devices. An answer to it would by implication be an answer to many of the questions raised by my commentators.

Intentional goal-directedness is an activity of functional consciousness, and I have already given a logical mechanistic account of that activity. Now functional consciousness consists of sequences and interrelations of immediate experiences. Hence our mind-body reducibility question becomes the question of whether immediate experiences are reducible to logical atoms. Immediate experiences are subjective events, so that our question "Are immediate experiences reducible to logical atoms?" is a computerized form of the traditional materialist-idealist question: "Are immediate experiences material events or subjective mental events?"

Let us now return to Laing's thought experiment (sec. 3.2) and imagine a robot made of artificial parts that, part by part, functions as you do, and whose parts are organized as your parts are. I have argued that this robot would be just like you with respect to desire structure, intentional goal-directedness, and functional consciousness. It follows from our recent analyses of conscious rule-following and free choice that the robot could have these capacities too, in just the way you possess them. Would your robot double have immediate experiences, and if so, would these immediate experiences be like yours?

By hypothesis, the only difference between you and your Doppelganger is a difference in the materials of which the logical parts are composed: hardware versus organic compounds. This difference does not seem relevant to the question of whether the robot would have immediate experiences. For both computer hardware parts and organic parts are reducible to inorganic materials. It even seems likely that computer and bio-engineering technology will develop a spectrum of logical atoms that range gradually from the artificial to the biological. Hence it is likely that your robot double would have immediate experiences, and that these immediate experiences would be like yours.

The preceding argument for logical materialism has considerable force, but it clearly does not settle the issue. One also needs direct evidence for reducibility or, on the negative side, strong reasons to believe that the reduction cannot be carried out, perhaps reasons to believe that immediate experiences are irreducible mental atoms. Since neither exists, we are limited to relevant reflections.

The prima facie case against reductionism is that immediate experiences seem very different in kind from material events. They are sometimes appearances of material entities, but still do not seem to be material entities. That is why the distinction between secondary properties and primary properties was drawn in the first place. These differences have seemed so great to many philosophers that they have held mind to be different in kind from matter; this is a common basis for subjective idealism and dualism.

I think the foregoing intuitions have considerable force. Running against them are two kinds of consideration, the relevance of a perspective or point of view, and a basic feature of reductions. As we have argued, immediate experiences and physical entities are known from two very different points of view, the inner and the outer. I am suggesting that when a sufficiently complicated computer is organized appropriately, it will be both conscious and self-conscious, can take both points of view, and will have immediate experiences. If this is so, it may be that the prima facie irreducibility of immediate experiences to logical atoms is only a reflection of the differences between these two points of view.

This position is supported by a basic feature of reductions, that the properties of an embedded subsystem are very different from the properties of the underlying system. We saw this in the case of the free-will problem: an indeterministic conscious free-choice process can be embedded in a deterministic system. Further support comes from the history of science. It seemed intuitively clear for many centuries that heat was a fluid and not an activity. But the kinetic theory of heat reduces heat to molecular activity.

Let me introduce the notion of "reduction distance" to help here. Most reductions cannot be carried out in detail. But one can give reductive definitions and, using these, derive the laws of the embedded subsystem from the laws of the underlying system. The proof of the gas law from the laws of mechanics is an example. A reduction is a kind of computation, and the complexity of computations can be measured relatively to a suitably chosen computer (Burks 1980a, p. 173). Hence we can define the *reduction distance* from an embedded subsystem to an underlying system as the complexity of the process of reducing the first system to the second.

Assume for purposes of argument that immediate experiences are ultimately reducible to the functioning of inorganic molecules. The reduction distance from experience to molecules is exceedingly large, many orders of magnitude larger than the reduction distance for the gas law. This estimate is supported directly by what is known in the life sciences and indirectly by what is known of evolution. Given the great reduction distance from immediate experience to inorganic matter, it is not surprising that the reduction seems counter-intuitive. To bridge the gap somewhat I have introduced the levels of logical atomism and of computer accounts of intentional goal-directedness, will and desire, and consciousness. I am suggesting that immediate experiences are events in the conscious functioning of a very complex intentional goal-directed system. This can be a logical mechanism.

For a further explication of my theory of mind, I will discuss some alternative views, using my own analysis of immediate experiences while recognizing that it is somewhat different from the analyses given by these alternative theories.

Consider first epiphenomenalism and Santayana's doctrine of essences. It is appropriate to call both of these theories "near-materialisms"; for they both give materialistic accounts of all mental entities except immediate experiences, and they deny these any causative role. Epiphenomenalism is the doctrine that immediate experiences are caused by neural events, but are not reducible to them and cannot affect them in any way. In other words, immediate experiences are mental "epiphenomena". Evolutionary epiphenomenalism incorporates this view into an evolutionary naturalism by saying that immediate experiences emerged at a certain stage in the evolutionary process and have continued thereafter. But it is very improbable that evolution would both produce and maintain such a pervasive and long-lived non-adaptive phenomenon. Moreover, as my account of immediate experiences shows, these experiences are events in consciousness and clearly do play a role in adaptation.

Santayana was a materialist and recognized the difficulty of fitting immediate experiences into a materialistic metaphysics. His solution was to say that immediate experiences are not events at all, but are "essences" or specific universals (cf. p. 422 above and p. 453 below). Thus on his account immediate experiences do not belong to the temporal realm of existents but to the atemporal realm of Platonic universals, and so the question of whether immediate experiences are material or mental is illegitimate. However, the entities I have identified as immediate experiences are clearly spatial-temporal events and they play a role in evolutionary adaptation. Immediate experiences have properties and so are instantiations of Platonic universals; but they are not Platonic universals.

Let us turn next to a traditional thought experiment designed to show that immediate experiences are logically separable from physical events, from which it is concluded that immediate experiences are irreducibly mental. This is the "reversibility argument" for idealism or dualism.

The imagined model is best explained by means of the following diagram. In the terminology of the previous section (p. 423), there are two actual objects with different actual secondary properties, blue and red. There are two percipients, each with an immediate experience (or appearance object) of the blue actual object, and similarly for the red actual object.

First Observer

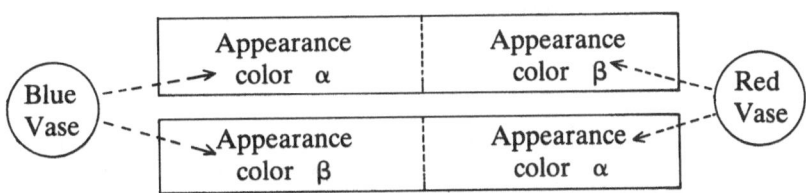

Second Observer

The two observers are normal and are seeing the vases under normal lighting conditions.

It seems logically possible that the subjective colors of their immediate experiences could be reversed, as in the diagram. In terms of the fourfold distinction of primary and secondary qualities of the previous section, the actual secondary property of blueness would produce an appearance secondary property of α for the first observer but an appearance secondary property of β for the second observer, whereas the actual secondary property of redness would have the reverse effects.

This thought experiment has been used to argue for a thesis basic to both idealism and dualism: immediate experiences are not reducible to bodily events and hence cannot be physical in nature. The argument runs like this. Although immediate experiences may be related to bodily events by a principle of psychophysical correlation, the reversibility model shows that immediate experiences are logically independent of bodily events and could exist even if there were no bodily events. Consequently, immediate experiences are not material or physical in nature; they belong to a separate metaphysical category of the mental.

On my reductive account, immediate experiences are embedded subsystems of physiological systems, and hence ultimately of physical-chemical systems. Embedding is a logical construction, involving logical constraints as well as logical alternatives. Moreover, by its very nature, an embedded subsystem must be simpler than the underlying system, and this puts an additional constraint on the embedding relation that holds between an appearance object and the corresponding actual object. These constraints seem especially compelling in the case of the appearance properties of pleasure and pain and the physiological correlates of satisfaction and frustration. For the pleasure-pain difference has arisen in evolution as a manifestation of the satisfaction-frustration difference. The opposite association, that of pleasure with frustration and pain with satisfaction, would not be evolutionarily stable.

I think the logical constraints involved in the operation of embedding cast doubt on the claim that the properties of immediate experiences or appearance objects are logically independent of the properties of the corresponding actual objects. For purposes of argument, however, I will grant the logical possibility of reversibility

But this admission does not yield the conclusion that immediate experiences or appearance objects are irreducibly mental. For it is logically possible for immediate experiences and appearance objects to exist without there being any underlying physiological and physical system.

More generally, while the states and laws of an embedded subsystem are defined in terms of the states and laws of an underlying system, it would be logically possible for the subsystem so defined to exist in the absence of the underlying system. It is logically possible, for example, that the gas of the gas law be an ultimate reality, the gas being a continuous fluid-like material with a certain temperature and exerting a certain pressure on its container. (Compare Descartes' view that space is filled with matter and that motion has a vortex character, i.e., is ultimately cyclical.) Similarly, the atoms of the Greek atomists and nineteenth century chemists could have been ultimate and irreducible, though in fact they were not.

Correspondingly, a person's immediate experiences might be physical and material in character without in fact being embedded subsystems of an underlying physiological body. This could be so for immediate experiences of pleasure and pain as well as for all others. If only immediate experiences did exist, the universe would constitute a materialistic version of solipsism or, if other persons existed, of phenomenalism. Immediate experiences would be material but would not be reducible to the world of physics and chemistry, because that world would not exist. They would float freely in their own space-time milieu, somewhat like Hume's impressions of sense and inner feeling.

Hence the reversibility argument fails to establish the idealist and dualist thesis that immediate experiences are irreducibly mental. Moreover, the theory of evolution constitutes a great difficulty for both idealism and dualism. Evolution proceeds almost continuously, in small steps; see the discussion of the teleological continuum in Burks 1984, sec. 3. This continuity is incompatible with the sharpness of the dualist's distinction between mind and matter, and its occurrence seems ad hoc on an idealistic or phenomenalistic view.

I conclude that immediate experiences are basically material events in a functionally conscious logical mechanism. This theory could be tested by simulating the evolutionary development of mind from biological processes, and by building and studying robots that possessed human mental powers. Such a robot might be constructed so that it had communication channels from its surface to various internal levels of its hardware, software, and mental activity, including the unconscious as well as the conscious. Then humans could see from the outside, and even influence, what the robot was experiencing from its inner point of view. Such experiments would provide much better empirical information about consciousness and self-consciousness and their relation to control than can be obtained from phenomenology, on the one hand, or from common sense or experimental or physiological psychology, on the other. Whether it would be proper to conduct such experiments is a moral question.

4. EVOLUTION AND INDUCTION

4.1 Causal laws.

The first two papers of this volume, those by Suppe and by Cohen, deal with my views on the foundations of induction, and so do the next six to be discussed. Moore and Uchii treat the concept of causal necessity. Clendinnen and Rosenkrantz criticize my presupposition theory of induction. Railton and Skyrms argue that probabilistic laws are basic while I hold that they are not. In answering these six writers, I will develop further my mechanistic philosophy of evolution, which will provide a basis for discussing the last paper in this volume, Audi's essay on values.

Edward Moore's "Actuality and Potentiality" is especially appropriate for this volume because my own concepts of causal necessity and dispositional probability were influenced by Peirce's "would-be's". There are important differences, however. Peirce held that a causal law is a limiting case of a probabilistic law ("matter is effete mind"), thereby committing himself to the corresponding conceptual thesis that causal necessity is a limiting case of dispositional probability. I believe this thesis is incorrect, even though my analysis of causal necessity does parallel my dispositional analysis of empirical probability (*CCR*, pp. 496–497).

Moore gives an excellent account of actualities and potentialities in Peirce's philosophy. He then argues successfully that a positivistic or pragmatic theory of meaning cannot account for the concept of potentiality; in this connection, he reviews the history of meaning-criteria from Ockham through Peirce to the logical positivists. I agree fully with Moore's thesis that Peirce's later realism is incompatible with the pragmatism of "How to Make Our Ideas Clear"; I will develop this point in due course.

Moore was primarily responsible for starting the new edition, *Writings of Charles S. Peirce*, at Indiana/Purdue University in Indianapolis, currently being edited by Max Fisch, Christian Kloessel, and others. This edition has two main merits over the first one (of which I edited the last two volumes): it is to be much longer (more than 20 volumes versus 8); and it is chronological, so that the reader will be able to see more clearly how Peirce's interests and views evolved.

Through his involvement with the new edition, Moore stimulated me to work on Peirce again, particularly on the nature of Peirce's realism. I will discuss three questions here: Did Peirce have a significant nominalist phase? What metaphysical role did he assign to final causes? Is his pragmatism compatible with his later realism? To answer these questions, one must distinguish four stages in Peirce's intellectual biography.

(I) From 1867–71, Peirce published his series on logic, in the *Proceedings of the American Academy of Arts and Sciences*; his review of John Venn's *The Logic of Chance*; his series on intuitive knowledge, in the *Journal of Speculative Philosophy*; and his review of Fraser's edition of *The Works of George Berkeley*. In these papers, he used his knowledge of the logic of relations and of Darwinian evolution to work his way from Kant's individualistic categories to a scientific community limit theory of inquiry, knowledge, and reality that was somewhat Hegelian in spirit.

(II) Starting soon thereafter, Pierce developed his empiricist or verifiability version of pragmatism, which culminated in his *Popular Science Monthly* series "Illustrations of the Logic of Science" of 1877–78, his "Note on the Theory of the Economy of Research" of 1879, and his Johns Hopkins University logic lecture of 1882 in which he said

This is the age of methods; and the university which is to be the exponent of the living condition of the human mind, must be the university of methods (Peirce 1931, 7.62).

During this period, Peirce focused on the logic of science and how to use it to improve research.

(III) Peirce's 1891–93 series on metaphysics in *The Monist* developed his highly speculative evolutionary and teleological cosmology. This was a hierarchy of three doctrines: tychism (the laws of nature are probabilistic), synechism (the evolution of laws and systems proceeds continuously), and agapism (final causes guide evolution toward a value goal). The social evolution of knowledge that developed in Stages I and II now became the last part of an evolutionary process beginning in a primeval chaos and advancing to better and better states of affairs ("Truth crushed to earth will rise again.")

(IV) From around 1900 until his death, Peirce outlined or sketched a grand synthetic extension of his philosophy. His system was to be based on a classification of all the sciences, together with a comprehensive semiotics developed from his earlier logic, classification of signs, and theory of inquiry. This grand philosophy was to encompass his earlier limit theory of knowledge and reality, his pragmatism, and his evolutionary cosmology. He expected it to have practical applications to science and engineering (compare the modern subject of operations research) and to have implications for the more specialized branches of philosophy, such as ethics and the philosophy of religion.

There was a rather nice and natural progression through these four stages of Peirce's philosophical interests and doctrines. Stage I concerned the social evolution of commonsense and scientific knowledge. Peirce's theory in this stage was a Hegelian-like transformation of Kant's transcendental logic. Stage II made explicit the role of the environment or external world in this evolutionary process: adaptation to the environment in the case of common sense, and observation and experimentation in the case of science. Stage III extended the time frame of this evolutionary process back to the infinite past. In Stage IV, Peirce projected a grand and comprehensive evolutionary philosophy based on a new logic-cum-semiotics and encompassing all branches of philosophy.

We discussed some aspects of Peirce's realism earlier, in connection with the relation of language to the environment (sec. 3.1). Let us now consider whether Peirce was a nominalist or a realist at each of these stages. "Nominalism" and "realism" are highly ambiguous terms. They are applicable to theories about things, theories about properties, and theories about laws. Although the three metaphysical categories of thing, property, and law are interrelated, the nominalism-realism issue takes a somewhat different form in each case. We will consider here the nominalism-realism issue with respect to laws, since in all four stages Peirce concentrated on the limiting law-like character of an evolutionary process.

In both Stage I and Stage II, Peirce treated universal laws (e.g., "if dry gunpowder is ignited it explodes") as contingent summaries of fact or Humean successions, expressible with material implication, and he treated probabilistic laws as assertions of *Actual* limiting relative frequencies. It is hard to decide whether to call this view nominalism or realism. It is more realistic than a relativistic or finitistic nominalism, but, as Peirce himself recognized in Stage IV, his theory of Stages I and II was less realistic than his later theory. For this reason I propose to call his view of law in Stages I and II *extensional realism*, as contrasted to a *modal realism* that he held in Stages III and IV.

Peirce's own terminology with respect to extensional realism was not consistent. In his review of Venn's *Logic of Chance*, he said that the actual limiting relative frequency view of probability was nominalistic. But in his review of Fraser's *Berkeley*, he stated the social-inquiry-limit definition of truth and reality and called it realism. Thus he came down on one side of the issue in the case of probabilistic laws, and on the other side in the case of universal laws!

In "How to Make our Ideas Clear" (Stage II), Peirce used material implication to symbolize a general law, analyzing the hardness of a diamond in terms of tests by scratching. He recognized that material implication did not capture counterfactual meanings, and said that it was an arbitrary matter of usage whether we say that an untestable diamond is hard or soft. But in Stage IV, he said that this earlier treatment of hardness was nominalistic and wrong, and that the correct view was realistic: The diamond's hardness was a potentiality or "would-be" (Peirce 1931, 5.403, 5.453, 5.457). At this stage he also held that probabilities were "would-be's" (*CCR*, sec. 8.4.2).

What I am calling Peirce's modal realism is the doctrine that both universal and probabilistic causal connections are dispositional. They are *"would be's"* and *"can be's"*, as contrasted to "the *will be's*, the actually *Is's*, and the *have been's* [that] only cover actuality" (Peirce 1931, 8.216–217). The idea of a "would-be" is a modal notion, so that the name "modal realism" for Peirce's realism of stages III and IV is appropriate. My own logic of causal statements and dispositional theory of empirical probability were stimulated by Peirce's modal realism (Burks 1951c, 1964; *CCR* chs. 3,4,6,7,8).

Thus the answer to the question "Did Peirce have a significant nominalist phase?" is that during Stages I and II he held the position I have called extensional realism. That position could be called nominalistic, as Peirce sometimes called it, but it asserts the reality of infinite limits, something a relativist or finitist about knowledge would deny, and so it is somewhat realistic.

Actually, the realism of Peirce's last two stages is a very much stronger doctrine than what I am calling modal realism. I will call it *final cause realism*, because the construal of Thirds as final causes is an essential component of it. For in *The Monist* series of 1891–93 and in subsequent writings, Peirce developed a theory in which Thirds operated as final causes, objective Platonic-Aristotelian forces guiding the evolutionary process toward an ultimate limit of "concrete reasonableness".

"Truth, crushed to earth, shall rise again." ... ideas are not all mere creations of this or that mind, but on the contrary have a power of finding or creating their vehicles, and having found them, of conferring upon them the ability to transform the face of the earth. (Peirce 1931, 1.217)

Peirce was here stating in flowery language his doctrine of agapism, or evolutionary love, according to which evolutionary processes were guided by objective values. I have discussed Peirce's final cause realism elsewhere and argued against it on the basis of my own logical mechanism (Burks, 1988b).

With this explanation of the metaphysical role that Peirce assigned to final causes in his later writings, we are ready to turn to the main question Moore discusses: Is Peirce's pragmatism of Stage II compatible with his modal realism of Stages III and IV? In Burks 1951c, I argued that his pragmatism and modal realism were not compatible, and this conclusion still seems correct to me.

Moreover, I think that Peirce recognized this problem tacitly in Stage IV. He broadened his semiotics greatly, and this implies a broader criterion of meaning than Stage II verificationism. Many of his arguments for the doctrines of Stages III and IV presupposed such a broader criterion of meaning. One example is his "neglected argument for the reality of God", which was an intuitive, abductive argument that had weight because, as a product of evolution, man's natural feelings should be given evidential weight in such cases (Peirce 1931, 6.452–493). Another example is his use of a design argument for the existence of God. Further, most of Peirce's arguments for modal realism and final cause realism were global, holistic, loose appeals to general evidence and explanatory power, not local, analytic, tight verifiability arguments. It is plausible to say that Peirce used a hierarchy of three criteria of meaning: intuitive, verificational, and general-explanatory. The meaning criterion of "How to Make Our Ideas Clear" was only the middle one of these three.

Thus Peirce had two successive theories of empirical meaning: a positivistic theory in Stage II and a broader theory in Stages III and IV. There is some similarity between Peirce's later theory and James' dual meaning criteria of scientific evidence and will-to-believe; and it is ironic that as he moved closer to James' pragmatism, Peirce renamed his view "pragmaticism" to distinguish it from James' form.

Let me turn now from history to analysis, and consider the new computer systems called "classifier systems" that we mentioned earlier in connection with computer simulation (p. 385). These systems are relevant to Peirce, because classifier systems capture some aspects of his ideas about concepts and pragmatic meaning, abduction (the logic of discovery), and some aspects of his evolutionary realism. On the other hand, these systems violate two of Peirce's basic principles. By being discrete, classifier systems violate his synechism (evolution is continuous in the strong sense of mathematical continuity). By being computational or logico-mechanical, classifier systems violate Peirce's agapism (evolution can be driven only by final causes).

Thus Peirce's views on evolution turn out to be an interesting mixture of the very advanced and the very traditional. On the one hand, he had a good understanding of Darwin's evolution, and saw how it could be extended backwards into pre-physics and forward to culture and science. He also had a good understanding of Babbage's analytical engine, and of its (limited) computing powers. But, on the other hand, Mendel's ideas had not yet been developed and the modern computer had not yet been conceived, so that there was no satisfactory mechanical or biological explanation of teleology. (Burks 1979a, 1984, and 1988b; Burks and Burks, 1988).

John Holland's *classifier systems* employ market economic principles for learning (induction) and genetic principles for discovery (what Peirce called "abduction"). For descriptions, see Holland 1975, 1985, 1986; Burks 1986c, 1988a; Holland and Burks 1987, Pending; Rick Riolo has written a general simulator. While these systems can be used for all kinds of computing, modeling, and simulation, it is convenient for our purposes to think of them as programs for intelligent goal-directed robots operating in problem-solving environments.

Input data, the results of internal computation, and output results are called *messages*. The instructions of a classifier system are conditional statements or inferential rules that react to messages that satisfy their conditions to produce new messages. Such an instruction is called a *classifier* because it divides the messages of the system into the general category of those that it accepts and the remainder. Note that a classifier is a universal statement, not a probabilistic one, though, as we shall see, probabilities enter into the processing of classifiers.

Conceptual generality is achieved computationally by using a richer alphabet for conditions than for messages. Thus if crimson is symbolized by "10" and burgundy is symbolized by "11", red can be symbolized by "1#", where # is satisfied by either 0 or 1. Similarly, "0#" could be used in a condition to represent two shades of green. In consequent terms of classifier conditionals, the "#" symbol is used to convey information from a message satisfying a condition to the newly created message; thus this message can combine old information with new.

A classifier system has a hierarchy of three levels: a basic classifier performance system, a market-economy learning procedure, and a genetic discovery procedure. The basic computation of classifiers is highly parallel, for at each major cycle every classifier is applied to every message to produce a new set of messages. Thus, at every cycle of robot computation, incoming messages and old internal messages are replaced by new internal messages and output messages.

Assume now that an environmental context is provided for the robot, a goal is defined, and the robot is programmed to recognize the goal. Because a classifier system is computationally universal, there will be a classifier program that will enable the robot to attain its goal. The learning procedure will enable the robot to improve its performance over successive runs.

In a basic classifier system, the size of the message set is unlimited; that is, every new message generated during a major cycle is carried over to the next major cycle, the old messages always being dropped. For the learning procedure, the size of the message set is limited, so that classifiers compete with one another to get their messages carried over for possible use at the next major cycle. This competition procedure is best explained as a recursion, with a general step for each major cycle and a terminating step when a goal is reached.

The selection of the messages to be carried over from one cycle to the next is made by means of a market-economy auction. Each classifier has a strength or capital. When a classifier produces a new message, it uses some of its strength to make a bid to have its message carried over. The auction system chooses (with some probabilistic variance) the messages with the highest bids and preserves them for the next major cycle. Each successful bidder then pays its bid amount to those classifiers that, in the preceding major cycle, produced the messages the successful bidder used to make its message. In turn, if the successful bidder's message is used in the next major cycle, the successful bidder will receive a payment or payments. Over successive runs, the strength of a classifier will vary according to its success and failure in producing messages that are used by other classifiers.

When the robot reaches its goal, a reward in the form of capital is distributed to those classifiers active at the last major cycle. Over repeated runs, these rewards gradually work back and increase the strengths of those classifiers that contribute useful messages earlier in the runs. Thus classifiers that produce messages that ultimately contribute to goal success increase in strength over successive runs, while classifiers that fail to make contributions gradually lose strength. This whole reward procedure is called the "bucket-brigade algorithm", because of the way rewards made on the successful completion of trial runs gradually work their way back to classifiers that contribute to the success early in the runs.

Thus the learning procedure introduces economic competition among classifiers. After repeated successful runs of a classifier program, the strength of a classifier is a measure of its contribution to the success of the whole program. The genetic procedure then extends this competition to include evolutionary competition, using the strengths developed by the learning procedure for this purpose.

The genetic procedure periodically chooses some of the weakest classifiers and replaces them with genetic combinations (offspring) of strong classifiers. Mutation and crossover are generally employed as the genetic operators. Mutation alters single letters, while crossover rearranges blocks of letters. When the blocks of letters that happen to be crossed represent meaningful concepts, this operation produces a new complex of concepts to be tested by the system.

The biological operations of mutation and crossover are, of course, probabilistic. Likewise, the auction system employs probability. Thus probability plays an essential role in processing classifiers. But, as we noted earlier, classifiers themselves are universal statements, not probabilistic statements.

The learning power of the bucket-brigade algorithm is limited by the potentialities inherent in the starting set of classifiers. The genetic algorithm removes this limitation by creating new classifiers from old ones. It can also generate new classifiers that are relevant to the environment by generalizing input and output messages.

The selection of internally generated messages for use in the system is based on the strengths of classifiers, and thus on their consequences over repeated runs. Likewise, the evaluation and selection of classifiers for reproduction and elimination is based on their consequences over repeated runs. Hence the system, like evolution, operates by an indirect logic, testing rules according to their consequences. Moreover, enumeration or repeated testing plays an essential role in these processes of learning and discovery. (Burks 1988a, secs. 3, 10).

Now we are in a position to see how classifier systems embody some of the key aspects of Peirce's analysis of concepts and his pragmatic theory of meaning. A classifier is a conditional statement or rule, very much like one of Peirce's "practical conditionals". According to pragmatism, the meaning of an empirical statement consists in its testable consequences; running a classifier system in an environment is a sophisticated way of testing the practical consequences of the classifiers in that environment. (pp. 399–405 above; CCR, sec. 4.2)

Classifier systems constitute a logic of discovery, of the evolutionary sort that Peirce was looking for under the name "logic of abduction" (Burks 1946c). However, it is clear from our analysis of evolution and our description of classifier systems that Peirce had neither the theoretical concepts nor the computing power needed to develop a logic of discovery that went beyond a corpus of heuristic advice.

Classifier systems illustrate an important feature of induction. As a classifier system evolves under the discovery procedure, the conditions and consequent terms of classifiers evolve along with their interconnections. Correspondingly, the empirical inquirer is searching for laws and theories that relate properties and functions to one another correctly. Hence inquiry must formulate causally or probabilistically relevant properties, in addition to establishing verified connections among them.

Put in terms of grammatical levels, inquiry must generate and validate useful concepts as well as promising hypotheses. The subtlety of concepts like *ellipse* for describing planetary motions or that of *heat* as a form of motion shows that creativity is needed not only for theories, but also for concepts. Induction by enumeration is important here too, for the more often a concept has worked in a given domain the more likely it is to have other applications (pp. 369, 399; Burks 1980a, sec. 4).

Finally, let me indicate briefly how classifier systems fit into Peirce's evolutionary realism. In Section 3.1, I outlined Peirce's theory of "generals" or universals, what he called "Thirds", in terms of a computer or robot using language to adapt to its environment. Thirds are generic universals of various interrelated kinds, occurring both in the mind and in nature: concepts and beliefs in the mind, properties and facts or laws in nature. Concepts are of properties, and beliefs are of facts and laws.

These sub-categories of Peirce's Thirds correspond to classifiers in several ways. The conditions and consequence of a classifier may express concepts, and the classifier itself may express a belief. Cooperating clusters of classifiers may express compound concepts or theories. The robot receives inputs from its environment and acts on its environment. Insofar as the robot is adapted to its environment, some of its concepts will match properties of the environment and some of its beliefs will be true of the environment. On the other hand, some classifiers play only an intermediate computational role in the calculations of the system.

We have been discussing Peirce's theory of Thirds or generic universals and their analogs in classifier systems. But the messages of a classifier system represent specific properties or universals, and they correspond to a different category in Peirce's system, that of a psychological First. According to Peirce, an immediate experience is a directly given specific quality (compare Kant's "given appearance"). Peirce's view here is close to the view of Santayana mentioned earlier, that an immediate experience is an "essence". Indeed, Santayana seems to have got this view from Peirce (p. 440; Fisch 1966, p. ix).

A Peircean Third is so-called because it consists of two or more specific qualities (Firsts) bound together by some rule. We will give two examples and their correlates in classifier systems.

A general color concept (Third) is a collection of specific colors (Firsts) bound together by a habit (rule or program). Peirce's doctrine here is derived from Kant, who said

Objects are *given* to us by means of sensibility and it alone yields us *intuitions*; they are *thought* through the understanding, and from the understanding arise *concepts*. (Kant 1781, A19)

To illustrate Peirce's construction of Thirds from Firsts, consider the color example given earlier. A specific shade of crimson ("10") and a specific shade of burgundy ("11") are both instances of a general concept of red represented by the condition "1#". This condition connects the two shades of red together according to the matching rule governing the generality operator "#".

Our second example of a Third and its correlate in classifier systems is that of a classifier itself. Consider a classifier with a single condition, which accepts each message satisfying its condition and transforms it into a new message. This classifier is a general rule defining a set of message pairs. In other words, it defines a message transformation, from a message satisfying its condition to the generated message, and thus expresses a law of message sequencing. In Peirce's terminology, it is a Third binding together all these pairs of Firsts.

We are now in a position to explain how classifier systems fit into or model Peirce's evolutionary realism. While Santayana was an evolutionary materialist, Peirce was an evolutionary panpsychist. Peirce held that mind is part of nature, and that mind-nature evolves probabilistically (tychism) and continuously (synechism) under the direction of Thirds qua final causes (agapism). As nature evolves, Thirds also evolve, by changing their membership of Firsts or of joint occurrences of Firsts; compare the traditional notion of "natural kind". Thus for Peirce natural properties and natural laws evolve together; moreover, as individual minds and the community of scientists adapt to an evolving nature, their concepts track natural properties and their beliefs track natural laws.

The analogy of this account of evolution to the operation of a classifier program is as follows. Consider a robot adapting to its environment under the direction of a classifier program. As the robot learns about and adjusts to its environment, its classifier program evolves under the direction of market economy and evolutionary discovery procedures. The strengths of its classifiers change, and so the sequences of messages generated change. Moreover, the classifiers themselves change: their conditions, their consequence terms, and the connections between these. Thus the classes of messages recognized and the classes of message sequences generated evolve as the whole system evolves. This process is analogous to the evolution of Thirds (general concepts and beliefs) in Peirce's account of evolution.

Furthermore, insofar as the robot adapts to its environment, both the conditions in its classifiers and the transformations made by these classifiers evolve to adaptive representations of the environment. Finally, when a classifier robot adapts to an evolving environment, its classifier conditions track natural properties and its classifiers track natural laws and other associations of environmental properties.

I have now concluded my discussion of Peirce's philosophy. Let us turn to Soshichi Uchii's paper on my logic of causal statements (*CCR*, chs. 6 and 7). Since probability plays such an essential role in classifiers, this shift may seem like one from the probabilistic to the causal or deterministic. But it should be kept in mind that a computer, when operating correctly, is a deterministic system, and the randomness needed is obtained in the form of pseudo-randomness (*CCR*, sec. 9.4.3).

Uchii compares my logic of causal statements (*CCR*, chs. 6 and 7) with some later logics for counterfactual conditionals, mainly those of Robert Stalnaker and David Lewis; I am pleased, of course, that on his analysis my theory comes off best. Uchii then suggests his own theory, which is an improvement on mine.

Uchii covers both formal and informal (application) issues. This is important, because in all formalization there is a choice between what is to be formalized and what is better treated informally. His paper constitutes a most excellent analysis of the problem of causal counterfactuals and an important contribution to its solution. I regret that his many papers in Japanese have not been available to most other writers on the subject.

Although formal logics for causal statements are intended as models for human causal and counterfactual usage, an important ultimate test of them will come when robots are made with sufficient intentional power and independence that they will have need for causal reasoning. The best way for me to comment on Uchii's formal contribution is to outline informally the kinds of situation I had in mind when constructing the logic of causal statements. My earlier comments on intentionality and free choice are relevant here (secs. 3.2 and 3.3), as is also my rule of completeness for causal subjunctives cited by Uchii (p. 192 ff).

The subjunctives of science and engineering are used to express claims concerning how a system would behave if it were started in some state, perhaps one different from its actual initial state. The speaker considers a system defined by states and laws. Each allowable starting state will produce a history, or a probabilistic tree of histories in the stochastic case; and each history is a possible world. Hence the logic of subjunctives is a modal logic. Note that the formal logic of subjunctives is essentially the same for a scientist attempting to predict the behavior of a system or understand its functioning as it is for an engineer attempting to achieve a certain goal.

In reflecting on how to obtain a certain kind of final state, the analyst may imagine changing only part of a starting state. In doing so, he may produce a causally impossible state, so that his subjunctive remains ambiguous until he specifies enough additional changes to make the starting state causally possible. This situation often arises in ordinary-language subjunctives. Lewis Carroll's barbershop paradox illustrates this point (*CCR*, sec. 7.4.2).

Because engineering adaptation and empirical science have evolved gradually out of commonsense knowledge and adaptation, our model of how subjunctives function in engineering and science applies also to ordinary discourse. Consider an intentional, free being B (human or robot) operating in an environment, having a local goal, and on the basis of its knowledge constructing a causal model of a relatively isolated, causally self-sufficient, local space-time subsystem of this environment that encompasses the possibility of the goal. Call this subsystem S. The being B envisages various choices it might make with respect to S and traces their consequences. For B to carry out a choice is for it to control a set of states of the subsystem S; thus it is using causal subjunctives and counterfactuals in its planning. Also, if another being reviews these choices after the fact and evaluates the wisdom or morality of the first being's choice, the second being is employing causally modal statements.

The system S is, of course, an embedded subsystem of the environment (cf. sec 3.4). As with embeddings generally, the set of allowable starting states should be restricted so that the embedded subsystem will be causally complete. If such is not the case, the embedded subsystem should be enlarged to include enough causally relevant factors (states and/or laws) to make it complete. The dictator example that Uchii mentions illustrates this point.

An example suggested by my late colleague, Charles L. Stevenson, shows how competition requires one to consider a much more complicated subsystem. Take the counterfactual "If he had pushed the switch, the light would have gone on". Normally, if you push a light switch you cause the on-off state of the light bulb to reverse. Thus a subsystem S based on the macrostates of the switches, wiring, light bulb, etc., is an adequate model for that counterfactual. But if the light is controlled by two switches and there is a person at each switch trying to thwart the other, then the causal consequence of each person's action depends on the micro-states of both of them, as well as on the macro-states of the switches, wiring, light bulb, etc. In the competitive case, then, a much deeper and more complicated subsystem S' must be used to evaluate the truth-status of the counterfactual "If he had pushed the switch, the light would have gone on".

In the ordinary use of language, we simplify discourse and thought by talking about common or standard situations (subsystems), not spelling out their characteristics unless a need arises to do so. This feature of ordinary usage is relevant to the symbolization of counterfactuals. Thus if there are special factors involved in a situation, as in the dictator case, they need to be made explicit. This latter entails a shift to a more comprehensive and complete subsystem, analogous to the shift that occurs when a default condition occurs in the use of computer software. For reasons of efficiency, the software is written to handle the most common cases; when an exception occurs, the user is asked to specify an additional condition (thereby specifying a more complex subsystem) before the software proceeds.

Thus in modeling discourse one should distinguish levels of discourse, very much as one should distinguish levels of structure in an organic system. The old problem of existential import illustrates the same point. When I was a graduate student, philosophers argued endlessly over whether or not "All S are P" implies the existence of S's. But in the standard case, when "All S are P" is asserted there are S's, and if there are not, the proper response to "Is every S a P?" is "There are no S's". If you ask "Are all the apples on the sideboard green?" and I find no apples there, I don't say "true" or "false", but "there are no apples there". I wonder if some of the problems of symbolizing subjunctives discussed by Uchii, such as whether elliptical causal implication (ec) is exportable, might involve, at least in part, levels of discourse.

Here is another case where an informal rule plays a key role in the meaning of a subjunctive. I agree with Uchii that the most important informal rule is a rule of completeness: in planning and evaluating goal-directed conduct, one should model the simplest complete causal subsystem that is adequate for the goal and the environment in which the goal is sought.

4.2 Evolution and the Presuppositions of Induction

The essays by Clendinnen and Rosenkrantz criticize my presupposition theory of induction. That of Skyrms does also, but in a different way, and it works out best to discuss his paper in connection with probabilistic laws. These critics and others may well wonder how my rather Kantian treatment of the fundamental concepts and principles of inductive logic can be reconciled with my view that biological evolution is a logical- mechanical process; this section is a good place to discuss that problem (pp. 470–474).

My presupposition theory of induction is a complex theory developed and defended over several chapters of *Chance, Cause, Reason* and completed in Chapter 10. It embodies three conceptual systems and associated doctrines: a pragmatic analysis of inductive probability and utility (ch. 5); a modal logic of causality (chs. 6 & 7); standard inductive logic (sec. 3.2) and a suggested extension of it to include causality (sec. 10.3).

The presupposition theory holds that man's inductive reasoning is based on an innate structure-program complex that includes the concepts of inductive probability ($\mathbf{P}$) and causal necessity ($\Box^c$) as well as rules for the application of these concepts. The concept of inductive probability has two basic aspects. First, it is governed by the rules of the calculus of inductive probability and of standard inductive logic, and, second, it involves a pragmatic commitment to apply these formal rules in conditions of uncertainty.

The concept of causal necessity has the formal properties of the logic of causal statements. Its application in inductive reasoning rests on standard inductive logic and presupposes three synthetic propositions about the general structure of nature. The first is the *causal uniformity principle*: If a causal connection between non-indexical properties holds in one region of space-time, it holds throughout space-time (*CCR*, p. 572). The second is the *causal existence principle*: Some space-time region is governed by quasi-local laws (*CCR*, p. 577). The third is the *limited variety principle*: The laws of nature are based on a finite set of non-indexical, monadic properties, each requiring a minimal space-time region for its exemplification (*CCR*, p. 633).

Taken as a whole, *Chance, Cause, Reason* constitutes an argument for the presupposition theory. One of the most basic lines of argument in it, the line most relevant to Clendinnen's comments, begins with the presentation of alternative formal models of inductive logic (sec. 3.2) and proceeds to what I call Hume's thesis that there is no non-circular justification of induction (sec. 3.3). Three typical alternative inductive logics are: standard (the rule of induction by simple enumeration is applied), random (the successive states of nature are treated as independent of each other), and inverse (the more often something happens, the less likely it is expected to happen). Each of these inductive logics is deductively consistent. Their existence establishes Hume's thesis that there is no non-circular justification of induction.

Suppose one has a finite corpus of observations about frequencies and wishes to make a probabilistic prediction or infer a probabilistic generalization. Because deduction does not yield the desired result, inductive inference is needed. But there are alternative inductive logics, and the probabilistic conclusion one draws depends on the logic used. Moreover, each of these inductive logics is deductively consistent, so that there is no deductive justification for using one rather than another. Finally, an inductive justification would be circular.

The inquirer will actually use standard inductive logic or a variant of it (see also my discussion of enumerative and eliminative induction in sec. 2.2 above). In simple models like Carnap's, standard inductive logic is a special case of Bayes' method of finding an unknown probability (*CCR*, sec. 3.2.4). But since the quantity of data from which the inference is made is finite, the predictions made will depend critically on the prior probabilities assumed. Human inquirers agree roughly in their probabilistic predictions, reflecting the fact that they instinctively assign prior probabilities more or less in accord with standard inductive logic. Thus from a foundational point of view the use of standard inductive logic presupposes some strong assumptions.

With this background, I am ready to discuss F. John Clendinnen's essay "Presuppositions and the Normative Content of Probability Statements". Although I am in considerable disagreement with it, it is a very interesting essay, and I appreciate his detailed study of my work.

Clendinnen focuses on the problem of the relation of truth to normative force in the three domains of mathematics, inductive logic, and morality. What I call "logical truth" (sec. 2.4) may be taken as a starting point for the statement of this problem. The truths of mathematics and logic clearly do have normative force. After combining five objects with seven objects, if one believes that the usual conditions of distinctness and continuing existence are satisfied, one *should* believe that there are twelve objects. Any *rational* person would believe so. Moreover, the truth of "5 plus 7 equals 12" is absolute and non-relative. In pure logic and mathematics, everything follows from a contradiction, and rational discourse breaks down.

The problem of the relation of truth to normative force in the domains of inductive logic and morality can now be put in this form: To what extent, and in what way, does the normative force of deductive logic and mathematics carry over to the truths of inductive logic and of morality? There are two questions here, which can be stated in parallel form: Are there logically consistent alternative inductive logics, and if so what do they look like? Are there logically consistent alternative moralities, and if so, what are they like?

Let us consider the inductive logic problem first. In his *Logical Foundations of Probability*, Carnap held that the normative force of deductive logic carried over to statements of inductive logic and gave them normative force. My alternative models of inductive logic, together with Hume's thesis that there is no non-circular justification of induction, show that this is not so (*CCR*, sec. 5.7.1). It follows that atomic inductive probability statements are not logically true-or-false.

I used this result against Carnap in my article in the Carnap volume (Burks 1963). Clendinnen suggests that Carnap accepted my criticism, but I am not sure he did. Carnap did change his position, but his new position is unclear; in any case, he gave no answer to my arguments for Hume's thesis (Carnap 1963, pp. 980–83). Moreover, the traditional method of logical formalization, which Carnap was using (and to which he contributed greatly), is inadequate for this type of problem. Though this method is often useful, it is not suitable here, because the normative force of inductive statements is pragmatic in nature and so involves the voluntaristic aspect of human nature.

The view of the ordinary language philosopher is that standard inductive procedures are reasonable by virtue of the meaning of "reasonable". Clendinnen correctly points out that that this argument is question-begging (p. 212). Standard inductive logic is the norm underlying the ordinary use of inductive expressions; for example, "probable", "likely", "confirm", and "evidence" (*CCR*, p. 136). Consequently, ordinary inductive probability assignments tend to correspond to those of standard inductive logic. But there is no analytical or deductive connection between the probability value obtained by applying an inductive rule and a commitment to *behave* in accordance with that probability value.

There is a similar connection in ordinary language between moral terms and "rational". Kant held that keeping one's promises is the rational thing to do. According to various ethical theorists it is *rational* to maximize one's own happiness, to seek perfection, to work for the greatest general good, or to abide by moral rules. Whatever the specifics of the matter, I think these views reflect the close connection in ordinary language usage between morality and rationality. But again, there is no deductive connection between the action enjoined by a moral rule and a commitment to behave accordingly.

Clendinnen raises two interrelated questions on this topic that I should answer. He asks whether an ethical *ought* can be derived from an ethical *is* (p. 213). And he asks me to explain what I mean by saying that inductive probability statements are normative (pp. 209, 217). Both questions involve the relation of norms and commitment, on the one side, to relevant facts and reasons, on the other. In my opinion the "ought" cannot be derived from the "is" in either case, but the explanation is somewhat different in the two cases. I will postpone the first question (to sec. 4.4, where general value issues are discussed) and answer the second question here by quoting from *Chance, Cause, Reason*.

According to the pragmatic theory, a system of atomic inductive probability statements has two interrelated but coequal aspects, an a priori aspect and a pragmatic aspect. The a priori aspect concerns the assignment of probability values to statements by formal rules. Probability values are and should be assigned according to the rules of the calculus of probability and standard inductive logic. Consequently, both individual atomic probability statements and systems of them have logically true-or-false components.

The pragmatic aspect of inductive probability concerns the application of these probability assignments. An atomic inductive probability statement does and *should* express a disposition to act in certain ways in conditions of uncertainty. A system of such statements does and *should* express a commitment to make one's probability assignments conform to the calculus of probability and standard inductive logic. The calculus of choice explicates the concepts of probability and utility as they are and *should be* reflected in the choices of a single subject. (*CCR*, p. 319)

The normative terms in this quotation have been italicized here, and Clendinnen's second question amounts to a demand for an explanation of what they mean. The calculus of choice is the foundation, since it contains the calculus of probability as an embedded subsystem. The rationale of the "ought" for these two systems is an extension of the rationale for deductive consistency, the value of employing a sufficiently complete formal system that enables one to establish interconnections by reasoning. If one makes a sufficiently complete set of choices or probability judgments, violation of the rules of these two formal systems may generate intransitive preferences, or a paradox of choice such as the sure-thing and variance paradoxes; it may even result in one's becoming the victim of a gambling "book".

The explanation of the use of "ought" in connection with standard inductive logic is somewhat different, and it was misleading of me to use "ought" in this context. As an instance of ordinary language usage, it was correct (*CCR*, pp. 136–137); it would, of course, been wrong to say "ought not". But I should have stayed on the level of philosophical theory and not descended to the practical level of ordinary normative discourse. For, according to the presupposition theory of induction, the human reasoner has a basic commitment to employ standard inductive logic, that is, an inductive logic that includes the rule of induction by simple enumeration (suitably qualified). By Hume's thesis, there is no non-circular justification of this commitment. "But standard inductive logic needs no justification. Indeed, our faith in it is so strong that no argument could cause us to abandon it" (*CCR*, p. 136). This commitment to inductive logic has an innate basis, concerning which we will say more when we discuss Rosenkrantz's paper.

Let us consider next Clendinnen's own position on the matter of justifying induction. He seems to accept my refutation of Carnap's a priori theory of inductive probability and also my refutation of the attempted justification of induction by an appeal to ordinary language usage. Now these refutations were based on Hume's thesis that there is no non-circular justification of induction. For if there were such a justification it would constitute a convincing argument for the a priori theory of inductive probability, and the ordinary language philosopher could simply say that ordinary usage reflects this justification.

Yet Clendinnen rejects Hume's thesis. He says that without a justification inductive language assertions lack normative force. Thus he finds my "account of the nature of the pragmatic component [of inductive probability] unsatisfactory" on the ground that "it provides no adequate basis for a normative interpretation of probability statements" (p. 209). But he is assuming that every normative statement needs a justification, even the most fundamental ones. In contrast, my view is that "the usual procedure for showing that one behavior pattern is more beneficial than any other involves the use of standard inductive logic . . . as a standard of evidence and probable predictive success", and that any attempt to justify this use is circular and question-begging (*CCR*, p. 134).

Clendinnen offers a justification of standard inductive logic (pp. 210, 218, 229–230). He argues, as did Hans Reichenbach, that we must assume the uniformity of nature because otherwise we could make no predictions. He means here the strong (extensional) uniformity of standard inductive logic (which includes enumerative induction), not the weaker uniformities of random and inverse inductive logic (and not the causal uniformity of my causal uniformity principle).

But one can make inductive predictions without using standard inductive logic. The existence of random inductive logic, inverse inductive logic, and other non-standard inductive logics shows this. I argued this point in considerable detail in *Chance, Cause, Reason* (sec. 3.3.1), and Clendinnen does not say how he would attempt to counter my arguments.

Clendinnen's type of argument was first advanced by Peirce and so is appropriately called the "Peirce-Reichenbach justification of induction". My own detailed rejection of it, which is also based on alternative inductive logics and Hume's thesis, is given in *Chance, Cause, Reason*, Section 3.4.2. Peirce put forward his justification of induction during Stage II of his thought, when he was an extensional realist and employed a verifiability theory of meaning (see sec. 4.1, above). I do not think Peirce's justification comports well with the quite general criteria of meaning he relied on in Stage IV of his thought, after he had become a modal realist.

I will add two comments relevant to Clendinnen's attempt to justify induction, and then move on to Rosenkrantz's attempt. These comments are on the alternatives to standard inductive logic. This logic uses the rule of induction by simple enumeration, and thus extrapolates the observed relative frequency of two properties occurring together. In contrast, random inductive logic treats the successive temporal states of nature as a random sequence. And inverse inductive logic employs an anti-enumeration rule: the probability of two properties occurring together tends *away from* the observed relative frequency with which they have occurred together in the past.

The first comment involves a comparison of standard and random inductive logics. Random inductive logic does not involve memory, for it makes what has happened in the past probabilistically irrelevant to what will happen in the future. But nature, as phenomenologically given, displays a great deal of uniformity over time as well as space. A defender of the Peirce-Reichenbach justification of induction might argue that the occurrence of such a highly organized and regular appearance is not accidental but must be accounted for, and the simplest explanation is that it results from the uniformity of the reality (i.e., nature) that produced the appearance. However, such a defense is question-begging, for the concept of "accident" is probabilistic, and its application here requires an inductive logic. On standard inductive logic, the extremely improbable does sometimes happen. On random inductive logic, the highly organized appearance of nature that one has is the result of chance!

The second comment involves a comparison of standard and inverse inductive logic with respect to infinite extension. An inductive logic model is constructed from properties (universals) and particulars (space-time events or individuals). A model of standard inductive logic can be based on either a finite or an infinite number of individual entities. But a model of inverse inductive logic can have only a finite number of individuals. The reason for this is worth looking into.

An infinite model of standard inductive logic is equivalent to a Bayesian inductive logic of the following sort. Bayes' method of finding an unknown probability applies to a mixture of independent processes. It consists in making repeated observations or performing successive experiments and extrapolating the observed relative frequencies. By what I call the "confirmation theorem", this method will eventually converge on the correct result. Expressed informally, this means that an investigator who repeatedly applies Bayes' method of finding an unknown probability to a *mixture of independent processes* will ultimately find the truth. In essence, Peirce and Reichenbach were presupposing the applicability of this Bayesian model to inductive inference when giving their justification of induction (*CCR*, sec. 2.5.3).

As we said, a standard inductive logic based on an infinite number of particulars is equivalent to Bayes' method of finding an unknown probability. This equivalence follows from de Finetti's theorem about exchangeability. An inductive logic is "exchangeable" when the probability of a statement about some individuals does not depend on what *particular* individuals are referred to by the formula. DeFinetti's theorem is: A complete probability assignment to an *infinite* process is exchangeable if and only if that process is a mixture of independent processes. Clearly, a standard inductive logic for a model with an infinite number of individuals makes an exchangeable complete probability assignment to an infinite process, and so a standard inductive logic with infinitely many individuals is equivalent to a mixture of independent processes.

It also follows from de Finetti's exchangeability theorem that an inverse inductive logic is necessarily finite in extension. For this logic employs an anti-enumeration rule, whereas any exchangeable process follows an enumeration rule (random inductive logic being a limiting case). An inverse inductive logic can be defined for any finite number of individuals, but its assignment of probabilities to universes is not extendable when more individuals are added to the model. Hence an inverse inductive logic cannot be defined for a model with infinitely many individuals (*CCR*, sec. 3.2.4).

The reason for this difference between inverse inductive logic and standard inductive logic is rooted in the very nature of randomness. By the law of large numbers, random universes become relatively more frequent as the number of individuals increases, and in the long run random universes prevail. As Skyrms points out (p. 287), almost all infinite random sequences have a limiting relative frequency.

Now standard inductive logic assigns higher probabilities to uniform universes than to random universes, and so is applicable to infinite models as well as to finite models (cf. p. 370 above). But inverse inductive logic assigns higher probabilities to random universes than to uniform universes, and so is applicable only to finite models, not to infinite models. It follows from these considerations that when Peirce and Reichenbach assumed an infinite model with a limiting relative frequency, they were already making a foundationally significant uniformity assumption.

Peirce and Reichenbach also argued that if there is no limit then there is nothing for scientific method to accomplish. But this argument is refuted by the existence of alternative inductive logics for finite models. For one can make inductive predictions with any of these models. Moreover, the scientist works with finite data and the applications of science are to finite cases. Hence, from a pragmatic point of view, inductive inference is finite and finite models are sufficient to prove Hume's thesis that there is no non-circular justification of induction.

Let us turn now to Roger Rosenkrantz's article, which also deals with my presupposition theory of induction. After providing a good short summary of my theory and the role that cellular automata models play in its formulation, he makes several specific criticisms of it. He then offers a justification of induction that, if valid, would render my presuppositions of induction superfluous.

I wish to thank him, not only for this paper, but for his careful review of my *Chance, Cause, Reason* (Rosenkrantz 1982). I will take up his specific criticisms first. His putative justification of induction is based on the claim that there are objective prior probabilities, and it makes a natural transition to the consideration of empirical probabilities in the next section.

Rosenkrantz is doubtful that my causal model of standard inductive logic will work, saying it "has a slight flaw. The different causal transition laws are not mutually exclusive" (p. 240). But the model is not organized in terms of causal laws; it is organized in terms of *possible causal systems*, and these ARE mutually exclusive (*CCR*, p. 626).

My causal model of standard inductive logic is summarized above (pp. 372). The model is admittedly speculative. However, as noted earlier, Uchii's papers on it contribute substantially to testing it, and as far as they go they show that it is workable (p. 373). I do not know of a better explanation of the difference between a universal statement about nature that is causally true and a universal statement that is contingently true (*CCR*, p. 617).

Rosenkrantz (p. 241) also argues – as does Skyrms (p. 313) – that my system cannot distinguish between the confirmation of a contingent universal and the confirmation of the corresponding causal universal. Scientists obviously make this distinction; the question is whether my causal models of standard inductive logic make it. Let me give an intuitive explanation of how they do, indeed, make this distinction.

The examples of contingent and causal universals that I cite (*CCR*, p. 617) are:

> (1) All objects that were released at a certain time and place did in fact fall [It is also the case that all the objects released were heavier than air],
>
> (2) All released objects heavier than air fall.

Consider now a causal model of standard inductive logic, and focus on the ACTUAL causal system of that model. Statement (2) is true in all universes of the actual causal system. In contrast, (1) is false in many universes of the actual causal system, even though it is true in the actual universe of this causal system. This difference between (1) and (2) induces a difference between the prior probabilities of (1) and of (2), which in turn is reflected in a confirmation difference between these two statements.

Furthermore, it seems to me that Rosenkrantz's alternative explanation of the difference between causal truth and contingent truth does not succeed. He presents his explanation in terms of the Ptolemaic-Copernican controversy (p. 242). But consider a Copernican orrery with the planets moving around a fixed earth along paths that are circles on circles on circles, etc. Now grasp the orrery by the earth and lift it off the table. It becomes a Ptolemaic orrery! These two orreries are formally equivalent and so have the same logical consequences. This defeats Rosenkrantz's explanation. It is true that the two orreries differ in simplicity and heuristic value (cf. *CCR*, p. 20). But the notions of simplicity and heuristic are not helpful here, for they are as complicated and problematic as the distinction between causal and contingent truth.

Rosenkrantz makes two comments on the holism-foundationalism issue that seem to me to point in opposite directions. In his review, he says "A suspicion dawns that the approach of the presupposition theory is altogether too holistic" (p. 332). Now, it is true that my presuppositions have a global character. Nevertheless, they constitute a *foundational* basis for inductive reasoning, somewhat analogous to logic as a foundational account of deductive reasoning.

On the other hand, in his paper for the present volume, Rosenkrantz suggests that the presupposition theory of induction "bars the road to inquiry", because the presuppositions cannot be criticized (p. 245). But to demand evidence for the presuppositions is to make a holistic-coherence demand that is not compatible with the presupposition theory. The basic argument I gave for that theory was that inductive inference presupposes an inductive method (*CCR*, sec. 3.3). If that is so, then the presuppositions cannot be linearly criticized by means of the inductive method they underlie.

Since Rosenkrantz has touched on the basic contrast between atomistic-foundational and holistic-coherent approaches to metaphysics and epistemology, let me comment on this contrast.

The absolute idealisms of Hegel and his followers, Henri Bergson's vitalism, and relativistic pragmatisms are paradigmatic examples of holistic, coherence philosophies. Greek atomism, Hume's philosophy, Kant's *Critique of Pure Reason*, and early logical positivism are good examples of foundational, reductionist philosophies. Both my presupposition theory of induction and my philosophy of logical mechanism are foundational, atomistic, and reductionist.

I have already defended my atomistic-foundational philosophy in connection with the contributions of Suppe, Cohen, Thompson, and Arbib. The concept of empirical truth is defined in terms of observation and experiment, and realism is analyzed in terms of adjustment to the environment. My theory of empirical truth and inductive verification is hierarchical, resting on many distinctions that do not play basic or significant roles in holistic-coherence theories: learning a language versus understanding it, logic of discovery versus logic of argument, and descriptive and explanatory versus normative and evaluative.

Holistic-coherent systems, whether actual or symbolic, have two distinguishing characteristics. These are: circularity (the whole depends on its parts), and mutual supportiveness (two or more constituents cooperate). Let us look at them from both a theoretical and a practical point of view.

Circularity and mutual influence seem incompatible with atomic foundationalism. But in principle, a holistic-coherent system can be reduced to a hierarchical system with feedback (Burks, 1988a). When this is done, the circularity in the system becomes temporal cyclicity, for though a feedback path is circular, transit around it involves delay. Similarly, mutual supportiveness involves non-linear interactions among the parts.

According to the philosophy of logical mechanism, all basic scientific knowledge is atomistic, foundational, and reductive in character. But discovery, understanding, applications, and human design procedures are matters of practice, and for these, holistic explanations and coherence-type arguments are necessary modes of discourse. They are required for gaining general knowledge of complex, non-linear systems, and they are especially important in dealing with teleological systems.

To understand or design a complex system one must work at many levels, and also back and forth between levels, sometimes working from the bottom up and sometimes from the top down. One proceeds by successive approximations. This is especially true of a holistic or coherent system, because of the mutual interdependence of whole and part. But it is true for any complicated system, even a formal axiomatic system, and reflects how the mind copes with complexity.

Embedded subsystems play an essential role in understanding complex non-linear systems, including holistic and coherent systems. An embedded subsystem is simpler than its underlying system, and is thus more useful for understanding and obtaining detailed knowledge of a system. The underlying system is more complex, usually too complex to yield detailed knowledge of the whole system. But the underlying system constitutes the precise reference standard against which models must ultimately be tested, and its atomic basis is often simpler than that of the embedded subsystem. For example, the human mind is very complex, yet in principle it is equivalent to a finite automaton, which may be constructed from logical *nan*'s (or *nor*'s) ـnd erasable bit memories.

Let us consider an example of a holistic system. What is the function of the heart? Its function is to pump blood through the circulatory system and thereby transport chemicals throughout the organism. In so doing, the heart contributes to the functioning and continued existence of the organism *and thereby to its own continuing existence.* The heart is an organ of the organism, and it depends for its operation not only on the operation of the whole organism but even on the chemicals it transports directly to itself. This explanation is self-referential, but it is not circular in a harmful way.

This request for a functional or teleological explanation presupposes a whole system of which the heart is a part, and the answer to the request relates the part to a larger subsystem (the circulatory system) and that in turn to the whole system (the organism). The answer also covers the feedback from the heart to itself via the circulatory system. Thus this holistic explanation involves the circularity of feedback and is relative to the purpose of the whole system.

Moreover, since there are alternative ways of achieving a certain function or goal, the functional question about the heart leads naturally to more general questions about the evolutionary design process. Why did heart-using organisms arise? What alternative subsystems are there for performing the function of the heart and circulatory system? Did evolution produce any of these? Under what circumstances might it have produced still others?

Here one is looking at the design of a hierarchical system in terms of alternative subsystems at various levels. It calls to mind an engineer designing a radically new type of automobile by combining one novel subsystem with many old subsystems, except that evolution works with a continuity design constraint. In both cases, the suitability of the design is tested in an environment. In the next section we will raise another evolutionary design question, one about the origin of sexual reproduction.

Let us now turn to the evolutionary basis of inductive reasoning. As Rosenkrantz says, it would be "viciously circular to claim inductive support for the presuppositions of induction" (p. 243). However, it is not circular to employ standard inductive logic to gather empirical evidence and draw inductive conclusions about biological evolution and the nature of the human species. This does involve a transition from the inner to the outer point of view, but there is no logical incoherence or vicious circularity in this transition. Rather, it is more like a case of self-reference, such as a Gödel formula referring to itself or a von Neumann self-reproducing automaton containing a description of itself.

We humans stand out in evolution for our conscious goal-directedness, language use, and sociality – together with the institutions, technologies, and cultures associated with these capacities. Human capacities have developed because of their adaptive value (Burks 1979a, 1984, 1986a). A person is born with the potential to develop these capacities under nurturing and education over a time span that is short compared to a normal life span. Thus a person begins life with an *innate structure-program complex* "which is the physiological basis of his innate information capacities, and as he learns from experience he acquires a new structure-program complex", his *acquired structure-program complex* (*CCR*, p. 612).

Consider first the inductive portion of man's acquired structure-program complex, and then its innate basis. My pragmatic theory of inductive probability holds that humans are disposed to reason inductively more or less in conformity with the rules of the calculus of inductive probability and standard inductive logic (*CCR*, p. 33). My presupposition theory of induction adds that humans reason in terms of causal models of induction based on the presuppositions of limited variety, causal existence, and causal uniformity (p. 372 above; *CCR*, sec. 10.6).

The presupposition theory of induction goes further and says that one's acquired capacity for inductive reasoning has an innate basis that involves a priori concepts of probability and causal necessity and the presuppositions of induction. This does not mean, of course, that a human is born with these explicit ideas and beliefs. Rather, the innate structure-program complex contains materials from which human inductive habits and practices develop as a result of training and interaction with the environment.

The general heredity-environment question is: How much human learning is owing to the inborn human learning system, and how much is owing to the educative contributions of environment and culture? Neurophysiology and computer simulation are a long way from being able to make a scientific study of man's innate structure-program complex. But, according to the philosophy of logical mechanism, a human is equivalent to a finite automaton. It follows that how the mind develops and functions is an empirical matter.

Kant's critical philosophy had two parts, one epistemological and positive, the other a critique of metaphysics. The first treated of what the inner point of view tells us about the nature of our minds, the epistemological apparatus presupposed in scientific inquiry. Kant held that the mind had an innate structure of categories and basic principles, including categories of causality, and that this structure was the logical foundation of science. My own view of causal necessity owes much to Kant, but Kant had no adequate concept of probability or inductive logic. Thus my presupposition theory of induction is a substantial modification and extension of Kantianism. (*CCR*, secs. 10.5, 10.6)

Kant's critical philosophy denies that there is any metaphysical explanation of the origin of mind, and leaves such issues to faith. Thus he believed that the innate structure of the mind had no natural explanation, that is, could not be fitted into the scientific picture constructed by the mind on the basis of empirical data. Here I depart from Kant completely, for I hold that the structure of the human mind has a natural and empirical explanation that can be found by using scientific method and the outer point of view. Kant took classical physics as a model of science, which is not adequate for explaining the origin and operation of mind. Darwin's theory of biological evolution, modern genetics, and computer logic are needed for this.

Peirce, influenced by Darwinian evolution, was the first to advance an evolutionary explanation of the origin of mind, and my ideas have been influenced by his. Peirce thought that people were born with general concepts and abilities that facilitate the learning process, including rough concepts of space and time, a feel for the nature of physical forces, and the ability to suggest plausible hypotheses about natural laws. He believed that these native endowments were the product of biological evolution.

... the mind of man is strongly adapted to the comprehensions of the world ... [so] that certain conceptions, highly important for such a comprehension, naturally arise in his mind....

How are we to explain this adaptation? The great unity and indispensableness of the conceptions of time, space, and force, even to the lowest intelligence, are such as to suggest that they are the results of natural selection. ... as that animal would have an immense advantage in the struggle for life whose mechanical conceptions did not break down in a novel situation (such as development must bring about), there would be a constant selection in favor of more and more correct ideas of these matters. (Peirce 1931, 6.417–418)

The human ability to guess plausible hypotheses is an example of this evolutionary adaptation.

We are therefore bound to hope that, although the possible explanations of our facts may be strictly innumerable, yet our mind will be able, in some finite number of guesses, to guess the sole true explanation of them. (*Ibid.* 7.219)

... man's mind has a natural adaptation to imagining correct theories of some kinds. (*Ibid.* 5.591)

We ought to give high authority to natural, instinctive conceptions of the mind, so far as they are of any practical utility. For natural selection, or whatever the principle of evolution may be, is there to adapt them to the welfare of the species. (*Ibid.* 7.409; see also 1.630, 2.753, 5.591, 6.416)

For Peirce, instinct is the starting point of learning and inquiry. In early life this innate knowledge is reflected in adaptive responses to the environment, and as the child evolves into an adult these adaptive responses develop into commonsense knowledge. Similarly, as culture and civilization have evolved, commonsense knowledge has gradually developed into scientific knowledge.

Side by side, then, with the well established proposition that all knowledge is based on experience, and that science is only advanced by the experimental verifications of theories, we have to place this other equally important truth, that all human knowledge, up to the highest flights of science, is but the development of our inborn animal instincts. (*Ibid.* 2.754)

When we theorized about molar dynamics we were guided by our instincts. Those instincts had some tendency to be true; because they had been formed under the influence of the very laws that we were investigating. But as we penetrate further and further from the surface of nature, instinct ceases to give any decided answers.... (*Ibid.* 7.508)

(See also IBID. 1.628–649, 2.753–754, 5.586, 5.591, 6.50, 6.476, 6.494–504, 7.38, 7.381, 8.97–99, 8.223.)

George Santayana held a similar view about mankind's belief in induction. He thought this belief was rooted in instinct or "animal faith", and called one of his books *Scepticism and Animal Faith*.

The evidence of data is only obvious; they give no evidence of anything else; they are not witnesses. ... If I hypostasise an essence into a fact, instinctively placing it in relations which are not given within it, I am putting my trust in animal faith, not in any evidence or implication of my actual experience. I turn to an assumed world about me, because I have organs for turning, just as I expect a future to reel itself out without interruption because I am wound up to go on myself.

Complete scepticism is accordingly not inconsistent with animal faith; the admission that nothing given exists is not incompatible with belief in things not given. I may yield to the suasion of instinct, and practise the arts with a humble confidence, without in the least disavowing the most rigorous criticism of knowledge or hypostatising any of the data of sense or fancy.... (Santayana 1923, pp. 99–100, 105)

Santayana did not work out the structure of the innate aspect of human inductive reasoning, nor did he consider the probabilistic aspect of induction. Peirce said more on the subject, enough to show the influence of Darwin on his thinking. My own formulation goes beyond Peirce in the development and application of the causal modalities and the theory of inductive probability.

Thus our human capacity for inductive reasoning is an outgrowth of evolution. It is evolution operating at the intentional, conscious, linguistic level. This level has produced explicit, organized scientific and mathematical knowledge. Of course, a foundational analysis of inductive reasoning is much more sophisticated than commonsense inductive reasoning (the manifestation of Santayana's animal faith). Similarly, a formalized Peano arithmetic is much more sophisticated than the use of counter board and abacus.

If my account of the biological basis of the presuppositions is correct, inductive reasoning should share some important features with evolution. I think it does. One such feature is the intertwined roles of repetition and probability, roles that are brought to the fore when Bayes' theorem is used for explaining and modeling induction. Another is the interactive evolution of the conceptual structure of science with the verification of theories. Inductive inquiry aims at discovering causally relevant properties as well as verifying the laws that connect properties. Thus concepts evolve along with theories (pp. 369, 399, 405; Burks 1980a, sec. 4).

Evolution and induction both employ the logic of indirect evaluation and selection, for they both involve the evaluation and selection of entities according to the properties of their productions or consequences (Burks 1988, secs. 3, 10). A logic of discovery involves the production of new ideas and hypotheses from old materials, and genetics works similarly. Evolution and induction are both learning processes, in which alternatives compete in a natural environment or in an environment of tests (Holland 1975, 1976, 1980, 1984, 1985, 1986; Burks 1986c, 1988a; Holland, Holyoak, Nisbett, Thagard 1986; Holland and Burks, 1987).

I will conclude this section with a critique of Rosenkrantz's own philosophy of probability and his attempt to justify induction. He has advanced an objective Bayesian theory of probability, using Bayes' theorem to explain the confirmation process, and arguing that there is an objective assignment of prior probabilities to the alternative hypotheses (Rosenkrantz 1981 and pp. 247–249 above). In this last respect, his theory differs from the subjective or personalistic theory of probability (Ramsey, De Finetti, and Savage), for the latter theory holds that prior probability assignments are personal and relative.

Rosenkrantz believes that he has established the objectivity of prior probabilities, and that by so doing he has justified induction. Actually, the justification problem is considerably more complicated than this. It includes three problems, which we will treat in sequence: (I) the finitude problem, (II) the prior probability problem, and (III) the causal necessity problem. Rosenkrantz does not address the first and third of these, so that even had he solved the second he would not have provided a justification of induction.

(I) *The Finitude Problem.* Bayesians use a potentially infinite probability process as a model of nature. In technical terms, this is a mixture of independent processes, with the prior probability distribution left to the investigator. The process is investigated by repeated sampling and the extrapolation of the results. When the prior probability distribution over hypotheses is specified, the result is what I can Bayes' method of finding an unknown probability (*CCR*, p. 76).

This inductive setup presupposes an infinite process and thus excludes inverse inductive logic by fiat. It also excludes random inductive logic, since it leaves the limiting frequency open, as something to be learned by induction. Hence it is essentially a formal version of the Peirce-Reichenbach putative justification of induction, against which I have argued at length (*CCR*, sec. 3.4.2).

The intuition underlying the Peirce-Reichenbach justification of induction is this: An investigator has nothing to lose by assuming that a sequence of experiments or observations has a limit and applying Bayes' method to find this limit. The defender of random inductive logic replies: Inquiry has a cost. The defender of inverse inductive logic replies: Bayes' method makes the wrong *kind* of prior probability assignment, favoring uniform universes over random universes, whereas the opposite should be done. One can, of course, ignore these considerations, but that is like saying that 5 plus 7 *is* 12, and we don't need Kant's Transcendental Aesthetic, or Peano's axioms, or quantification theory, or anything else, to ground it.

Even if one accepts Bayes' method of finding an unknown probability, there remain basic alternatives. As de Finetti showed, Bayes' method is equivalent to assuming that the investigative sequence is exchangeable (see p. 371 above). But one might want to make a stronger assumption, such as the symmetry of the prior probability assignment, as Carnap did for the inductive methods of his continuum. This brings us to the next justificatory problem.

(II) *The Prior Probability Problem.* The subjectivist or personalist in probability theory says that differences in prior probability assignments to alternative hypotheses do not matter, for over a sufficiently long run of observations two disputants will come to agree. This is generally satisfactory in practice. But from a foundational point of view it is question begging, because, for each finite quantity of evidence, there are prior probability disagreements so great that the evidence will not resolve them (*CCR*, pp. 91, 530).

Rosenkrantz has treated this problem in his interesting book *Inference, Method and Decision*, and he summarizes his position briefly above (pp. 248–249). He makes two claims, that there is an "objective" prior probability assignment to alternative hypotheses, and that this assignment is a "cognitive analog" of the Dutch "book" notion of betting consistency. It seems to me that neither of these claims is correct. I will discuss the second claim first, since my criticism of it is independent of whether or not the first claim is true.

My calculus of choice, described above (pp. 358–360), is an axiomatic system of rules of choice for probabilistic situations. These choice rules yield the standard class of utility functions. Since the use of utility theory protects one from having a betting book made against oneself, these rules provide a foundation for the book theorem. Now there is obviously a very large *cognitive* gap between the rules of the calculus of choice and inductive rules for predicting the future on the basis of memories of the past. For this reason, no solution of the prior probability problem could be a natural cognitive extension of the Dutch "book" notion of betting consistency (cf. *CCR*, ch. 3).

Consider next Rosenkrantz's purported solution to the problem of choosing the prior probabilities in Bayes' method of finding an unknown probability. There are several alternative hypotheses, perhaps a continuum of them, and the total prior probability of unity must be distributed among them. This problem has long been the Achilles' heel of the a priori and personalistic theories of inductive probability; it is also a difficulty for what I call the positivistic theory of empirical probability (*CCR*, secs. 5.7 and 8.3.3). There are many different logically possible prior probability assignments, most of which do not lead to standard inductive logic.

It was originally proposed to solve this problem by means of "the principle of indifference": assign the same prior probability to each alternative under consideration. This principle is in need of justification, and thus to assume it is question-begging. But even if the principle of indifference is accepted, the problem remains. For there are many different ways of formulating the alternatives, and an equal distribution of prior probabilities under one formulation often yields an unequal distribution under a logically equivalent formulation.

This fact is illustrated by the well-known von Kries paradox. Consider a uniform distribution of prior probabilities for a specific density that is known to lie in the interval from 1 to 2. Transform it to an a prior probability distribution for specific volumes over the interval from 1 to 1/2. The resulting distribution is non-uniform.

For example, a uniform distribution for density implies

(D) Probability (the specific density is between 3/2 and 2) = 1/2.

But a uniform distribution for specific volume implies

(V) Probability (the specific volume is between 2/3 and 1/2) = 1/3.

Now, it follows from the definition of specific volume (it is the reciprocal of specific density) that the statements within the parentheses of (D) and (V) are logically equivalent. Therefore, (D) and (V) are contradictory.

Rosenkrantz seems to hold that there is a strengthened form of the principle of indifference that gives an assignment of prior probabilities that is invariant under any reformulation of the alternatives. He bases this claim on E. T. Jaynes' theory of entropy and information in physics. Rosenkrantz believes that the prior probabilities yielded by this form of the principle of indifference are "objective" and should be accepted by every rational person. Also, he thinks that this constitutes a justification of induction.

As we have already noted, a justification of induction must solve the finitude problem and the causal necessity problem, as well as the prior probability problem. *But Rosenkrantz has not solved the prior probability problem, for whatever the virtues of the new principle of indifference, it is not logically true.* It is not logically true because there are other logically consistent ways of assigning prior probabilities to the alternatives.

Standard inductive logic is a norm underlying our use of "rational", but this fact about language usage shows only that rationality presupposes standard inductive logic, not that true statements of standard inductive logic are logically true in the strict sense (*CCR*, sec. 3.3.2). Carnap stated a formalism that yields standard inductive logic, and the new principle of indifference may be part of an improved formalism that yields the same result (see p. 369 above). My own view is that one's assignments of prior probabilities to alternative hypotheses is rooted in one's innate structure-program complex, and hence is a result of evolutionary learning.

(III) *The Causal Necessity Problem.* What is the nature of causal necessity and how are causal necessity and inductive probability related? This problem is the focus of *Chance, Cause, Reason,* and my presupposition theory is one answer to it. It is argued there that causally necessary statements are not reducible to extensional (non-modal) statements, so that an account of the inductive verification of the latter statements does not cover the former. Until one has an adequate account of the inductive verification of causally necessary statements AND a justification of this process, one does not have a justification of induction.

Rosenkrantz does not address this problem. Nor do the subjectivist (personalist) or the frequentist. Consequently, there is a big gap in each of these theories of induction.

4.3 Probabilistic law and empirical probabilities.

Railton's "Taking Probability Seriously" addresses empirically verifiable probabilistic laws, and he says little about inductive probability. He favors a dispositional or propensity interpretation of empirical probability, but seeks one that is more objective and less inductive than my dispositional theory. I appreciate very much his careful analyses and his stimulating suggestions.

Skyrms' "Presuppositions of Induction" presents an alternative, more Humean, theory of presuppositions than mine. He seems to accept what I call Hume's thesis that there is no non-circular justification of induction (pp. 285, 297, 309), for which reason I did not discuss his paper in the previous section. But because the notion of probabilistic resilience plays a crucial role in his theory, it is appropriate to discuss it in the present section. His theory of probabilistic resilience is more Humean than my somewhat Kantian dispositional theory. I thank him for his interesting and suggestive paper as well as for his excellent review of *Chance, Cause, Reason* (Skyrms 1980b).

The notion of probabilistic law and its relation to causality is a large subject. I limit my comments to four topics: the interrelations of the probabilistic and causal modalities, some questions about determinism and indeterminism, the problem of verifying empirical probability statements, and the nature of probabilistic explanations.

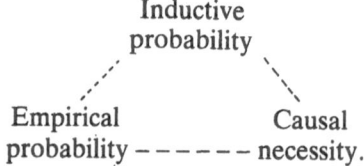

There are two key probabilistic concepts and one key causal concept, shown in the preceding triangle

There are many different theories about causality and these two kinds of probability, as well as about their interrelations.

Comparisons among these theories are complicated by the fact that each theory employs its own terminology, and this terminology often reflects some aspect of the theory. The concepts of inductive probability, epistemic probability, a priori probability, pragmatic probability, subjective probability, personal probability, and Carnap's probability-sub-1 are similar, although the theories that employ these terms vary widely. Likewise, the concepts of empirical probability, statistical probability, frequency probability, objective probability, physical probability, probabilistic law, dispositional probability, propensity probability, probabilistic resiliencies, and Carnap's probability-sub-2 are similar, but the users of these different terms often disagree.

For me, the basic or foundational concept of inductive logic is inductive probability. True atomic inductive probability statements are neither logically nor empirically true. They belong to a third category, that of *inductive statements*. The nature of these statements is described by my pragmatic and dispositional theories of inductive probability (*CCR*, p. 656, chs. 5, 8).

My dispositional theory of empirical probability reduces empirical probability to inductive probability.

An empirical probability theory is dispositional in nature. It asserts that the system underlying the disposition is such that each possible observable outcome has a certain inductive probability (degree of prediction). In the context of alternative theories and a prior probability assignment, these probabilistic predictions may be used to confirm or disconfirm the theory: a finite portion of the actual process produced by the system is observed and Bayes's theorem is applied to the result. (*CCR*, pp. 518–519)

My colleague Peter Railton quotes me correctly as being dissatisfied with the frequency interpretation of physical probability and proposing a dispositional analysis of physical probability. He then says "I find myself in agreement with Burks' positions thus abstractly stated, but I would like to explore here a disagreement over the *kind* of dispositional analysis to be given of physical probability statements" (p. 251). He goes on to suggest that he takes physical probability "more *seriously* more *realistic*, if you will – than *epistemic* interpretations permit" [italics added] and seems to include my dispositional theory of empirical probability in the latter category. Let me comment on the three italicized words.

"Epistemic" is not quite the right word here. It fits a Humean view of probabilistic dispositions, like that of Skyrms, which Railton also has in mind (pp. 265–266) and on which I will comment in a moment. But "constitutive" is a better word to apply to my dispositional theory than "epistemic", for my view is more Kantian than Humean (*CCR*, sec. 10.5.2).

Railton's phrase "more realistic" is perhaps appropriate, in the sense that Locke was more realistic than Kant, and Kant more realistic than Hume. But empirical probability statements are about nature, and so my view is also realistic, unless one wants to argue that acceptance of Hume's thesis (that there is no non-circular justification of induction), which is an epistemological doctrine, precludes realism, which is a metaphysical doctrine. And, of course, I am more realistic than most philosophers in my commitment to the causal modalities.

"Serious" is not the right word, either, for the eighty pages or so I devote to the issue reflect considerable thought about the problem over the years. *Chance, Cause, Reason* contains extended discussions of the traditional theories of physical probability, namely, frequency theories. (Frequency theories are, to the best of my knowledge, the only fully worked out accounts of the view that probabilistic laws are basic). I rejected frequency theories only after careful consideration.

Moreover, I take seriously the problem of *verifying* empirical probability statements. This is the third main question of *Chance, Cause, Reason* (p. 656): What is the general nature of an empirical probability theory? How are empirical probabilities related to inductive probabilities? Why is the traditional calculus of probability applicable to empirical probabilities?

The dispositional theory of empirical probability is my answer to this question. It reduces probabilistic laws to inductive probabilities, and so I conclude that "the concept of inductive probability is epistemologically more fundamental than the concept of empirical probability" (*CCR*, p. 517).

People do use probabilistic laws and explanations based on them, and this is an important part of scientific and engineering practice. One can, of course, just accept these practices as facts and pursue the matter no further, in which case one's philosophy of probability does not matter. But an important problem in the foundations of induction remains. Knowledge of empirical probabilities is probabilistic, and this probabilistic knowledge needs to be accounted for. Moreover, the sense of probability involved here is that of inductive probability, so that empirical probabilities need to be related to inductive probabilities.

My reason, then, for not accepting the widely held thesis that probabilistic laws are basic is epistemological, negative, and somewhat personal: the only satisfactory *foundational* account of how empirical probability statements are verified makes empirical probabilities depend on inductive probabilities.

Railton does not realize that this is the reason I say that "all basic laws are causal, and there are no basic probabilistic laws" (*CCR*, p. 657). Instead, he thinks it is that I am a determinist. Clendinnen also thinks I am a determinist. I am not a determinist, and I never say that I am; on the other hand, I also omit to say explicitly that I am not a determinist. Let me clarify my position here.

The essential point is that determinism and probabilism are not the only alternatives – an indeterminism in which events happen arbitrarily is a third alternative. The three basic alternatives are best explained in terms of cellular automata (*CCR*, sec. 9.2.3). Under *determinism* each state of the system has a unique successor state. Under *indeterminism* at least one system state (and typically many) has two or more successors. Under *probabilism* there is a probabilistic distribution on the successors to any given system state. Note that probabilism is indeterminism *plus* a probability distribution over alternative successor states. Hence when I say "all basic laws are causal and there are no basic probabilistic laws" (*ibid.*), I am not committed to determinism but only to the arbitrary form of indeterminism.

These are the pure cases. The actual alternatives are more complicated, involving local laws of various kinds. There are four main alternatives, given below, ranging from one extreme (determinism) to the other (probabilism).

(1) *Determinism* = nature is completely governed by causal laws of a quasi-local and uniform character.

(2) *Classical indeterminism* = nature is for the most part governed by causally uniform and quasi-local laws, *but* some events are accidental and not predetermined. Lucretius's indeterminism is an example.

(3) *Causal-probabilistic indeterminism* = there are both basic causal laws and basic probabilistic laws. This is the working view of most physicists, since classical physics and relativity theory are causal, whereas quantum mechanics is probabilistic, and there is as yet no satisfactory unification of the two.

(4) *Metaphysical probabilism* = probabilistic laws are basic, and causal laws are reducible to them. (Peirce's tychism is a combination of this doctrine with the theory that probabilistic laws have evolved over time).

"Near-determinism" as defined in *Chance, Cause, Reason* (p. 577) embraces both (2) and (3).

Since the development of quantum mechanics in this century, determinism has been on the defensive. More than fifty years ago John von Neumann gave a "proof" that there are no hidden variables in quantum theory, i.e., that quantum mechanics cannot be embedded in a deterministic system. Since then the issue has been debated back and forth, but recent experimental results strongly support von Neumann's conclusion (Shimony 1988). Though there are still problems in the interpretation of quantum mechanics, this experimental evidence is a conclusive argument against determinism, and hence for indeterminism. Let us see how this conclusion relates to my views on induction.

The presupposition theory of induction is stated so as to allow for near-determinism as well as determinism (*CCR*, pp. 649, 577, 589). Later in this section I will reformulate the causal uniformity and causal existence presuppositions so that they explicitly cover objective probability laws as well as laws of nature (which involve causal necessity).

There still remain some important problems about the relation of the two kinds of laws, problems that need to be solved before one has an adequate theory of probabilistic laws. Since basic probabilistic laws are expressed as empirical probability statements, some of these problems are stated in my third main question (p. 480 above). Other problems can be expressed in terms of embedding.

We have seen how probabilistic laws can be embedded in deterministic systems (pp. 434–437). Can the reverse be done, that is, can causal laws of nature and deterministic systems be embedded in probabilistic systems? The limiting probabilities of 1 and 0 are kinds of necessity and impossibility (*CCR*, p. 272), but this fact is not sufficient to establish the embedding of a deterministic system in a probabilistic system, or to show how the logic of causal statements is related to the calculus of probability.

Although my presupposition theory of induction is compatible with the view that probabilistic laws are basic, this is not true of my dispositional theory of empirical probability, nor of my unified theory of probability, causality, and induction. I should have pointed out these incompatibilities in *Chance, Cause, Reason* and emphasized that my reason for them was not a commitment to determinism. My reason, as given there and repeated a moment ago, was my inability to find a satisfactory account of the inductive verification of basic probabilistic laws, and I still think that is an unsolved problem.

As we noted a few paragraphs back, a probabilistic law involves both a set of possible consequences and a probability distribution over them. The question at issue now concerns the nature of this concept of probability. Is it objective, perhaps some improved frequency concept; or is it inductive, resting ultimately on standard inductive logic?

My foundational approach to empirical verification starts from induction by simple enumeration. Quantum mechanics should in principle be derivable from raw empirical data by a Bayesian verification process. The gap between the levels of theory and data is tremendous, of course, for quantum theory is very mathematical and sophisticated whereas induction by simple enumeration is quite primitive. Perhaps the gap between the two levels is unbridgeable. But repetition and probability are essential features of biological evolution, which over a long continuous process produced humans and then science. Moreover, the reduction distance from modern higher mathematics to basic logic and arithmetic is very great, and yet it has been bridged.

Of course, even on my view there is a sense in which probabilistic laws are quite basic. They are expressed by empirical probability statements. The epistemology of standard inductive logic is presupposed by and constitutive of empirical inquiry. Relative to this epistemology, probabilistic laws are basic. Thus, metaphysically speaking, probabilistic laws are basic, but from the inner point of view epistemology underlies metaphysics.

This completes my remarks on the interrelations of determinism, indeterminism, and probabilism. I turn now to the problem of verifying empirical probability statements, discussing Railton's views first. Let me note in making this transition that some of the metaphysical and epistemological issues underlying these two problems are the same. Both problems concern the structure of systems: embedding and reductive relations within a hierarchical system, and knowledge and action relations within such a system (the extent to which a subsystem can learn about the whole and modify it). Thus, as we will see in a moment, probabilistic dispositions involve the interrelations of three levels: observed frequencies, properties underlying a disposition, and lower-level microscopic systems.

Railton characterizes his alternative dispositional analysis of physical probability as

> a *propensity interpretation* according to which physical probability is not to be deductively analyzed into any sort of epistemic probability, but instead is viewed as an objective property of physical systems. Physical probability, on this interpretation, is a measure of the dispositional tendencies of indeterministic physical systems to yield certain outcomes. One way these chance dispositions manifest themselves is in the relative frequencies obtained in sequences of trials.... But a propensity is not reducible to its manifestations or to the expectations it warrants, just as dispositions in general are not so reducible. Quantum mechanics seems to afford us plausible examples of propensities. (pp. 266–267)

However, he does not actually formulate such a theory of physical probability. He only states a desired characteristic of such a theory, namely, that an empirical probability statement refer to an objective property of an indeterministic or probabilistic system. Frequency theories of empirical probability satisfy this objectivity requirement, but, as Railton recognizes, they fail to give an adequate account of how empirical probability statements are verified.

Actually, Railton's objectivity requirement is satisfied by the referent of a term in my dispositional analysis of an empirical probability statement, namely, what I call the *property* or *system* underlying a probabilistic disposition (*CCR*, pp. 491f., 518–19, 528 – cf. 442f.). "System" is used here in the sense of a system structure or complex of properties. The system underlying a disposition is an objective structure of physical properties that manifests itself in relative frequencies, but it is not reducible to these manifestations.

Thus, what underlies a probabilistic disposition is a relatively self-sufficient complex system, "a set of physical properties that is complete with respect to properties of a certain kind and a certain level of specificity, together with the causal laws governing these properties".

This complex system is in principle describable by a theory that is complete in an appropriate sense. The theory does not yield unique predictions, however, for the underlying system is indeterministic with respect to the observable relative frequencies. (*CCR*, pp. 493, 500, 517, 546)

My dispositional analysis of empirical probability employs my analysis of the relation of an embedded subsystem to its underlying system, including the fact that the deterministic vs. indeterministic status of an embedded subsystem may be different from that of its underlying system (CCR, sec. 9.4.1). An empirical probability involves a hierarchy of three systems. The observable relative frequencies are on the top level, the system underlying the probabilistic disposition is on the middle level, and the system in which the latter is embedded is on the bottom level. The middle level system is probabilistic, but the bottom level system may be deterministic.

Take a sequence of random coin tosses as an example. The actual finite sequence of coin tosses constitutes the top level. The underlying system of potentially infinite coin tosses constitutes the middle level. The basic mechanical microsystem constitutes the bottom level. The states of the middle-level probabilistic system are clusters of states of the underlying deterministic system (see pp. 434–436 above). The bottom-level system contains the middle-level probabilistic system as an embedded subsystem, but the bottom-level system is itself a deterministic system. Thus it satisfies Railton's injunction (p. 267) "needless to say, fully deterministic systems do not have probabilistic propensities in this sense since they have no *chance* dispositions", while yet containing a probabilistic subsystem.

Whatever the form of one's theory of objective or physical probability may be, the theory needs to account for the inductive verification of objective probability assertions, what I call "empirical probability statements". Railton does not address this question. My answer to it is that the verification of all empirical theories (causal as well as probabilistic) is best modeled by means of inductive probability. As I stated earlier, that is why I concluded that inductive probability is more basic than empirical probability, and that probabilistic laws are accordingly not ultimate.

Let me turn now to Brian Skyrms' subjective theory of probability and his correlated account of objective probabilities, which he calls "resiliencies" (his paper in the present volume, and Skyrms 1977, 1978,

1980a). Railton discusses Skyrms views on subjective probability and resiliences with apparent sympathy, and so I think I will be answering both of them on the issue of objective probability.

There is a substantial philosophical gap between my dispositional theory of empirical probability and what Skyrms and Railton have in mind. However, my positivistic theory of empirical probability (*CCR*, sec. 8.3.3), which I suggested as a second best, is much closer to their views. A good way to respond to their comments is to consider the positivistic theory and some variants of it.

The positivistic theory lists several basic aspects of empirical probability: a priori, empirical, pragmatic, and subjective. In any particular case, the empirical aspect refers to an objective physical property that is manifested by relative frequencies; in this respect, then, the positivistic theory satisfies Railton's objectivity requirement.

Consider the prior probability assignment presupposed by a Bayesian verification of an empirical probability statement. The positivistic theory treats this assignment subjectively and relatively, as do the subjective and personalistic theories. But this approach leaves the verification of empirical probability statements unexplained. For in any actual case there is only a finite amount of data, so that agreement between two investigators on posteriori probabilities presupposes a certain amount of agreement on the prior probabilities. This is the main weakness of the positivistic theory, and it is a strong argument for the dispositional theory of empirical probability (*CCR*, pp. 529–31).

This is also a grave difficulty for subjective and personal theories, and for those of Rosenkrantz (see p. 476 above) and Skyrms. It is a mystery on subjective theories of probability how there can be any objective or empirical probabilities, for by their very nature objective probabilities cannot be known subjectively. Empirical probability statements refer to something in nature, not something in the mind.

De Finetti, a subjectivist, attempted to solve the problem of objectivity by means of the notion of exchangeability. Because Skyrms seems to follow him in this attempt, it is important to evaluate it. We mentioned de Finetti's ideas earlier (secs. 2.2, 4.2), but for present purposes it is best to restate them.

Consider some domain of individual events and properties, and a probability assignment to all possible statements about that domain. For a probability assignment to be exchangeable means that the probability of a statement does not depend on which particular individuals the statement is about. Every independent process with a specific given probability is exchangeable, as is every probabilistic mixture or weighted combination of such processes.

Carnap dealt with the same notion in terms of isomorphism. Two formulas are isomorphic if one can be transformed into the other by a systematic renaming of the individuals referred to. For example, QaQb~Qc is isomorphic to Qa~QbQc. In his inductive logics, Carnap assigns equal probabilities to isomorphic universe descriptions; thus all his inductive logics are exchangeable (*CCR*, p. 125).

(Markov chains are simple examples of non-exchangeable processes. The events in a Markov chain are ordered, and the order is essential. These are called chains because the probability that an event or link in the chain has a certain property depends on the properties of the previous link, that is, its immediate predecessor. Since interchanging individual names in statements changes the predecessor relation, a Markov chain is not exchangeable.)

De Finetti proved an important theorem about probability assignments to infinite processes. (An infinite process is one with infinitely many individuals, though only finitely many properties or individual states). The theorem is: An infinite probabilistic process is exchangeable if and only if that process is a probabilistic mixture of independent processes (*CCR*, sec. 3.2.4).

We can now state and evaluate de Finetti's (and Skyrm's) attempt to use the notion of exchangeability to give a subjective probability account of objective probability. He proposed to do this by analyzing a statement such as

(U) This is an independent process of unknown probability

into

(E) This is an exchangeable process.

The equivalence is acceptable. But it does not show that on the subjective theory of probability (U) and (E) are empirical statements, verifiable or refutable by repeated sampling. It establishes only the formal equivalence of (U) and (E). (*CCR*, sec. 8.4.3).

Significantly, De Finetti's proposed analysis of (U) into (E) suffers from the fact that both statements are ambiguous with respect to the prior probability distributions involved. This is an important point, because prior probabilities are essential to the verification process and are the Achilles' heel of any attempt by the subjective theory of probability to account for the objectivity of empirical probabilities. (*CCR*, sec. 8.4.3).

De Finetti had in mind the classic model of induction studied originally by Peirce and Reichenbach, and in more sophisticated forms by later writers. In this model the investigator assumes that the process is a random, independent sequence of states (events), but does not know the limiting relative frequency of the process. He or she learns this frequency by repeated sampling. The traditional rule of induction by simple enumeration is a special case of this model, involving the sequential correlation of two properties. This correlation may be statistical, as in the case of a probabilistic law.

(Usually only single properties are considered, but most results are easily extended to cover any finite number of individual states. Indeed, we will see in a moment that Carnap's inductive logics are classic models of induction in a different guise.)

Skyrms emphasizes the classic model, titling the main part of his paper "Learning from Experience". I discuss the model under the name, "Bayes' method of finding an unknown probability", and state the confirmation theorem for it (*CCR*, sec. 2.5).

Now, a mixture of independent processes is not fully defined until prior probabilities are assigned to the alternative hypotheses. It is true that the prior probability assignment does not matter in the limit, and (as a consequence) is often downplayed. Peirce and Reichenbach clearly did so (*CCR*, sec. 3.4). I do not think that de Finetti and Skyrms treat it adequately. But since verification is always based on a finite amount of data, from a foundational point of view the prior probability distribution must be considered.

When prior probabilities are taken into account, the old difficulty reappears. Consider the empirical hypothesis

(I) This is an independent process of probability 0.6,

which is investigated in the context of competing hypotheses. Now for any finite set of observations of the process, there is a prior probability assignment to the alternative hypotheses such that the posterior probability of (I) is less than one-half relative to those observations. Hence, if prior probability assignments are subjective and relative, statement (I) is not really verifiable, for empirical verification is public in nature.

Thus de Finetti's attempt to use exchangeability to give a subjective account of empirical probability fails. This failure undermines Skyrms' theory of resiliencies or probabilistic dispositions.

De Finetti's exchangeability results do contribute to our understanding of inductive models and of the process of verifying empirical probability statements. For this reason, Skyrms is justified in making a careful study of exchangeability and its extensions to Markov chains and ergodic processes generally. Let us expand on this point a little before continuing our discussion of the prior probability problem.

The relevance of de Finetti's exchangeability theorem to induction may be illustrated with Carnap's continuum of inductive methods or logics. Each of these logics satisfies the condition: two isomorphic state descriptions have the same prior probability. Hence each logic is exchangeable. It follows by de Finetti's theorem that each inductive logic of Carnap's continuum is equivalent to a probabilistic mixture of independent processes, and hence to a classical model of induction (*CCR*, sec. 3.2).

In a finite model, the number of universe descriptions isomorphic to a given universe description D is a measure of the *extensional randomness* of D. For example, Qa~QbQcQd~Qe is more uniform than QaQbQcQdQe. The *extensional uniformity* of a state description may be defined as the inverse of its randomness. Carnapian inductive logics favor extensional uniformity over extensional randomness, for in each finite model the 'higher the uniformity of a universe description the higher its prior probability.

There are inductive logics that reverse this, favoring random universes over uniform ones. Inverse inductive logic is an example (*CCR*, pp. 103, 116, 127). This is an exchangeable logic in which the predicted probability tends away from the observed relative frequency. Thus in inverse inductive logic an anti-enumerative rule holds, in contrast to the rule of simple enumeration of the classical models (mixtures of independent processes).

The question naturally arises: Could there be an inverse inductive logic with an infinite number of events? De Finetti's theorem provides the answer: since inverse inductive logic is exchangeable but is not a mixture of independent processes, it cannot be infinite. Skyrms has an interesting discussion of Diaconis's formal result on this matter (p. 295).

These considerations establish that de Finetti's exchangeability theorem is relevant to inductive logic, even though the theorem does not help solve the problem of prior probabilities that besets subjective and personalistic accounts of empirical probability and also my positivistic theory of empirical probability. My dispositional theory of empirical probability was designed so as to solve this problem, but let us look here for another possible solution.

Note that both the classic models of induction and Carnap's inductive logics can be used either locally, to model the verification of empirical probability statements, or globally, as models of comprehensive inductive logics. My positivistic theory of empirical probability employs local models, while my pragmatic theory of inductive probability employs global models of standard inductive logic. In these global models, approximate prior probabilities are assigned to possible universes.

This suggests that one solution to the prior probability problem for empirical probabilities is to combine the positivistic theory with the pragmatic theory into a single theory, which I will call the *pragmatic-positivistic theory of probability*. According to the pragmatic theory, humans do employ the rules of standard inductive logic in making inferences from empirical data. This practice implies sufficient public agreement on prior probabilities to explain the amount of agreement that exists on posterior probabilities. Thus the pragmatic theory solves the prior probability problem of the positivistic theory. ·

This suggested pragmatic-positivistic theory of probability fits in with the doctrine of innate capacities that I advocated in the last section, the use of a crude form of standard inductive logic being an inborn capacity. The natural use of induction is included in what Santayana called animal faith. Explicit statements of inductive rules are conceptual formulations of habitual inductive practices.

Skyrms may be sympathetic to this pragmatic-positivistic theory of probability. He begins his paper with the remark that there is good scientific practice (method), including

logically consistent ways of assimilating evidence which count as bad science, or unscientific or outright mad. To this extent, at the very least, scientific method will have its presuppositions. (Whether we classify them as a special kind of knowledge is another question.) An examination of the presuppositions of scientific induction is nothing other than an analysis of scientific method itself. (p. 285)

However, rules of inference can be formulated as statements.

Moreover, as I argued against Carnap, if one reasons inductively according to the rules of standard inductive logic, then one is presupposing certain general principles. The finitude of the language reflects limited variety. To make inductive inferences that are invariant over space and time, weighing extensionally uniform universe descriptions more heavily that extensionally random ones, is to reason in accord with a uniformity of nature principle (Burks 1951e).

In passing, I would like to comment on Skyrms' terms "unscientific" and "outright mad". Of course, standard inductive logic is definitive of good "science", and pragmatically we all employ it. Hence anyone who tried to follow an alternative inductive logic would be "outright mad" (*CCR*, secs. 3.3.2, 5.6.1).

One last point needs to be made about my proposed pragmatic-positivistic theory of probability. As so far specified, it does not account for the causal modalities. My logic of causal statements characterizes the deductive interrelations of causal and counterfactual statements (see the paper by Uchii in this volume) and shows that the causal modalities are all-or-none in nature. In contrast, the probabilistic modalities are matters of degree – compare the difference between determinism and probabilism.

To fill this gap, I extend the pragmatic-positivistic theory to incorporate the appropriate causal materials. For example, the uniformity of nature and causal existence principles are extended to cover probabilistic laws as well as causal laws (cf. *CCR*, pp. 572, 577).

Uniformity of law principle: If a causal connection or *objective probability connection* holds between non-indexical properties in one region of space-time, then it holds throughout space-time.

Existence of law principle: Some space-time region is governed by causal and/or *probabilistic* laws to a substantial degree.

I think that both this pragmatic-positivistic theory of probability and my earlier dispositional theory of empirical probability give good explanations of objective probability, better than any foundational alternatives I know of. But, they still analyze objective probabilities by means of inductive probabilities. Hence neither satisfies Railton's and Skyrms' desire for an account of objective probability that is independent of inductive probability. I am sceptical, however, that there is such an account.

This is an opportune place to discuss probabilistic explanations. I do not agree with Railton that the determinism issue is relevant to what model of probabilistic explanation is best (pp. 256–261). Explanations, theories, and verifications are all practical in that their complexity is limited by what humans can grasp and manipulate, by themselves or with the aid of equipment. The value of a probabilistic explanation does not depend on whether or not the system depicted is embedded in a deterministic system. Laplace's dictum "probability is a measure of our ignorance" should be replaced by "probability simplifies knowledge".

Deductive explanations and predictions are all-or-none, whereas probabilistic explanations and predictions are matters of degree and typically involve distributions. These are complicating factors. Thus to the question "Why did this person die from lung cancer?" an informed physician might reply that smoking was the most likely cause, but that lung cancer was common in the man's family and this factor may have contributed to his death. Because they involve distributions of events rather than single events, probabilistic predictions are more difficult to verify than deductive predictions. Probabilistic predictions of rare events are especially difficult to verify because of the very large samples required.

Teleological systems have traditionally been regarded as holistic, and often as only understandable in the context of a larger holistic system such as the great chain of being or an evolutionary process (Burks, 1988b). Whether or not this is so, teleological systems are generally complex and hence good candidates for probabilistic explanations. Almost any evolutionary explanation is probabilistic or potentially so, and we will give two examples from evolutionary genetics.

The first example is R. A. Fisher's classic theorem connecting fitness to genetic variance. This theorem states that the rate of increase of fitness in a species at any time is equal to its genetic variance in fitness at that time (Fisher 1929, pp. 37, 50–51). A greater variance in the gene frequencies of a population produces a greater variance in the trait distribution of that population, which in turn helps the population to adjust to a wider range of environments.

Today, with powerful computers and simulation techniques, statistical studies of evolutionary processes constitute a fruitful area of research. Fisher's theorem could be illustrated on a computer, and it could be simulated in a context that would be more realistic than Fisher's mathematics could capture. We give next a probabilistic explanation that is somewhat hypothetical and that is now being tested by a combination of computer simulations and the collection of empirical data.

This example concerns the evolutionary explanation of the origin and stability of sexual reproduction. Evolution involves successions of reproducing and competing organisms interacting with a changing environment. New genomes are produced by a stochastic process, using genetic operators such as mutation and crossover. Although reproduction was asexual or clonal to begin with, sexual or aclonal reproduction soon developed. Why and how this happened, and why sex continues as a fundamental mode of reproduction – these are interesting questions. The final answers are not yet in, but a promising hypothesis illustrates very well the nature of probabilistic explanation and prediction.

Since sexual reproduction involves the cooperation of two parents, one to produce eggs and the other to fertilize them, it seems less efficient than asexual reproduction, done by a single parent. And so while sexual reproduction could begin by chance, an economic explanation is needed to account for its persistence and development. W. D. Hamilton has suggested an ingenious answer to this question (1982; Hamilton and Zuk, 1982).

All large metazoa are sexual. All carry inside themselves, or are from time to time invaded by, a very large number of small living things, such as bacteria, viruses, protozoa, and yeasts. Many of these smaller organisms are parasites, harmful to their larger hosts. These parasites generally reproduce asexually. Hamilton's hypothesis is that the sexuality of large organisms is connected biologically to the existence and potential for rapid multiplication of these small internal parasites.

The host and its parasites are in evolutionary competition, so that the rate at which each one evolves is crucial to its success. The rate of evolutionary change or learning depends on several factors, two of which are central to Hamilton's hypothesis. The first is the rate of reproduction of a species, the number of generations per unit time interval. The second factor is the amount of change in a species per generation.

A multicellular organism has many cells of different kinds, organized hierarchically. Such a structure requires more time to develop than a single cell, and hence the life span and regeneration cycle of a large organism is much larger than those of a small organism. For example, some parasites in elephants reproduce tens of thousands of generations within one life span of the host.

The amount of change in a species per generation depends on the genetic operators employed in reproduction, such as mutation and crossover. Since these operators are probabilistic, this explanation of sex is probabilistic. Moreover, the probabilities (e.g., the mutation rate) may be low, so that relatively rare events may play a significant role in the process. Now, the sexual mode of reproduction mixes two different genomes, whereas the asexual mode varies only one genome. For this reason, sexual reproduction brings about much more genetic change per generation than does asexual.

In particular, sexual reproduction employs crossover, which operates at a higher grammatical level than mutation. Mutation changes the equivalent of a letter in a genetic sequence, possibly generating a new "word" in a genetic program. Crossover juxtaposes part of the genetic sequence of one parent with part of the genetic sequence of the other parent, thereby generating new combinations of "words" in the genetic program of thee offspring. Looked at conceptually, mutation produces new "concepts" to be tested by the environment, whereas crossover produces new combinations of concepts already tested by the environment. The importance of crossover for sexual reproduction is confirmed by the fact that the longer the life span of a species, the stronger the role of crossover.

As we have noted, because of its physiological complexity a large host organism cannot reproduce as often as its much smaller internal parasites. Hamilton's hypothesis on the origin and continuance of sex is that the greater rate of genetic change produced by sexual reproduction enables the host to compensate for the faster rate of reproduction of its parasites. In other words, by employing sex the larger host compensates for its longer life span.

The relevance of parasites to an organism's vitality is shown by the size of its immune system, which protects the organism from its parasites. In the case of humans, the number of cells in the immune system is comparable to the number of cells in the brain, though of course brain cells are much more complicated than the cells of the immune system.

This hypothesis about the origin and stability of sex is now being tested by a combination of empirical observation and probabilistic computer simulations. Such studies are very complex, since they must take account of many environmental factors and organism characteristics. Different probability distributions need to be tested. The strength of the various forces of interaction and the stability or robustness of the mechanisms and processes must be considered. There are many related questions: What are the conditions under which sex arises and develops? What is the role of partially sexual modes of reproduction in evolution? Do asymetrical roles in sex always emerge early? How did the immune system evolve in relation to sex? And so forth.

Hamilton's general hypothesis about the origin and stability of sex seems essentially correct to me. But whatever its truth status, it illustrates very well the nature of probabilistic explanations. In probabilistic systems, low probabilities may play significant roles – for example, statistical noise may prevent a system from "stalling" and keep it searching for new possibilities. Inputs and outputs, as well as internal computation, may involve probabilistic distributions.

Probabilistic models, explanations, and predictions look quite different from their deductive counterparts. Yet both kinds can be carried out on a deterministic computer, for pseudo-random numbers can be used to simulate bona-fide randomness (*CCR*, sec. 9.4). This shows once again that there are interesting interrelations between probabilistic and deterministic systems, and that a unified theory about both types of systems and their interrelations is needed.

4.4 Evolution, Science, and Values.

I turn finally to Robert Audi's "Scientific Objectivity and the Evaluation of Hypotheses". He discusses the question: Can scientific judgments be objective and factual, or must they incorporate moral or normative value considerations as well?

This is an important foundational question, of the same general kind as the question of the separability of mathematical knowledge from empirical knowledge (sec. 2.4). I appreciate very much Audi's raising it, for his doing so gives me an opportunity to round out my views by discussing values and morals.

Those who advocate the view that science must mix values and/or morals with its theories and facts do not usually make clear what they are claiming. They do not start from a stated view of values and

morals. Nor do they adequately distinguish the tentative acceptances of
theoretical science from the total commitments of engineering and applied science.

Science has, of course, value and moral aspects, for it is a cooperative human enterprise. On the one hand, scientists qua humans
exhibit human nature in its many variations, including not only
morality but immorality. On the other hand, there are many social and
normative questions associated with scientific research. What research
should be done, and why? Is work on nuclear weapons, germ warfare,
and recombinant DNA ethical? Which scientific results should be
applied, and under what conditions? Should society apply scientific
theories that have not been rigidly established?

Prima facie, these value and moral issues concerning science seem
separable from the results of science, and hence not ultimately destruc-
tive of its objectivity. To show that this is indeed so, Audi analyzes the
general concept of objectivity and then applies the result to scientific
objectivity as a special case. I think his analysis is convincing, but I
would like to treat the subject differently, in terms of my philosophy of
logical mechanism, as a general reduction question. This approach will
also enable me to comment further on evolution and to sketch briefly
my views on the foundations of values and morals.

Science is a very complex holistic system, involving observation,
reasoning, and action. The thesis that scientific results must be partially
value-laden is an irreducibility thesis, to the effect that it is not possible
even in principle to separate the value and moral constituents of the
scientific enterprise from its empirical and logical constituents. The underlying issues here concern human nature and the nature of scientific
communities.

According to the person=robot thesis of the philosophy of logical
mechanism, a finite deterministic automaton can perform all natural
human functions (sec. 3.2). Now it seems to me that an ideal scientific
robot could, in principle, be organized into relatively independent
divisions, as follows. There would be an input system to process sensory inputs and send them to a central computer. Correspondingly,
there would be an output system to execute instructions received from
the central computer. The central computer would have two parts.

The first part would do the information processing proper. It would
have the knowledge and skills of deductive logic and mathematics, on
the one hand, and the knowledge and skills of inductive logic and em-

pirical science, on the other. In terms of my foundational classification of statements, this part of the central computer would enable the robot to discover and verify empirically true-or-false statements, developing and using mathematics as needed. It would receive information from the robot's sensors, process this information, and send orders for actions and observations to the robot's outputs.

The second part of the central computer would contain the goals, desires, and driving forces of the system, and would control the computations and actions of the rest of the system.

Empirical science would be developed by the first (information-processing) part of the central computer, while morality and values would be associated with the second (control) part. The robot could be designed so that its main purpose was to gain scientific knowledge. The control part of the central computer would direct the activities of the information-processing part of the robot to this end, but would not bias the truth-values of its results. Hence this computer architecture would allow the content of science (and mathematics) to develop independently of morals and normative values.

An opponent might counter that, since scientific results must contain implicit value and/or moral judgments, such a division of labor between the parts of a computing system is not possible. But this proposed centralized computer organization does look feasible, and so the burden of proof belongs to the philosopher who thinks it is not.

An opponent might also object that the suggested robot model is irrelevant because human scientists are not organized in that way. Rather, human interests and drives are distributed among information-processing capabilities, an organization that may prevent the individual scientist from separating facts from values. This is a fair point, but one must also consider whether the interaction of scientists can ameliorate such biases. Thus we need to discuss human nature and the nature of the scientific community.

The distributed computer organization of humans is a consequence of the step-by-step design procedure of evolution. Sensory and action abilities, information-processing powers, and also drives, interests, and emotions have evolved together in continuing interaction. This evolution has occurred in the small-step manner of the teleological continuum (Burks 1984).

Similarly, the growth of an individual takes place in small increments. This fact applies to beliefs as well as to other aspects of individual development, and hence is relevant to the present question of whether scientific results can be value-free. People come to believe things by a stepwise process that is driven by their desires and interests, and that is limited by the partial character of their information, their ingrained habits, and their opportunities.

Human interests, desires, and emotions restrict the nature and quantity of evidence they gather, accept, and retain. Humans make heavy use of authority, and their associations influence which authorities they rely on. Belief is necessarily based on partial information, which is often influenced by one's emotions; one may even come to believe what one wants to believe.

Thus many human judgments on factual matters are infected by value or moral considerations. Two questions then arise: How fundamental is this characteristic of human nature? To what degree can scientific procedures overcome its effects on inquiry? As we mentioned earlier (p. 413), the relevant features of human nature, both individual and social, are rooted in the fundamental nature of the evolutionary process, to which we now turn.

In its basic logic, an evolutionary process consists of iterations of a two-stage cycle involving organisms and their genetic programs. The cycle consists of a reproduction stage alternating with a selection stage. During the reproduction stage, a new generation of organisms is produced from the old generation by means of a genetic mechanism that involves copying with variation and/or mixing. During the selection stage, the produced organisms compete in an environment, some organisms being selected for use in the next cycle and others being rejected.

Thus the reproduction stage of an evolutionary cycle works with the materials given to it by the selection stage of the preceding cycle, produces new designs that are tested by the environment, and passes the results to the next cycle. This quite general account is abstracted from organic evolution, but there are interesting analogues in scientific inquiry and in computer systems. Consider, for example, a Bayesian model of inquiry in which one hypothesis is chosen over alternatives on the basis of its predictions. Genetic programs may be viewed as hypotheses that are evaluated in terms of their environmental consequences and then used to generate new hypotheses. Both the evaluation procedure and the generation procedure of evolution are used in a new computing system for learning and discovery with which I am involved. This system is based on rules that compete with one another to have their computations used, the most successful rules being mixed periodically with one another to yield new rules. (Burks 1984, 1988a, 1988b; Holland 1976, 1986; Holland and Burks 1987, and pending)

Starting from simple organisms, and operating over a very long time span, this basic evolutionary logic has produced humans. Looked at holistically and abstractly, evolutionary biological history is a "teleological continuum" that ranges from simple direct-response organisms at the beginning to intentional, conscious, goal-directed humans at the present (p. 402). From our human perspective this is progress. But though evolution has produced consciousness and intentionality from very simple, direct-response organisms (and originally from quite elementary forms of existence), this advance is not the result of some regulative goal plan, elan vital, or final causality. Rather, the logic of evolution is indirect and probabilistic in its operation: it selects genetic designs indirectly according to their comparative suitability to the environment, and it employs chance in the construction of new genetic plans from old. This *indirect nature of evolutionary development* is important, for it enables a mechanistic system that is not inherently goal-directed to develop goal-directed systems.

Let us now look at how the logic of evolution, operating in a suitable environment, has resulted in certain fundamental characteristics of organisms, and ultimately of human nature. Consider various genetic programs operating in an environment. The property of self-reproducing viable copies is by its very nature a self-perpetuating property. For this reason, the ability to reproduce appeared early in evolution and led to the development of drives and desires. (Copying is associated with the reproduction stage of evolution, whereas viability is associated with the selection stage.)

Now whether or not an organism survives and produces viable offspring depends on three kinds of factors: the sensing and acting abilities of the organism; its internal processing capabilities; and its drives to develop, reproduce, and care for its offspring. The success of an organism depends on the relation of all of these factors to those of other organisms and to the properties of the environment.

It will be convenient to have names for the three types of factors entering into the operation of an organism. Let us call its input and output abilities *bodily skills*. These include some processing abilities, for sensing and acting organs are often intelligent terminals. We will call the remaining computing capacities *reasoning abilities*. And we will call the organism's drives, desires, and powers of will its *volitions*.

Traditional philosophers, pragmatists excluded, have generally neglected the first kind of factor and have lumped the other two kinds into single faculties of Reason and Will. But such terminology connotes much more unification than actually exists. Information processing is distributed throughout the organism. "Will" is also distributed in organisms, both architecturally and functionally. A human has many volitions, and these often compete. The conflicts between short-term interests and long-term interests are important, especially for understanding individual conduct and social behavior. In the relatively rare well-organized person, volitions are arranged hierarchically and are highly unified.

The pluralistic organization and functioning of bodily skills, reasoning abilities, and volitions result from the step-by-step nature of evolution. For as evolution proceeds through the teleological continuum, factors of all three kinds develop together. Moreover, all these factors are the product of the indirect nature of evolutionary development. Genetic plans that produce superior bodily skills, internal reasoning abilities, and volitions that work well together tend to win out over others.

The question before us is whether scientists, operating in a scientific community, can separate facts from values – a question that takes us into the realm of morality and ethics. For the rule, "be honest", presupposes that an individual or institution can distinguish the truth from what it desires. This distinction is important, because truth has adaptive value. Hence we must return to the topic of morality and consider its evolutionary basis (cf. pp. 427–431).

To understand how social institutions, including morality and law, arose in the evolutionary process, we need first to discuss competition and cooperation among organisms, and to distinguish two kinds of interest within an organism.

Since the materials of the environment are limited, organisms compete with each other for the fulfillment of their interests. They may also cooperate toward this end. Cooperation may bring additional benefits, but it also allows an organism to act as if it were going to cooperate and then to make a gain by failing to do so. At the human level this form of behavior is called "cheating", but the concept is quite naturally generalized to cover non-intentional forms of behavior that follow the same pattern and from which human cheating evolved.

Cooperation and cheating have developed throughout evolution. These modes of behavior became explicit and very complicated with the development of human language and various social control mechanisms, including ethics and government. Humans employ sophisticated cheating strategies, which are often successful.

So far we have treated the interests of an organism as unified, whereas organisms usually have many competing interests. One distinction between kinds of interest is particularly relevant to the emergence of social institutions. This is the distinction between what I call "kin interests" and "direct interests".

Kin interests are interests in having viable and successful descendants and relatives, and are thus indirect. In contrast, *direct interests* are interests in the organism's own growth and development and in non-reproductive success. In the case of intentional beings, the latter category includes personal and spiritual achievements.

Satisfaction of kin interests requires maturation and hence partial satisfaction of direct interests. But these two kinds of interests can conflict in the same organism, and often do. Parents seriously harming their own children is an example of such a conflict.

Both kinds of interest result from the indirect and cyclic nature of evolutionary development: a genetic program produces an organism that is tested by the environment and, according to its success, produces further genetic programs; this cycle is iterated indefinitely. The kin interests of an organism are directly associated with its genetic program, the constituents of which have an interest in surviving and spreading. The direct interests of an organism are the interests of the organism per se, apart from its kin interests.

"Direct interest" could be called "organism interest", or, in the case of humans, "self- interest". However, the former term is better because the interests of an organism include kin interests as well as direct interests. Of course, an organism also has interests in the satisfaction of interests of non-kin, but I will not say much about these here.

There are two primary modes of organism reproduction, asexual and sexual, and a logical difference between them is relevant to the origin of social institutions. In *a*sexual or parthenogenetic reproduction, the genetic program of an offspring is a copy of the genetic program of the parent, usually with minor variations (mutations). In sexual reproduction, the genetic program of an offspring is a combination of portions of the parents' genetic programs, with the operations of mutation and crossover playing the dominant roles.

Two consequences flow from this difference. First, there is much more variation in genetic programs under sexual than under asexual reproduction. If one thinks of a genetic program as a hypothesis to be tested in an environment, sexual reproduction produces more variation in the hypotheses generated than does asexual reproduction. Whether this is an adaptive advantage depends, of course, on the nature of the starting hypotheses in relation to the properties of the environment. For example, if a genetic program is already well fitted to a static environment, the cost of trying different programs may exceed the gain.

The second consequence concerns the extent and degree of genetic relatedness in a population, and is relevant to the emergence of social groups. This point can be illustrated by imagining a species whose members have all descended from one "Adam" and "Eve" pair whose genetic programs were very different.

In the *a*sexual case, there will be two extended families, one descended from Adam and one from Eve. Within each family genetic relatedness, and hence kin interests, will be very strong. As a consequence, cooperation with other members of the family will serve kin interests as well as organism interests. This factor will be a strong force for cooperation and the development of sociality within the extended family. But genetic relatedness between families will be extremely weak, and so the motives for cooperation will be weak. It would be natural for the two families to feud, like the Hatfields and the McCoys.

In the sexual case (with outbreeding), there will be one extended family in which everyone is related to everyone else to some degree. Hence there will be a force for cooperation throughout the species, though its strength will not be as great as in the asexual case.

Consider now the difference between asexual and sexual reproduction in a population. Other things being equal, the effect of sexual reproduction with outbreeding is to expand the size of kin groups and so also of social groups. It is then natural for enforcement mechanisms to develop in such a way that individuals that are only weakly connected by kin interest cooperate with each another. This cooperation in turn leads naturally to enforcement procedures for bringing about the cooperation of totally unrelated individuals.

It is worth noting that sexual reproduction requires a new kind of cooperation, cooperation between mates. But it still involves competition, because the interests of the mates are different. This difference applies to kin interests as well as to direct interests, for parents may compete over who will raise the offspring (Dawkins 1976, ch. 9). Thus the interaction of competition and cooperation in a population with sexual reproduction is extremely complicated.

To me, the most interesting extended families occur among insect species and homo sapiens, and thus on two very disparate evolutionary levels. The extended families of some insect species have substantial social structures, including queens, nursemaids, workers, soldiers, farmers, and dairymaids. These social systems are based on instinctive habits, not on culture, language, and intentionality as in the case of humans. Still, the similarities between insect and human social organizations seem remarkable in view of the vast difference in complexity between the two species.

Most social insect species belong to the order hymenoptera, and an important factor in the origin of their social systems is the different genetic structure of male and female. The males are haploid (with a single genetic program), while the females (queen and workers) are diploid (with a double genetic program). It follows logically that a female worker is more closely related to her sister workers than a (female) queen is related to her offspring.

This difference in kin interest makes it more profitable genetically for a female to take care of her sisters than to have offspring. Consequently, there is an adaptive value to a social system based on a queen, a few males, and many worker females. I think insect sociality illustrates the importance of kin interests in the evolution of sociality. This is another example of a probabilistic explanation (cf. pp. 492–495), for sociality, with worker sterility, has evolved eleven times independently among hymenoptera and only once among other insects (in the termites).[8]

I think kin interests are the evolutionary, voluntaristic roots of most sociality, even though many other factors are obviously needed for kin interests to operate in such a way that sociality develops. Especially important at higher evolutionary levels is the transfer of skills, habits, and practices between generations by a training process in which younger individuals learn from older kin. This requires that generations overlap and have sufficient contact opportunities for such learning to take place.

The learning may be imitative or instructional. Moreover, these practices are themselves passed directly from one generation to the next. As evolution proceeds, the transfer of skills and knowledge from one generation to the next becomes institutionalized into educational activities of various kinds. Thus evolution has developed modes for the *direct* transfer of information between generations to supplement *indirect* genetic transfer.

In our human case, sociality is also based on language use and advanced intelligence and skills. In the presence of these and other factors, cooperation among members of extended families led to wider social groups, such as tribes and states, and thence to cultures. The educational process of transferring information now becomes general cultural transfer and extends to non-kin. Let me speculate on how these advances might have come about.

Even though an extended human family is bound together by kin interests, there are normally some conflicts between the direct interests and the kin interests of members of the extended family. These conflicts tend to become stronger as extended families become larger and the average level of kin-relatedness decreases. Then direct interest and kin interest both favor the development of social mechanisms to control such conflicts.

As the extended family evolved into communities, tribes, and small states, these social control mechanisms evolved into the control mechanisms of these larger groups, such as religion, morality, law, and enforcement institutions associated with them. In this way direct interests and kin interests led to the development of *social interest*, or motivation to cooperate and keep agreements with others under suitable circumstances. Thus kin interest is an important evolutionary root of human social organization, morality, and culture, and ultimately of science. (Burks 1986b, Lecture 6)

The social control mechanisms that evolve in this way are necessarily imperfect. Kin interests often conflict with direct interests, and the latter may then prevail over the requirements of morality or law. A more important limitation on the effectiveness of morality and law is the fact that the cooperative pull of kin interest is very weak in large groups. Consequently, individuals often pursue their direct and/or kin interests in violation of morality or law. Such control institutions as the courts and police have arisen to counter this tendency by means of force. But of course, the members of these institutions are in an advantageous position to pursue their own individual and/or kin interests by violating the rules they are created to enforce.

Moral and legal violations occur on various levels of a social system. For example, an individual may follow the mores of a sub-group and in doing so break the mores of the larger society, as in organized crime. Or an individual may break the laws of the society with which he or she is most closely identified, to further either a direct interest or a kin interest. To be a member of a group one must appear to accept its morality, but this allows considerable deviation from it in practice.

As we saw when discussing the nature of free choice earlier (p. 430), the concept of an evolutionarily stable strategy is relevant here. Consider these two limiting cases: at one extreme almost everyone is honest, while at the other extreme almost no one is honest. In the first case it would be very easy to cheat, and the payoff for cheating would be great; the amount of cheating, then, would naturally increase. At the other extreme there would be social chaos, and that would produce a strong force for the emergence of control mechanisms, such as morality and law. The actual situation will fall somewhere in between, that is, there will be a mixture of immorality and morality and a mixture of lawlessness and lawfulness.

Imagine that the alternatives are simplified into a continuum from the two extremes of perfect moral-legal conformity to moral-legal chaos. Then the actual state of society will be represented by some point on this continuum. If the relevant circumstances are more or less fixed, this point will be a "stable point" in the dynamic sense of that term. At this stable point there will be a balance of the forces of immorality, illegality, etc., and the forces of moral, legal, and ideological control. Thus by the very nature of human nature, morality and law are only partly followed.

Generally, this stable point will not be the point of maximum utility, for the interests of the members of a society are directed toward their own fulfillment, and the state of affairs resulting from these interacting forces will not in general maximize utility. This would be so even if the governing moral-legal system should be directed toward maximizing utility. The stable point of a society may evolve, subject to the constraints of the society's history and its environmental resources.

As a background for my analysis of moral and value judgments, I will now recapitulate my views on morality, its origin, and its degree of success.

The moral code of a society is a semiotic control system for guiding and regulating the actions of its free and intentional members so as to contribute to a stable social system of opportunities for goal achievement (Burks 1986b, pp. 86, 88). It evolved as a means of mediating among conflicting interests.

Biological evolution is both competitive and cooperative. The direct interests of organisms are fulfilled by cooperation as well as by competition. The kin interests of organisms constitute additional forces for cooperation among kin and lead to control mechanisms within extended families. In the case of humans, these have evolved into the moral and legal control practices and institutions of societies.

However, because of the strengths of direct interests and kin interests, and their limited scopes, morality and law are necessarily imperfect. Consequently, in any stable society there are significant amounts of immorality and lawbreaking.

The preceding account of the evolutionary origin and consequent nature of morality is only a sketch. Much more needs to be said about it at the analytic level. Empirical data and computer models are needed to develop and establish it, and to learn about the roles played in the evolution of morality by various environmental circumstances and distributions of competing organisms, small as well as large, internal as well as external (cf. sec. 2.3 and pp. 492–495, above). Nevertheless, this account provides a sufficient foundation for discussing the basic nature of value and moral statements and for extending my earlier classification of statements (p. 386) to cover them. I will add two new kinds of statements to the classification: empirical value statements and deontic statements.

Empirical value statements are reports on the successes and failures of goal-directed systems. They are statements about the success or failure of goal-seeking activities, or about the fulfillment or frustration of drives and desires. They apply to any organism or organization of organisms of the teleological continuum (p. 402) and to any equivalent artificial goal-directed systems. The achievements and failures of both kin interests and direct interests are possible subjects of empirical value statements. Thus biological statements about the fitness of organisms or groups of organisms belong to this class of statements.

I think that empirical value statements constitute a well-grounded basic category of statements, distinguishable from other empirically true-or-false statements. Because morality is a human institution, the empirical value statements of most concern to it are those about humans, though statements about other living things also play a role (as in Jainism). Statements about the satisfactions and frustrations of desires, as well as reports on immediate feelings of pleasure and pain, are empirical value statements. The drives and desires involved may be unconscious or conscious. Assertions of individual or social utility are empirical value statements. Judgments about good and bad are empirical value statements if they have an evaluative rather than an obligatory or moral sense.

Since we are concerned with the separability of science from values and moral norms, it is natural to ask whether empirical value statements can be the subject of scientific study, particularly when they are about humans. Being empirical statements about the operation of logical-mechanical systems, they are verifiable in principle. However, the relations between desires and goals, on the one hand, and behavior and its internal consequences, on the other, are very complex. Moreover, data about the internal states of these systems is generally not accessible. All of this is especially true for humans, in whom much of what is relevant occurs below the level of consciousness. In practice, then, it is generally very difficult to verify empirical value statements.

Deontic statements express the rules and injunctions of a morality. They state obligations and rights, that is, what a free and intentional member of a society should do or is morally entitled to. Let us consider next whether deontic statements are reducible to any of the other types of statements in my classification.

Empirical statements characterize how reality is, whereas deontic statements say how it should be. If the rules of a moral system were universally followed, some ideal state of the universe would be attained, such as maximizing utility in a society or fulfilling the rights of all its members. But this is a goal, not an achievement; indeed, on my account of human nature this is a goal that can be only partly achieved in practice. Hence deontic statements are not empirical value statements, and so are not empirically true-or-false.

The relation of deontic statements to rationality and logic is more complex. Reasoning is an important aspect of morality, and in ordinary usage "rational" is closely identified with morality. In this sense, the moral choice is "the rational choice" and an immoral act is "unreasonable". It is often said that maximizing utility in conditions of uncertainty is "the rational thing to do". But this rule does not determine a moral choice, for the utility of a strategy depends on the utilities of its possible consequences.

We have seen that morality is a social control system of a group or society, involving a commitment by members of the group to a moral code. In practice, as we have noted, an adherent sometimes pretends to follow the code while deviating from it. The moral group may vary in size from small to large, may include all humans, and often involves animals indirectly. Moral agents may, and often do, belong to more than one group, and these groups may have conflicting moralities. Thus a morality has a scope of application, the group of people for which it is the morality.

A morality also has a content, its structure of moral rules used to regulate the activities of the individuals of the group. These rules need not be applied to people outside the group, and often are not. For example, the moral rules applied by members of an ethnic or social group to one another are often very different from the rules they apply to outsiders. Similarly, the moral rules applied to the enemy in war are different from the rules applied to one's countrymen and allies. Thus a morality involves loyalty to a specific group and rests on the moral motivations of the individuals in that group.

A moral system has several basic functional aspects: intentional rule-application and moral reasoning; non-cognitive aspects of emotive meaning and imperative force; an intuitive or immediate-habitual aspect; a naturalistic or evolutionary basis; an ideal aspect connected with further development; and perhaps others (Burks 1986b, Lecture 6). Thus moralities have a common structure. But the contents of moralities vary, and so there are many different moral systems, each logically consistent internally.

It is clear from these considerations that deontic statements are not logically true-or-false in my sense of the term (p. 391). Logic doesn't dictate the content of a morality, its scope of application, or the degree to which it is followed. There is nothing illogical in following a different morality or in systematically violating a morality.

I conclude that there are four basic types of statements or modes of truth. These are, in order of increasing dependence: logically true-or-false, inductive, empirically true-or-false (with empirical value statements as a special case), and deontic.

We have now separated out values and morals from empirical science on a theoretical level, first in our conception of a robot that has separate subsystems for information processing and for goal-direction and motivation, and then in our classification of basic types of statements. But these results are not sufficient to show that scientific conclusions can be objective. For it may be that this theoretical separation can never be made in practice. It may be that human abilities, beliefs, and volitions develop together step by step in such a way that value considerations or deontic beliefs will always interfere with the objectivity of scientific conclusions. If so, then the subjectivity claim that Audi has putatively disproved will actually be true.

There are as yet no formalisms or computer simulations of the actual way beliefs and volitions develop. To help focus on what is needed, we will review two formalisms that treat this subject matter in an idealized way and indicate the important issues they omit. We will then suggest how computer simulation might be used to study these issues. The two formalisms are: Bayes' method of finding an unknown probability extended to ideal models of nature (pp. 367–370) and the calculus of choice (pp. 358–360).

As we noted earlier in this section, a Bayesian procedure in which one hypothesis is chosen over alternatives on the basis of its environmental consequences illustrates the indirect nature of evolutionary development. But this model is incomplete in that it does not take account of the cost of making observations or of processing them. Moreover, a Bayesian model assumes that the logic of discovery has already done its work, so that the investigator has the appropriate concepts and statement compounds of them to test. Furthermore, it treats concepts as precise, whereas all empirically applicable concepts are vague, and there is an epistemological cost of reducing vagueness. (Burks 1946b; 1978a, pp. 408–409; 1980a, pp. 173–174, 177–184)

Bayesianism gives no account whatever of motivation. The calculus of choice adds this factor to the calculus of probability, for it treats preferences (of consequences and mixed acts) along with partial beliefs (subjective probabilities). Under the assumption that a subject obeys its axioms in making a complete set of choices in conditions of uncertainty, it allocates degrees of belief to statements, and utilities to consequences, so that the subject is maximizing utility.

But the calculus of choice is also unrealistic in the amount of global planning it presupposes. For its concept of maximum utility is based on a detailed consideration of a very large class of strategies, in other words, of a large class of logically possible universes. For economic reasons, only a few strategies can be considered at one time, and these only for a few levels; compare a checkers or chess playing program. For practical reasons, actual intentional strategies evolve as they are executed. Preferences also evolve, and conflicts among them wax and wane. (*CCR*, sec. 5.6.2)

Thus both of these formalisms presuppose that humans are rational beings with rational preferences and give no account of how beliefs and goals develop interactively step by step. This development process is evolutionary in character and should be studied from that point of view.

Evolution works by indirection, not by means of explicit strategies. It produces comparative solutions, perhaps locally maximal in some sense, but not globally maximal. A good theory of scientific or epistemological "convergence" should take evolution as a model rather than a search for an unknown probability by repeated sampling, as Peirce and Reichenbach did.

The practical limitations of intentional conduct, and of evolutionary processes generally, can be illuminated by a different kind of formalism, that used in the computational search for the maximum value of a function. Consider a multidimensional function that is defined by an algorithm for calculating the value of the function from any given set of values of the variables. Suppose one wishes to find the (approximate) maximum of the function over some finite range. For simple cases one can choose a reference grid, calculate the value of the function at every point of the grid to the desired degree of accuracy, and pick the maximum. Unfortunately, most functions met with in practice are too complicated for this procedure to be feasible.

A common computational search algorithm used to solve such a problem is aptly named "hill climbing". It may be illustrated with a function of a single variable. The computer chooses a value of the variable at random, calculates the value of the function and its derivative at that point, and then moves in the direction of increasing slope to calculate the next function value. This method will find the maximum of a well-behaved function if that function has a single, global peak. But it will generally fail if the function has one or more local maxima which are smaller than the global maximum, for the calculation may get stuck on one of these "false peaks". Hence hill climbing can find a local maximum but not a global one.

Let us examine the relevance of this hill-climbing functional maximization algorithm, first, to biological evolution, and, second, to human growth.

Evolution can be thought of as a search process that tests alternative designs (organisms) and selects from them according to their suitability in relation to the environment. These designs result from genetic programs interacting with the parents and the environment. The representation of an evolutionary system in a function space is complicated, but for our purposes it suffices to associate the variables of the multidimensional space with the alleles of a genetic program, the possible genes at a given allele constituting values of the variable. Each combination of genes determines a point in the coordinate system, and the function value at that point represents the adaptiveness or fitness of that genetic program. Of course, the fitness depends on the nature of the environment and on the fitness of competing organisms. These may change, but for the purpose of this illustration let us assume that they are fixed.

Evolution involves very many local competitions, and its multidimensional adaptive landscape will have many false peaks. The mutation operator in genetics is a local search operator, trying neighborhood positions in the search space, and hence sometimes moving up a gradient of the adaptive landscape (hill climbing). Hence the evolving organism may climb a false peak. However, the crossover operator of genetics searches the space of possibilities in a more global way, causing the system to jump from one area of the search space to another. In other words, mutation results in small jumps in the adaptive landscape, whereas crossover results in large jumps.

Searching for the maximum of a function is relevant to individual decision-making in the following way. Think of the individual as a conscious, intentional robot operating in a set of possible universes based on a cellular automaton framework (see pp. 414, 449). At each moment the robot is in some individual state, from which it can "move" by choice to any of several possible individual states. Associated with each such state is an evaluation function measuring the robot's apparent degree of satisfaction, or personal utility. Within its economic limitations, the robot moves so as to maximize its apparent utility.

Human beliefs and desires evolve together interactively. At each step a choice is made on the basis of a small amount of information and limited consideration of alternatives. Because of this step-by-step evolution, changes in belief interact with changes in desires and goals. It has long been accepted that what one wants is influenced by what one believes. I think it is also the case that one can come to believe what one wants to believe, over a period of time. Similarly, social groups and whole societies may come to believe what they want to believe.

Thus individual decision-making, like evolution, can be viewed as a search for the maximization of a function. The multidimensional representation of such a function may appropriately be called an "adaptive landscape". Moreover, except for radical changes, the step-by-step development of belief and desire in a person's life can be viewed as an attempt to find the maximum of a function by means of hill climbing.

Revolutionary changes in a person's life are radical jumps from one point of the adaptive landscape to another in an attempt to find a better hill to climb. The long-term goal is to reach a high peak, but distant peaks are not visible to the searcher. For the most part, people do local hill climbing. The notion of being stuck on a false peak in the adaptive landscape seems to fit some personal phenomena, such as phobias, compulsions, "hang ups" and obsessions, and some forms of suicide.

We are now ready to address the question raised at the opening of this section: *can* scientific judgments be objective and factual, or *must* they be distorted by moral and value considerations? The use of the modals "can" and "must" are essential here, for many putative scientific judgments are biased by moral and value considerations. Moreover, science has often been subverted to serve some economic, political, or military purpose, and will probably be so subverted again.

Thus the important questions are these: Does the institution of science operate so as to separate empirical truth from values and morals? If so, how does it accomplish this? In particular, how does it overcome natural human tendencies to let desire dominate belief?

Peirce's final-cause theory includes an answer to the first of these questions: objective values are final causes leading evolution inevitably towards an ultimate limit of "concrete reasonableness" (p. 447). But Peirce's theory does not explain how this result is accomplished. The philosophy of logical mechanism must answer all these questions, and do so in terms of the efficient causality of biological, cultural, and scientific evolution.

Because scientists are humans and the social sciences are about humans, I will begin with the natural sciences (the physical and the biological sciences) and treat the social sciences later. The biological sciences do study humans, but mainly as organisms, though the line between the biological and the social sciences cannot be sharp because of the teleological continuum.

The essence of my account of how the natural sciences arrive at value-free conclusions is as follows. Science has evolved the goal of discovering and verifying empirical statements about nature, statements not biased by value and moral considerations. And science has developed reliable methods for achieving this goal.

Thus there are two interrelated dimensions of science, its methods, and its results, the former changing more slowly than the latter. Scientific results are based on reliable methods of observation and experiment. An experiment must be repeatable, so that any qualified scientist can check the claimed result. Theories and models guide data collection and are influenced by the outcomes. By contributing rigor, the use of mathematics and computation in science helps to make empirical conclusions value-free. The long run adaptive value of scientific applications has a similar effect.

Competition in both experiment and theory contributes to the independence of scientific results from values and morals. The tendency of humans to cling to their beliefs is mitigated in science by generational change. This latter is especially important for theories that seem to conflict with common sense, such as non-Euclidean geometry, the relativity of simultaneity and the interchangeability of space with time, the variable curvature and possible finitude of space, and some quantum phenomena. Thus science advances by changes of generation, as does evolution generally.

I think the foregoing considerations establish that the natural sciences can reach value-free, objective conclusions. The natural sciences have developed this goal, and it is being achieved by the continuing institutionalized competition among scientists and interaction between experiment and theory. Let us turn, then, to the social sciences.

As a transition, we pause to examine the consequences of controlled competition in legal proceedings of the English type, since the outcomes of these proceedings are intended to reflect values and norms as well as empirical facts. A trial must come to a definitive empirical conclusion in a single case (e.g., Is the accused guilty or not?) subject to many economic and legal constraints (e.g., It is better that ten guilty persons escape than that one innocent suffer). In contrast, both the social and the natural sciences use probability estimates, and they can wait indefinitely for definitive answers.

Although the judicial system has these practical constraints, in the area of specific human conduct it can sometimes gather better data than the social sciences. Where the stakes are high, much research can be done, and the procedures of depositions, direct examination, and cross-examination, together with the threat of punishment for perjury, enable the judicial system to probe more deeply into human conduct than a social scientist can. For an example, see the account of the ENIAC trial in Burks and Burks, 1988. Had it not been for this trial, the question of who invented the electronic computer could never have been settled.

Of course, even here the data were limited to what people really did, thought, and said; they did not include verifiable information about motives and goals. This circumstance illustrates the limitations of the social sciences. Individuals are complex non-linear systems, groups of individuals are of course more complex, and interacting groups are still more complex. Most of what goes on in individuals and groups occurs below the level of behavior, and, indeed, below the level of consciousness.

Thus the social sciences cannot gather the most important data about the extremely complex non-linear systems they are studying. It is accordingly very difficult for the social sciences to develop deep theories about their subject matter and to have a strong interaction between theory and experiment. As a consequence, their theories are much less robust than those of the natural sciences.

Since the social sciences study human behavior and culture, in which morality and ideology play important roles, these sciences impinge more directly on human interests than do the natural sciences. The results of these sciences may have social and ideological implications that people do not like, and this may produce opposition to the inquiry itself. Renaissance society had difficulty reconciling heliocentric astronomy with its earth-centered ideology. Similarly, contemporary society may have difficulty reconciling the ideological implications of genetic evolution with social and moral views that are rooted historically in transcendental and final-cause accounts of teleology.

For these reasons, it is much easier for values and morals to intrude into the results of the social sciences than into those of the natural sciences.

5. THE PHILOSOPHY OF LOGICAL MECHANISM

The organization of these responses is quite naturally derived from that of the contributors' papers. To put my replies in perspective, and to bring out their potential unity, I conclude with a summary of my philosophical point of view.

I have dealt at length with *man, society*, and *science*, three tightly interrelated systems central to philosophy. There are two very different ways of analyzing these complex systems, and philosophers disagree as to which way is the more fundamental. Most of what I have said in these replies can be organized around this classic contrast.

Man, society, and science are *holistic-coherent systems*. A holistic system is highly integrated, so that its operation depends critically on the interrelated functioning of its parts. The corresponding concept for an epistemological system is "coherence", which involves beliefs that support each other and perhaps also desires interacting with beliefs. These are the senses in which "the whole is greater than the sum of its parts" – the operation of the whole is a non-linear feedback function of the operations of its parts.

All living things are goal-directed, and biological evolution has gradually produced more and more complicated goal-directed systems. Man and, derivatively, society and science, are intentional goal-directed systems. Humans are consciously and freely goal-directed. Goal-directed systems seem to operate under the direction of final causes, as well as in accordance with efficient or mechanical causes.

There are two dominant philosophical traditions concerning the ultimate nature of holistic goal-directed systems and correlated coherent systems. The main tradition is that holism and coherence are ultimate and irreducible: a holistic system cannot be understood in terms of the interrelated operation of parts governed by physical laws; correspondingly, coherent knowledge systems cannot be analyzed by means of formal logical structures.

The opposite and minority tradition is that most if not all holistic and coherent systems are in principle reducible. The most prominent form of this view originated with the Greek atomists, who held that humans are compounds of material atoms operating in accord with mechanical laws. However, this atomistic, mechanical tradition could never give an adequate account of the most unique human abilities: explicit reasoning, intentional goal-directedness, consciousness, and the inner point of view.

Also, the philosophical tradition of mechanism has never provided an adequate explanation of living things and their goal-directedness; nor, more basically, of how biological evolution works. This failure underlies its further failure to account for the advanced human abilities, since these have been produced by an evolutionary process that started from a purely physical system.

My responses to the essays in this volume have been guided by an enriched form of atom-compound theory, which I call "the philosophy of logical mechanism". It augments traditional atom-compound materialism with concepts from logic (broadly conceived), computers, and recent evolutionary genetics. Using this enriched base, the philosophy of logical mechanism gives a reductive account of many holistic-coherent systems. Let me indicate briefly the nature of these reductions.

A digital computer or robot is a compound of primitive building blocks. As a hardware object, the computer or robot is a compound of logical switches, memory elements, input and output devices, and communication elements. It may have input and output devices capable of acting on itself, and it may be self-repairing or even self-reproducing.

For a computer or robot to be efficient it needs to have a well-organized structure. Its atomic parts are typically organized into a hierarchy of many levels, with dynamic feedback cycles both within and between levels. The system will involve energy or drive. A hardware system may be governed by deterministic laws, probabilistic laws, or some combination of these. A deterministic computer is a kind of deductive system.

Communication involves language, and efficient organization requires control. Language also involves atoms and compounds: primitive characters make up words, instructions constitute programs, and data are organized into data structures. Computers operate by means of a hierarchy of programming languages. These logico-linguistic powers enable computers to reason and to simulate.

Computer simulation is a natural extension of the logical formalization of various kinds of reasoning: deduction, induction, decision theory. Until recently formalization was limited to organizing knowledge already acquired by humans. We are now on the threshold of computer systems that will learn from experience and discover new knowledge.

Modern computing systems, viewed as both hardware and soft-ware, and in terms of both constituents and organization, are the basis for an alternative model for complex systems. A *hierarchical-feedback system* is an ordered, dynamic structure of many levels. The atoms of one level are the building blocks of the compounds on the next higher level. There are dynamic interconnections, including cyclic feedback connections, within and between levels.

Many complex non-linear systems of engineering and physics are hierarchical-feedback systems. The laws of these systems can be deterministic, probabilistic, or a mixture of both, and the type of law can change from one level to the next. The dynamical system of a sequence of random coin tossings is a hierarchical system that is probabilistic or chaotic at the macrolevel but deterministic at the mircrolevel.

Let me next summarize what it means to reduce one system to another. To say that a system is reducible to some underlying system is to say that the first system can be embedded in the second. The states of an embedded subsystem are sets of states of the underlying system, and the resultant laws of the embedded subsystem are simplifications of the laws of the underlying system.

For example, the gas law describes an embedded subsystem of a very complicated mechanical system of bouncing hard atoms. Consider next the internal principles operating in the conscious human process of free decision-making. These principles are very different from the laws governing nature, but the former are reducible to the latter in the ways we indicated in our discussion of mind.

Reductions may be practical, or they may be only theoretical, depending on the complexities of the systems and the embedding relation involved. The gas law, despite its limitations, illustrates the practical value of a reduction. Biochemistry and biophysics are subjects based on practical embeddings. Philosophy deals with theoretical reductions.

Organisms are reducible to hierarchical-feedback systems, mixed analog and digital. The languages employed are genetic programs for self-construction and chemical languages for internal communication.

Organic systems are based on fluid and evolving building blocks, structures that are only partially stable. The holism of an organism derives from dynamic cyclical feedback, including the contextually dependent amplification of small changes. The coherence of a human knowledge system has a similar basis.

The philosophy of logical mechanism holds that the holistic-coherent systems of man, society, and science are reducible to hierarchical-feedback systems. The fundamental issue here is the nature of the human person, for if humans are logico-mechanical, clearly society and science are.

Humans are unique in being conscious, self-conscious, free, and intentionally goal-directed. The philosophy of logical mechanism holds that consciousness is organizational in nature: it is a relatively efficient goal-directed control system, having immediate feelings, and capable of free choice. At least in principle, there could be a robot that performed all natural human functions and operated internally in the general way a human does. It would have desires and interests as well as reasoning abilities. This robot would be intentionally goal-directed and capable of making free choices in a society of similar robots. It would be self-conscious as well as conscious, and could take the inner point of view.

In my writings, I have presented several kinds of arguments for this logico-mechanical philosophy: arguments that man is equivalent to a finite deterministic automaton, reductive accounts of our unique abilities, and a logico-mechanical account of the evolutionary process that has produced these abilities.

NOTES

*I am grateful to the National Science Foundation for research support (Grants SE 82–18834, DCR 83–05830, and IRI 86–10225), to Alice Gantt for her typing of the manuscript, and to Alice Burks for her editorial assistance.

1. Since there will be very many references to it, they will be made with the abbreviation *CCR*, rather than in the usual way.

2. Holland's classifier systems are examples. See the references to Holland, Burks 1986c and 1988a, Holland *et al.* 1986, and Holland and Burks 1987 and pending.

3. It should be noted that for the infinite case a universal generalization has zero prior probability and hence is not confirmable. Jaakko Hintikka has shown how to avoid this (Hintikka 1968 and his other papers referred to therein).

4. See, for example, the proceedings of a conference on the physics of computation, published in the *International Journal of Theoretical Physics*, 21, 1982, nos. 3/4, 6/7,

and 12; the proceedings of a conference on cellular automata and physics, published in *Physica* 10D. 1984, nos. 1–2; Langton 1986 and the proceedings of his interdisciplinary workshop "Artificial Life: The Synthesis and Simulation of Living Systems", *Artificial Life*, Addison-Wesley, New York, 1989.

5. More generally, the question of whether a logical structure or a formal model fits some natural system is empirical, not logical. This principle applies to the claim that a particular formal language is a good model of some aspect of a natural language.

6. Though we can have immediate experiences of our own mental sequences, we cannot abstract the concept of causal necessity from these experiences (*CCR*, sec. 10.2.2).

7. For the concept of an evolutionary stable strategy see Dawkins 1976, ch. 5, and Maynard Smith 1982, chs. 5–7.

8. Dawkins 1976, ch. 10. William D. Hamilton developed the probabilistic explanation of the role of the haploid-diploid difference in the evolution of hymenoptera sociality. R. L. Trivers and H. Hare explained the adaptive value of workers biasing the sex ratio in favor of females.

REFERENCES

Allais, Maurice and Ole Hagen (eds.): 1979. *Expected Utility Hypotheses and the Allais Paradox*. Reidel, Dordrecht, Holland.

Axelrod, Robert: 1984. *The Evolution of Cooperation*. Basic Books, New York.

Brandt, Richard: 1979. *A Theory of the Good and the Right*. Oxford University Press, Oxford.

Burks, Alice and Arthur Burks: 1988. *The First Electronic Computer: The Atanasoff Story*. University of Michigan Press, Ann Arbor.

Burks, Arthur: 1946a. 'Laws of Nature and Reasonableness of Regret'. *Mind* **55**, 170–172.

Burks, Arthur: 1946b. 'Empiricism and Vagueness'. *The Journal of Philosophy* **43**, 477–486.

Burks, Arthur: 1946c. 'Peirce's Theory of Abduction'. *Philosophy of Science* **13**, 301–306.

Burks, Arthur: 1949. 'Icon, Index, and Symbol'. *Philosophy and Phenomenological Research* **9**, 673–689.

Burks, Arthur: 1951a. 'A Theory of Proper Names'. *Philosophical Studies* **2**, 36–45.

Burks, Arthur: 1951b. 'The Logic of Causal Propositions'. *Mind* **60**, 363–382.

Burks, Arthur: 1951c. Introduction to Peirce Selections. *Classic American Philosophers*, pp. 41–53, 113. Max Fisch *et al.* (eds.). Appleton-Century-Crofts, New York.

Burks, Arthur: 1951d. 'Reichenbach's Theory of Probability and Induction'. *The Review of Metaphysics* **4**, 377–393.

Burks, Arthur: 1951e. Review of Rudolf Carnap's *Logical Foundations of Probability*. *The Journal of Philosophy* **48**, 524–535.

Burks, Arthur: 1953a. 'The Presupposition Theory of Induction'. *Philosophy of Science* **20**, 177–197.

Burks, Arthur: 1953b. 'Justification in Science'. *Academic Freedom, Logic, and Religion*, pp. 109–125. Morton White (ed.). University of Pennsylvania Press, Philadelphia.

Burks, Arthur: 1953c. Review of Carnap's *The Continuum of Inductive Methods*. *The Journal of Philosophy* **50**, 731–734.

Burks, Arthur: 1955a. 'On the Presuppositions of Induction'. *The Review of Metaphysics* **8**, 574–611.

Burks, Arthur: 1955b. 'Dispositional Statements'. *Philosophy of Science* **22**, 175–193.

Burks, Arthur: 1963. 'On the Significance of Carnap's System of Inductive Logic for the Philosophy of Induction'. *The Philosophy of Rudolf Carnap*, pp. 739–759. P. A. Schilpp (ed.). Open Court, LaSalle, Illinois.

Burks, Arthur: 1964. 'Peirce's Two Theories of Probability'. *Studies in the Philosophy of Charles Sanders Peirce*, Second Series, pp. 141–150. Edward Moore and Richard Robin (eds.). University of Massachusetts Press, Amherst.

Burks, Arthur: 1967. 'Ontological Categories and Language'. *The Visva-Bharati Journal of Philosophy* **3**, February, 25–46. University of Visva-Bharati, Santiniketan, West Bengal, India.

Burks, Arthur: 1968. 'The Pragmatic-Humean Theory of Probability and Lewis' Theory'. *The Philosophy of C. I. Lewis*, pp. 415–463. P. A. Schilpp (ed.). Open Court, La-Salle, Illinois.

Burks, Arthur (ed.): 1970. *Essays on Cellular Automata*. University of Illinois Press, Urbana. Burks, "Von Neumann's Self-Reproducing Automata", is reprinted in von Neumann 1986, pp. 491–552.

Burks, Arthur: 1972. 'Logic, Computers, and Men'. *Proceedings and Addresses of the American Philosophical Association* **46**, 39–57.

Burks, Arthur: 1975a. 'Models of Deterministic Systems'. *Mathematical Systems Theory* **8**, 295–308.

Burks, Arthur: 1975b. 'Logic, Biology, and Automata – Some Historical Reflections' *International Journal of Man-Machine Studies* **7**, 297–312.

Burks, Arthur: 1977. *Chance, Cause, Reason – An Inquiry into the Nature of Scientific Evidence*. University of Chicago Press, Chicago.

Burks, Arthur: 1979a. 'Computer Science and Philosophy'. *Current Research in Philosophy of Science*, pp. 399–420. Peter Asquith and Henry Kyburg (eds.). Philosophy of Science Association, East Lansing, Michigan.

Burks, Arthur: 1979b. 'Computers and Control in Society'. *Nature and System* **1**, 231–243.

Burks, Arthur: 1980a. 'Enumerative Induction versus Eliminative Induction'. *Applications of Inductive Logic*, pp. 172–189. L. Jonathan Cohen and Mary Hesse (eds.). Oxford University Press, Oxford.

Burks, Arthur: 1980b. 'From ENIAC to the Stored Program Computer: Two Revolutions in Computers'. *A History oi Computing in the Twentieth Century*, pp. 311–344. Nicholas Metropolis, Jack Howlett, and Gian-Carlo Rota (eds.). Academic Press, New York.

Burks, Arthur: 1980c. 'Man: Sign or Algorithm? A Rhetorical Analysis of Peirce's Semiotics'. *Transactions of the Charles S. Peirce Society* **16**, 279–292.

Burks, Arthur: 1981. 'Programming and Structural Changes in Parallel Computers'. *CONPAR 81: Conference on Analyzing Problem Classes and Programming for Parallel Computing, Nurnberg, June 10–12. 1981, Proceedings*, pp. 1–24. Wolfgang Handler (ed.). Springer-Verlag, Berlin.

Burks, Arthur: 1984. 'Computers, Control, and Intentionality'. *Science, Computers, and the Information Onslaught*, pp. 29–55. Donald Kerr, Karl Braithwaite, N. Metropolis, David Sharp, and Gian-Carlo Rota (eds.). Academic Press, New York.

Burks, Arthur: 1986a. 'An Architectural Theory of Functional Consciousness'. *Current Issues in Teleology*, pp. 1–14. Nicholas Rescher (ed.). University Press of America, New York.

Burks, Arthur: 1986b. *Robots and Free Minds*. College of Literature, Science and the Arts, The University of Michigan, Ann Arbor. Includes Burks 1972 and 1979b.

Burks, Arthur: 1986c. 'A Radically Non-von Architecture for Learning and Discovery'. *CONPAR 86: Conference on Algorithms and Hardware for Parallel Processing, Aachen, September 17–19, Proceedings*, pp. 1–17. Wolfgang Handler (ed.). Springer-Verlag, Berlin.

Burks, Arthur: 1988a. 'The Logic of Evolution, and the Reduction of Holistic-Coherent Systems to Hierarchical-Feedback Systems'. *Causation in Decision, Belief Change in Statistics*, pp. 135–191. William Harper and Bryan Skyrms (eds.). Kluwer Academic Publishers, Dordrecht, Holland.

Burks, Arthur: 1988b. 'Teleology and Logical Mechanism'. *Synthese* 76 (1988) 333–370.

Burks, Arthur, H. H. Goldstine, and John von Neumann: 1946. *Preliminary Discussion of the Logical Design of an Electronic Computing Instrument*. Institute for Advanced Study, Princeton, 2nd edition, 1947, pp. vi + 42. Reprinted in William F. Aspray and Arthur W. Burks (eds.), *Papers of John von Neumann on Computers and Computer Theory*. MIT Press, Cambridge, Massachusetts, 1986, pp. 97–142.

Burks, Arthur and J. B. Wright: 1953. 'Theory of Logical Nets'. *Proceedings of the Institute of Radio Engineers* 41, 1357–1365.

Burks, Arthur and I. M. Copi: 1956. 'The Logical Design of an Idealized General-Purpose Computer'. *Journal of the Franklin Institute* 261, March, 299–314, and 261, April, 421–436.

Burks, Arthur and Hao Wang: 1957. 'The Logic of Automata'. *Journal of the Association for Computing Machinery* 4, April, 193–218, and 4, July, 279–297.

Burks, Arthur and Alice Burks: 1981. 'The ENIAC: First General-Purpose Electronic Computer'. *Annals of the History of Computing* 3, 310–389. Comments by John V. Atanasoff, J. G. Brainerd, J. Presper Eckert and Kay Mauchly, Brian Randell, and Conrad Zuse, together with the authors' responses to these comments, pp. 389–399.

Burks, Arthur and Alice Burks: 1982. Comment on two historical pieces by John W. Mauchly, *Annals of the History of Computing* 4, 254–256. Reply to comment by Byron E. Phelps, *ibid.*, pp. 285–287.

Carnap, Rudolf: 1950. *Logical Foundations of Probability*. 2nd edition, 1962. University of Chicago Press, Chicago.

Carnap, Rudolf: 1952. *The Continuum of Inductive Methods*. University of Chicago Press, Chicago.

Carnap, Rudolf: 1963. 'Replies and Systematic Expositions'. The *Philosophy of Rudolf Carnap*, pp, 859–1013. P. A. Schilpp (ed.). Open Court, LaSalle, Illinois.

Cliff, R., R. Freitas, R. Laing, G. von Tiesenhausen: 1982. 'Replicating Systems Concepts: Self-Replicating Lunar Factory and Demonstration'. *Advanced Automation for Space Missions*, pp. 189–335. R. Freitas and W. P. Gilbreath (eds.). NASA Conference Publication 2255.

Codd, E. G.: 1968. *Cellular Automata*. Academic Press, New York.

Cohen, L. Jonathan: 1970. *The Implications of Induction*. Methuen, London.

Cohen, L. Jonathan: 1977. *The Probable and the Provable*. Clarendon Press, Oxford.

Cohen, L. Jonathan: 1980. 'What has Inductive Logic to do with Causality?'. *Applications of Inductive Logic*, pp. 156–171. L. Jonathan Cohen and Mary Hesse (eds.). Oxford University Press, Oxford.

Cohen, L. Jonathan and Mary Hesse (eds.): 1980. *Applications of Inductive Logic*. Oxford University Press, Oxford.

Dawkins, Richard: 1976. *The Selfish Gene*. Oxford University Press, Oxford.

Dawkins, Richard: 1982. *The Extended Phenotype*. W. H. Freeman, San Francisco.

Elster, Jon: 1979. *Ulysses and the Sirens – Studies in Rationality and Irrationality*. Cambridge University Press, Cambridge.

Fisch, Max (general editor): 1966, Preface to the Fifth Printing. *Classic American Philosophers*. Appleton-Century- Crofts, New York. (The Peirce section was edited by Arthur Burks.)

Fisher, R. A.: 1929. *The Genetical Theory of Natural Selection*. Reprinted in 1958 by Dover Publications, New York.

Frankena, William: 1973. *Ethics*, 2nd edition. Prentice-Hall, Englewood Cliffs, New Jersey.

Fredkin, E. and T. Toffoli: 1982. 'Conservative Logic'. *International Journal of Theoretical Physics* **21**, 219–253.

Hamilton, W. D.: 1982. 'Pathogens as Causes of Genetic Diversity in Their Host Populations'. *Population Biology of Infectious Diseases*. R. M. Anderson and R. M. May (eds.). Springer-Verlag, Berlin.

Hamilton, W. D. and Marlene Zuk: 1982. 'Heritable Fitness and Bright Birds: A Role for Parasites?'. *Science* **218**, 384–386.

Hintikka, Jaakko: 1968. 'Induction by Enumeration and Induction by Elimination'. *The Problem of Inductive Logic*, pp. 191–216. I. Lakatos (ed.). North Holland, Amsterdam.

Holland, John: 1975. *Adaptation in Natural and Artificial Systems – An Introductory Analysis with Applications to Biology, Control, and Artificial Intelligence*. University of Michigan Press, Ann Arbor.

Holland, John: 1976. 'Adaptation'. *Progress in Theoretical Biology* **4**, pp. 263–296. R. Rosen and F. M. Snell (eds.). Academic Press, New York.

Holland, John: 1980. 'Adaptive Algorithms for Discovering and Using General Patterns in Growing Knowledge Bases'. *International Journal of Policy Analysis and Information Sciences* **2**, 217–240.

Holland, John: 1984. 'Genetic Algorithms and Adaptation'. *Adaptive Control in Ill-Defined Systems*, pp. 317–333. O. G. Selfridge, E. L. Rissland, and M. A. Arbib (eds.). Plenum Press, New York.

Holland, John: 1985. 'Properties of the Bucket-Brigade Algorithm'. *Proceedings of an International Conference on Genetic Algorithms and Their Applications, July 24–26*, pp. 1–7. John J. Grefenstette (ed.). Carnegie-Mellon University, Pittsburgh.

Holland, John: 1986. 'Escaping Brittleness: The Possibilities of General-Purpose Learning Algorithms Applied to Parallel Rule-Based Systems'. *Machine Learning II*, forthcoming. R. S. Michalski, J. G. Carbonell, and T. M. Mitchell (eds.).

Holland, John, Keith Holyoak, Richard Nisbett, and Paul Thagard: 1986. *Induction:*

Processes of Inference, Learning and Discovery. MIT Press, Cambridge, Massachusetts.

Holland, John and Arthur Burks: 1987. 'Adaptive Computer System Capable of Learning and Discovery'. United States Patent 4,697,242.

Holland, John and Arthur Burks: pending. 'Method of Controlling a Classifier System'. Patent applied for.

Hume, David: 1777. An Enquiry Concerning Human Understanding. In *Enquiries*, L. A. Selby-Bigge (ed.), 2nd edition. Oxford University Press, Oxford, 1902.

Kant, Immanuel: 1781. *Critique of Pure Reason*. N. K. Smith (ed.). Macmillan, London, 1933.

Kolata, Gina: 1986. 'What Does It Mean to be Random?'. *Science* **231**, 1068–1070.

Laing, Richard: 1975. 'Some Alternative Reproductive Strategies in Artificial Molecular Machines'. *Journal of Theoretical Biology* **64**, 63–84.

Laing, Richard: 1977. 'Automaton Models of Reproduction by Self-Inspection'. *Journal of Theoretical Biology* **66**, 437–456.

Langton, Christopher: 1986. 'Studying Artificial Life with Cellular Automata'. Physica **22D**, pp. 120–149.

Leibniz, Gottfried: 1714. (Paul and Anne Schrecker, eds. and translators.) *Monadology and Other Philosophical Essays*. Bobbs-Merrill, Indianapolis, Indiana, 1965.

Lewis, C. I.: 1946. *An Analysis of Knowledge and Valuation*. Open Court, LaSalle, Illinois.

Margolus, N.: 1984. 'Physics-Like Models of Computation'. *Physica* **10D**, 81–95.

Maynard Smith, John: 1982. *Evolution and the Theory of Games*. Cambridge University Press, Cambridge.

Nelson, R. J.: 1982. *The Logic of Mind*. Reidel, Dordrecht, Holland.

Parker, De Witt: 1931. *Human Values*. Harper Brothers, New York.

Peirce, Charles Sanders: 1931. *Collected Papers*. Vols. 1–6 edited by Charles Hartshorne and Paul Weiss, 1931–35. Vols. 7–8 edited by Arthur W. Burks, 1958. Harvard University Press, Cambridge, Massachusetts. References will be made in the standard form: "3.417" means "vol. 3, par. 417".

Peirce, Charles Sanders: 1982. *Writings of Charles S. Peirce, A Chronological Edition*. Max H. Fisch, Christian J. W. Kloesel, Edward C. Moore, *et al.* (eds.). Indiana University Press, Bloomington. Vols. 1–3 published, others to follow.

Ramsey, Frank: 1931. *The Foundations of Mathematics*. R. B. Braithwaite (ed.). Harcourt Brace, New York.

Rosenkrantz, Roger D.: 1977. *Inference, Method and Decision*. Reidel, Dordrecht, Holland.

Rosenkrantz, Roger D.: 1980. Review of *Chance, Cause, Reason. Philosophy of Science* **47**, 329–332.

Santayana, George: 1923. *Scepticism and Animal Faith*. Charles Scribner, New York.

Shimony, Abner: 1988. 'The Reality of the Quantum World'. *Scientific American* **258**, 46–53.

Skyrms, Brian: 1980a. *Causal Necessity*. Yale University Press, New Haven, Connecticut.

Skyrms, Brian: 1980b. Review of *Chance, Cause, Reason. Theory and Decision* **12**, 299–309.

Stevenson, Charles: 1944. *Ethics and Language*. Yale University Press, New Haven, Connecticut.

Suppe, Frederick (ed.): 1974. *The Structure of Scientific Theories*. University of Illinois Press, Urbana. 2nd edition, 1977, with an afterword.

Toffoli, Tommaso: 1977. 'Cellular Automaton Mechanics'. Ph.D. Thesis, Logic of Computers Group, The University of Michigan, Ann Arbor. Summarized in 'Computation and Construction Universality of Reversible Cellular Automata'. *Journal of Computer and System Sciences* **15**, 213–231.

Uchii, Soshichi: 1972. 'Inductive Logic with Causal Modalities: A Probabilistic Approach'. *Philosophy of Science* **39**, 162–178.

Uchii, Soshichi: 1973. 'Inductive Logic with Causal Modalities: A Deterministic Approach'. *Synthese* **26**, 264–303.

Uchii, Soshichi: 1977. 'Induction and Causality in a Cellular Space'. *PSA* **2**, 448–461. Peter Asquith and Frederick Suppe (eds.). Philosophy of Science Association, East Lansing, Michigan.

von Kries, Johannes: 1886. *Die Principien der Wahrscheinlichkeitsrechnung – Eine Logische Untersuchung*. J. C. B. Mohr, Freiburg in Briesgau.

von Neumann, John: 1966. *The Theory of Self-Reproducing Automata*. Edited and completed by Arthur W. Burks. University of Illinois Press, Urbana.

von Neumann, John: 1986. *Papers of John von Neumann on Computers and Computer Theory*. William Aspray and Arthur Burks (eds.). MIT Press, Cambridge, Massachusetts.

Wiener, Norbert: 1948. *Cybernetics, Control and Communication*. Wiley, New York.

Zeng-yuan, Yue and Zhang Bin: 1985. 'On the Sensitive Dynamical System and the Transition from the Apparently Deterministic Process to the Completely Random Process'. *Applied Mathematics and Mechanics* (English edition) **6**, 193–211. Shanghai University of Technology. Available from J. C. Balzer AG Scientific Publishing Co., Basel Switzerland.

Zeigler, Bernard: 1976. *Theory of Modelling and Simulation*. John Wiley, New York.

BIBLIOGRAPHY OF ARTHUR W. BURKS

Books

1. With John von Neumann and H. H. Goldstine. *Preliminary Discussion of the Logical Design of an Electronic Computing Instrument*. Princeton, Institute for Advanced Study, 1946. Second edition, 1947, pp. vi + 42.
2. Editor of the last two volumes of *Collected Papers of Charles Sanders Peirce*. Cambridge: Harvard University Press, 1958.
   Vol. VII, *Science and Philosophy*, pp. xiv + 415.
   Vol. VIII, *Reviews, Correspondence, and Bibliography*, pp. xii + 352.
3. Edited and completed John von Neumann's *Theory of Self-Reproducing Automata*. Urbana: University of Illinois Press, 1966, pp. xix + 388.
4. *Essays on Cellular Automata*. Editor and contributor. Urbana: University of Illinois Press, 1970, pp. xxvi + 375.
   My contributions are: The Introduction, pp. xi-xxvi; "von Neumann's Self-Reproducing Automata," pp. 3–64; reprint of my "Programming and the Theory of Automata," (1963), pp. 65–83; and reprint of my "Towards a Theory of Automata Based on More Realistic Primitive Elements" (1963), pp. 84–102.
5. *Chance, Cause, Reason – An Inquiry into the Nature of Scientific Evidence*. Chicago: University of Chicago Press, 1977, pp. xvi + 694. Paperback edition, 1979.
6. *Robots and Free Minds*. Ann Arbor: College of Literature, Science, and the Arts, University of Michigan, 1986, pp. vi + 97.
   This contains items 41, 48, 57 below and three unpublished lectures.
7. With William Aspray, edited *Papers of John von Neumann on Computers and Computer Theory*. Cambridge: MIT Press, 1987, pp. xviii + 624.
   The introductions are all by me, and my "von Neumann's Self-Reproducing Automata" (from item 4 above) is reprinted at pp. 491–552.
8. With Alice R. Burks, *The First Electronic Computer: The Atanasoff Story*. Ann Arbor: University of Michigan Press, 1988, pp. xii + 387.
9. *The Philosophy of Logical Mechanism*, edited by Merrilee Salmon. Dordrecht, Holland: Kluwer Academic Publishers, 1989.
   This is a festschrift on my philosophy, with replies by me.

Articles

1. "Peirce's Conception of Logic as a Normative Science." *The Philosophical Review 52* (March, 1943) 187–193.
2. With Paul Weiss. "Peirce's Sixty-Six Signs." *The Journal of Philosophy 42* (July, 1945) 383–388.
3. "Laws of Nature and Reasonableness of Regret." *Mind 55* (April, 1946) 170–172.
4. "Super Electronic Computing Machine." *Electronic Industries 5* (July, 1946) 62–67, 96.
5. "Empiricism and Vagueness." *The Journal of Philosophy 43* (August, 1946) 477–486.

6. "Peirce's Theory of Abduction." *Philosophy of Science 13* (October, 1946) 301–306.
7. "Electronic Computing Circuits of the ENIAC." *Proceedings of the Institute of Radio Engineers 35* (August, 1947) 756–767.
8. "Icon, Index, and Symbol." *Philosophy and Phenomenological Research 9* (June, 1949) 673–689.
9. With I. M. Copi. "Lewis Carroll's Barber Shop Paradox." *Mind 59* (April, 1950) 219–222.
10. "The Logic of Programming Electronic Digital Computers." *Industrial Mathematics 1* (1950) 36–52.
11. *Classic American Philosophers.* Ed. with Max Fisch *et al.* New York: Appelton-Century-Crofts, 1951, pp. vii + 493. Chapter I: "Charles Sanders Peirce," pp. 41–113.
12. "Reichenbach's Theory of Probability and Induction." *The Review of Metaphysics 4* (March, 1951) 377–393.
13. "A Theory of Proper Names." *Philosophical Studies 2* (April, 1951) 36–45.
14. "The Logic of Casual Propositions." *Mind 60* (July, 1951) 363–382. Abstract in *The Journal of Symbolic Logic 15* (March, 1950) 78.
15. Review of Rudolf Carnap's *Logical Foundations of Probability. The Journal of Philosophy 48* (August, 1951) 524–535.
    Review of Carnap's *The Continuum of Inductive Methods. Ibid. 50* (November, 1953), 731–734.
16. "The Presupposition Theory of Induction." *Philosophy of Science 20* (July, 1953) 177–197.
17. With J. B. Wright. "Theory of Logical Nets." *Proceedings of the Institute of Radio Engineers 41* (October, 1953) 1357–1365.
18. "Justification in Science." *Academic Freedom, Logic, and Religion,* pp. 109–125. Edited by Morton White. Philadelphia: University of Pennsylvania Press, 1953.
19. With Don Warren and J. B. Wright. "An Analysis of a Logical Machine Using Parenthesis-Free Notation." *Mathematical Tables and Other Aids to Computation 8* (April, 1954) 53–57 and frontispiece.
20. With Robert McNaughton, C. H. Pollmar, Don Warren, and J. B. Wright. "Complete Decoding Nets: General Theory and Minimality." *Journal of the Society for Industrial and Applied Mathematics 2* (December, 1954) 201–243.
21. "On the Presuppositions of Induction." *The Review of Metaphysics 8* (June, 1955) 574–611.
22. "Dispositional Statements." *Philosophy of Science 22* (July, 1955) 175–193.
23. With Robert McNaughton, C. H. Pollmar, Don Warren, and J. B. Wright. "The Folded Tree." *Journal of the Franklin Institute 260* (July, 1955) 9–24 and *260* (August, 1955) 115–126.
24. With I. M. Copi. "The Logical Design of an Idealized General-Purpose Computer." *Journal of the Franklin Institute 261* (March, 1956) 299–314, and *261* (April, 1956) 421–436.
25. With Hao Wang. "The Logic of Automata." *Journal of the Association for Computing Machinery 4* (April, 1957) 193–218 and *4* (July, 1957) 279–297.
26. "The Logic of Fixed and Growing Automata." *Proceedings of an International Symposium on the Theory of Switching, 2–5 April 1957.* Cambridge: Harvard University Press, 1959, Part I, pp. 147–188.

27. With James Alexander and Donald Flanders. "Arithmetic Unit for Digital Computer." United States Patent 2,936,115, May 10, 1960.
28. "Computation, Behavior, and Structure in Fixed and Growing Automata." Original paper and discussion in *Self-Organizing Systems*, pp. 282–311, 312–314. Edited by Marshall Yovits and Scott Cameron. New York: Pergamon Press, 1960.
    A revised version in *Behavioral Science 6* (January, 1961) 5–22.
29. Material on logic and computers. *Computer Handbook*, pp. 14–13 to 14–23. Edited by Harry Huskey and G. A. Korn. New York; McGraw Hill, 1962.
30. With J. B. Wright. "Sequence Generators and Digital Computers." *Recursive Function Theory* (Proceedings of Symposia in Pure Mathematics, Vol. V), pp. 139–199. Providence, Rhode Island: American Mathematical Society, 1962.
    Abstract and discussion in *Proceedings of the International Conference on Information Processing, UNESCO, Paris 15–20 June 1959, p. 425*. Paris UNESCO, 1960.
31. With J. B. Wright. "Sequence Generators, Graphs, and Formal Languages." *Information and Control 5* (September, 1962) 204–212.
32. "Programming and the Theory of Automata." *Computer Programming and Formal Systems*, pp. 100–117. Edited by P. Braffort and D. Hirschberg. Amsterdam: North-Holland Publishing Company, 1963.
33. "Toward a Theory of Automata Based on More Realistic Primitive Elements." *Information Processing 1962, Proceedings of IFIP Congress 62*, pp. 379–385. Edited by C. M. Popplewell. Amsterdam: North Holland Publishing Company, 1963.
34. "On the Significance of Carnap's System of Inductive Logic for the Philosophy of Induction." *The Philosophy of Rudolf Carnap*, pp. 739–759. Edited by P. A. Schilpp. La Salle, Illinois: Open Court, 1963.
35. "Peirce's Two Theories of Probability." *Studies in the Philosophy of Charles Sanders Peirce*, Second Series, pp. 141–150. Edited by Edward C. Moore and Richard S. Robin. Amherst: University of Massachusetts Press, 1964.
36. "Cellular Automata." In Russian. *Theory of Finite and Probabilistic Automata*, pp. 100–111. Edited by M. A. Gavrilov. Moscow: Nauka, 1965.
37. With Richard Laing. "The Propositional Calculus." This is a revision of my material in *Computer Handbook* (1962). *System Engineering Handbook*, pp. 41–1 to 41–11. Edited by Robert Machol *et al.* New York: McGraw-Hill, 1965.
38. "Automata Models of Self-Reproduction." *Information Processing 7* (May, 1966) 121–123. Information Processing Society of Japan, Tokyo.
39. "Ontological Categories and Language." *The Visva-Bharati Journal of Philosophy 3* (February, 1967) 25–46. Visva-Bharati (University), Santiniketan, West Bengal, India.
40. "The Pragmatic-Humean Theory of Probability and Lewis' Theory." *The Philosophy of C. I. Lewis*, pp. 415–463. Edited by P. A. Schilpp. La Salle, Illinois: Open Court, 1968.
41. "Logic, Computers, and Men." *Proceedings and Addresses of the American Philosophical Association 46* (1972–73) 39–57.
42. "Cellular Automata and Natural Systems." *Cybernetics and Bionics* (Proceedings of the 5th Congress of the Deutsche Gesellschaft für Kybernetik, Nürnberg, March 28–30, 1973), pp. 190–204. Edited by W. D. Keidel, W. Händler, and M. Spreng. Munich: R. Oldenbourg, 1974.
43. "Models of Deterministic Systems." *Mathematical Systems Theory 8* (1975) 295–308.

44. "Logic, Biology, and Automata – Some Historical Reflections." *International Journal of Man-Machine Studies 7* (1975) 297–312.

45. Review of Charles Peirce, *The New Elements of Mathematics* (2600 pp.), edited by Carolyn Eisele. *Bulletin of the American Mathematical Society 84* (1978) 913–918.

46. Preface, *Values and Morals, Essays in Honor of William Frankena, Charles Stevenson, and Richard Brandt*, pp. vii-xvii. Edited by A. Goldman and J. Kim. Dordrect, Holland: Reidel, 1978.

47. "Computer Science and Philosophy." *Current Research in Philosophy of Science*, pp. 399–420. Edited by Peter Asquith and Henry Kyburg. East Lansing, Michigan: Philosophy of Science Association, 1979.

48. "Computers and Control in Society." *Nature and System 1* (December, 1979) 231–243.

49. "Enumerative Induction versus Eliminative Induction." *Applications of Inductive Logic*, pp. 172–189. Edited by L. Jonathan Cohen and Mary Hesse. Oxford: Oxford University Press, 1980.

50. With Richard Brandt. Preface to William Frankena, *Thinking About Morality*, pp. vii-xii. Ann Arbor: University of Michigan Press, 1980.

51. "From ENIAC to the Stored Program Computer: Two Revolutions in Computers." *A History of Computing in the Twentieth Century*, pp. 311–344. Edited by Nicholas Metropolis, Jack Howlett, and Gian-Carlo Rota. New York: Academic Press, 1980.

52. "Man: Sign, or Algorithm? A Rhetorical Analysis of Peirce's Semiotics." *Transactions of the Charles S. Peirce Society 16* (1980) 279–292.
     Also published in *Image and Code*, pp. 57–70. Edited by Wendy Steiner. Ann Arbor, Michigan: Michigan Studies in the Humanities, Horace H. Rackham School of Graduate Studies, University of Michigan, 1981.

53. "Programming and Structural Changes in Parallel Computers." *CONPAR 81: Conference on Analyzing Problem Classes and Programming for Parallel Computing, Nürnberg, June 10–12, 1981, Proceedings*, pp. 1–24. Edited by Wolfgang Händler. Berlin: Springer-Verlag, 1981.

54. With Alice R. Burks. "The ENIAC: First General-Purpose Electronic Computer." *Annals of the History of Computing 3* (October, 1981) 310–389. Comments by John V. Atanasoff, J. G. Brainerd, J. Presper Eckert and Kay Mauchly, Brain Randell, and Conrad Zuse together with the authors' responses to these comments, pp. 389–399.

55. With Alice Burks. Comment on two historical pieces by John W. Mauchly, *Annals of the History of Computing 4* (July, 1982) 254–256. Reply to comment by Byron E. Phelps, *ibid.*, pp. 285–287.

56. "Computers, Control, and Intentionality." *Science, Computers, and the Information Onslaught*, pp. 29–55. Edited by Donald Kerr, Karl Braithwaite, N. Metropolis, David Sharp, Gian-Carlo Rota. New York: Academic Press, 1984.

57. "Computers and Metaphysics." (Distinguished Senior Faculty Lecture, University of Michigan, 26 Oct., 1982). Chinese translation by Hu Yao-Ding. *Zhexue Yicong (Philosophy Translations)*, number 3, 1984, pp. 16–23. Chinese Academy of Social Sciences, Beijing.

58. "Digital Machine Functions" (pp. 89–97) and two lectures on numerical mathematical methods (pp. 211–215 and 369–372). In *The Moore School Lectures: Theory and*

Techniques for Design of Electronic Digital Computers. Lectures given at the Moore School of Electrical Engineering, University of Pennsylvania, 1946, and originally edited by George W. Patterson and issued as reports in 1947 and 1948. Edited by Martin Campbell-Kelly and Michael R. Williams. Cambridge: MIT Press, 1985.

59. "An Architectural Theory of Functional Consciousness." *Current Issues in Teleology*, pp. 1–14. Edited by Nicholas Rescher. New York: University Press of America, 1986.

60. "A Radically Non-von Neumann Architecture for Learning and Discovery." *CONPAR 86: Conference on Algorithms and Hardware for Parallel Processing, September 17–19, Proceedings*, pp. 1–17. Wolfgang Händler et al. (eds.). Berlin: Springer-Verlag, 1986.

61. "Technology, Intentionality, and Consciousness." *Acta Psychologica Sinica 19* (1987) 158–166. In Chinese, with English abstract.

62. With John H. Holland. "Adaptive Computing System Capable of Learning and Discovery." U.S. Patent 4,697,242. September 29, 1987.

63. With Alice R. Burks. Letter to the editor on who invented the electronic computer. *Physics Today 46* (1987) 110–112.

64. "The Logic of Evolution, and the Reduction of Coherent-Holistic Systems to Hierarchical-Feedback Systems." *Causation in Decision, Belief Change and Statistics*, pp. 135–191. Edited by William Harper and Bryan Skyrms. Dordrecht, Holland: Kluwer Academic Publishers, 1988.

65. "Teleology and Logical Mechanism." *Synthese 76* (1988) 333–70. Item 61 is drawn from an earlier version.

66. With Alice R. Burks. "Peirce's and Marquand's Ideas for Electrical Computers." *Grazer Philosophische Studien* 32 (1988) 3–36.

This is derived from Appendix A of *The First Electronic Computer: The Atanasoff Story*.

In process

With John H. Holland. "Method of Controlling a Classifier System." U.S. Patent Application. Filed May 7, 1987.

REPRINTS AND TRANSLATIONS

1. "The Folded Tree" (1955). Russian translation in *Cybernetics Collection*, Vol. 2, pp. 53–82. Edited by A. P. Ershov, O. B. Lupanov, A. A. Liapunov, I. A. Poletayev, A. I. Prokhorov. Moscow: Foreign Language Publishing House, 1961.

2. *Logical Design of an Electronic Computing Instrument* (1946). Reprinted (with deletions) in *Datamation 8* (September, 1962) 24–31 and 8 (October, 1962) 36–41.

3. "The Logic of Automata" (1957). Reprinted in Wang Hao, *A Survey of Mathematical Logic*, pp. 175–223. Peking: Science Press, 1962. Distributed by North-Holland Publishing Company, Amsterdam.

4. *Logical Design of an Electronic Computing Instrument* (1946). Reprinted in *John*

von Neumann, Collected Works, Vol. V, pp. 34–79. Edited by A. H. Taub. Oxford: Pergamon Press, 1963.

5. "Theory of Logical Nets" (1953). Reprinted in *Sequential Machines – Selected Papers*, pp. 193–212. Edited by Edward F. Moore. Reading, Massachusetts: Addison-Wesley, 1964.

6. "Theory of Logical Nets" (1953). Russian translation in *Cybernetics Collection*, Vol. 2, pp. 33–57, 1962. Translated by B. A. Trakhtenbrot. (Trakhtenbrot had published an earlier translation in a less well-known publication in 1958.)

7. *Logical Design of an Electronic Computing Instrument* (1946). Russian translation in *Cybernetics Collection* (Kiberneticheskii Sbornik), Vol. 9, pp. 7–67, 1964.

8. "On the Presuppositions of Induction" (1955). Reprinted as Phil-42 of Bobbs-Merrill Reprint Series in Philosophy, 1969.

9. "Reichenbach's Theory of Probability and Induction" (1951). Reprinted as Phil-43 of Bobbs-Merrill Reprint Series in Philosophy, 1969.

10. *Logical Design of an Electronic Computing Instrument* (1946). Reprinted in part in *Perspectives on the Computer Revolution*, pp. 37–46. Edited by Z. W. Pylyshyn. Englewood Cliffs, New Jersey: Prentice-Hall, 1970.

11. "The Propositional Calculus" (1965). Translated into Russian by A. V. Shileiko and published by Soviet Radio Corporation in 1970.

12. *Logical Design of an Electronic Computing Instrument* (1946). Reprinted in *Computer Structures: Readings and Examples*, pp. 92–119. Edited by C. G. Bell and Allen Newell. New York: McGraw Hill, 1971.

13. von Neumann's *Theory of Self-Reproducing Automata* (1966). Translated into Russian by Vadim L. Stafanuk. With a preface by Victor I. Varshavsky (pp. 5–6) and a preface to the Russian edition by A. W. Burks (pp. 7–14). Moscow: Peace (Mir), 1971, p. 382.

14. *Essays on Cellular Automata* (1972). Reprinted by the Library of Computer and Information Sciences, Riverside, New Jersey, 1971.

15. "A Theory of Proper Names" (1951). Reprinted in *New Readings in Philosophical Analysis*, pp. 72–77. Edited by H. Feigl, W. Sellars, and K. Lehrer. New York: Appleton-Century-Crofts, 1972.

16. *Logical Design of an Electronic Computing Instrument* (1946). Reprinted in part in *The Origins of Digital Computers, Selected Papers*, pp. 371–385. Edited by Brian Randall. Berlin: Springer-Verlag, 1973.

17. Von Neumann's *Theory of Self-Reproducing Automata* (1966). Translated into Japanese by Hidetoshi Takahashi. Tokyo, Japan: Iwanami Press, 1975.

18. *Logical Design of an Electronic Computing Instrument* (1946). Reprinted in *Computer Design Development – Principal Papers*, pp. 221–259. Edited by Earl Swartzlander. Rochelle Park, New Jersey: Hayden Book, 1976.

19. "The Logic of Causal Propositions" (1951). Reprinted in *The Nature of Causation*, pp. 255–276. Edited by Myles Brand. Urbana: University of Illinois Press. 1976.

20. "The Logic of Causal Propositions" (1951). Italian translation in *Leggi di natura, modalità, ipotesi – La logical del ragionamento contro fattuale*, pp. 181–203. Edited by Claudio Pizzi. Milan: Giangiacomo Feltrinelli, 1978.

21. Chinese translation of Ch. 11 ("Chance, Cause, and Reason") of *Chance, Cause, Reason*. Translated by Chang Li Ping, checked by Wang Yu Tian. *National Scientific*

Philosophical Problem Translations, no. 4, 1984, pp. 82–86. Chinese Academy of Social Sciences, Beijing.

22. Chinese translation of "Computers and Control in Siciety" (1979). Translated by Kao Di. *Zhexue Yicong* (Philosophy Translations), no. 5, 1985, pp. 63–67.

SUBJECT INDEX

533

NAME INDEX